Meta-Analysis in Stata:

An Updated Collection from the Stata Journal

Second Edition

Meta-Analysis in Stata:

An Updated Collection from the Stata Journal

Second Edition

TOM M. PALMER, collection editor
Department of Mathematics and Statistics
Lancaster University
Lancaster, UK

JONATHAN A. C. STERNE, collection editor
School of Social and Community Medicine
University of Bristol
Bristol, UK

H. JOSEPH NEWTON, *Stata Journal* editor
Department of Statistics
Texas A&M University
College Station, TX

NICHOLAS J. COX, *Stata Journal* editor
Department of Geography
Durham University
Durham City, UK

A Stata Press Publication
StataCorp LP
College Station, Texas

Published by Stata Press, 4905 Lakeway Drive, College Station, Texas 77845
Typeset in LaTeX 2_ε
Printed in the United States of America

10 9 8 7 6 5 4 3 2 1

ISBN-10: 1-59718-147-1
ISBN-13: 978-1-59718-147-1

Library of Congress Control Number: 2015950607

Contents

4 Multivariate meta-analysis: metandi, mvmeta   211

5 Individual patient data meta-analysis: ipdforest and ipdmetan   265

6 Network meta-analysis: indirect, network package, network_graphs package   309

7 Advanced methods: glst, metamiss, sem, gsem, metacumbounds, metasim, metapow, and metapowplot   401

Introduction to the second edition

We are delighted that this second edition of *Meta-Analysis in Stata* reflects the continuing innovations in meta-analysis software made by the Stata community since the publication of the first edition in 2009. This new collection of articles about meta-analysis from the *Stata Technical Bulletin* and the *Stata Journal* includes 27 articles, of which 11 are new additions.

The main Stata meta-analysis command `metan` has been widely used by researchers and, according to Google Scholar, has to date been cited by over 300 articles (adding the citations for Bradburn, Deeks, and Altman [1998], Harris et al. [2008], and its listing on the Statistical Software Components archive). We hope that this collection will facilitate the widespread use of both the existing and new commands.

The new articles reflect recent methodological developments in meta-analysis and provide new commands implementing these methods. The second edition extends the structure of the first edition by including parts on multivariate meta-analysis, individual participant data (IPD) meta-analysis, and network meta-analysis.

Part 1 is concerned with fitting meta-analysis models. It additionally includes the article by Kontopantelis and Reeves (2010) describing the `metaan` command, which provides additional estimators for random-effects meta-analysis and can report alternative measures of heterogeneity.

Part 2 remains unchanged from the first edition.

Part 3 is concerned with investigation of bias. It additionally includes the article by Crowther, Abrams, and Lambert (2012) describing the `extfunnel` command, which can be used to examine the impact of a hypothetical additional study on a meta-analysis by augmenting the funnel plot with statistical significance or heterogeneity contours.

Part 4, which addresses multivariate (multiple outcomes) meta-analysis, discusses a substantial update to the `mvmeta` command for multivariate outcome meta-analysis as described by White (2011). The update includes multivariate meta-regression and additional postestimation reporting features, such as I^2 statistics for each outcome.

Part 5 is a new collection of commands for IPD meta-analysis. The article by Kontopantelis and Reeves (2013) describes the `ipdforest` command, which performs IPD meta-analysis using either hierarchical linear or logistic regression and can provide a forest plot. A two-stage approach to IPD meta-analysis is described by Fisher (2015) and implemented in the `ipdmetan` command. The command can incorporate studies reporting both IPD and study-level (aggregate) data and has options to fine tune the forest plots in such settings.

Part 6 includes three new articles on network meta-analysis, which is a major recent development in meta-analysis methodology (Bucher et al. 1997, Caldwell, Ades, and Higgins 2005; Salanti et al. 2008; Salanti 2012). The first article, by Miladinovic et al. (2014), concerns comparisons of treatments in the absence of direct evidence between them (so-called indirect comparisons). The second article, by White (Forthcoming), presents the `network` suite of commands for network meta-analysis, which is centered around fitting network meta-analysis models with the multivariate normal approach using `mvmeta`. Third the article, by Chaimani and Salanti (Forthcoming), describes the `network_graphs` package of graphical commands for network meta-analysis. These commands have been designed to work with the same data structures as those provided by the `network` suite.

Part 7 includes articles on various advanced meta-analysis methods. New articles include that by Crowther et al. (2013), which provides the `metasim`, `metapow`, and `metapowplot` commands. These estimate the probability that the conclusions of a meta-analysis will change given the inclusion of a hypothetical new study and are based on the methodology of Sutton et al. (2007). Stata 12 and 13 introduced the `sem` and `gsem` commands for structural equation modeling. These commands are very flexible and allow a wide range of constraints to be placed on the parameters in the model. Palmer and Sterne (Forthcoming) describe how these features enable these commands to fit fixed- and random-effects meta-analysis models, including meta-regression and multivariate meta-analysis models. Cumulative meta-analysis was discussed in the first edition by Sterne (1998). Through their `metacumbounds` command, Miladinovic, Hozo, and Djulbegovic (2013) automate the use of the "ldbounds" package for R (Casper and Perez 2014). This command implements trial sequential boundaries for cumulative meta-analyses for controlling the type I error of the meta-analysis.

Information about user-written commands for meta-analysis can be obtained by typing `help meta` in Stata. In addition to this, Stata maintains a frequently asked questions on meta-analysis at

http://www.stata.com/support/faqs/statistics/meta-analysis/

We hope that this second edition of articles about meta-analysis repeats the success of the first edition and continues to encourage users to implement the latest methods for meta-analysis in new Stata commands.

Tom M. Palmer and Jonathan A. C. Sterne
August 2015

1 References

Bradburn, M. J., J. J. Deeks, and D. G. Altman. 1998. sbe24: metan—an alternative meta-analysis command. *Stata Technical Bulletin* 44: 4–15. Reprinted in *Stata Technical Bulletin Reprints*, vol. 8, pp. 86–100. College Station, TX: Stata Press. (Reprinted in this collection on pp. 3–28.)

Bucher, H. C., G. H. Guyatt, L. E. Griffith, and S. D. Walter. 1997. The results of direct and indirect treatment comparisons in meta-analysis of randomized controlled trials. *Journal of Clinical Epidemiology* 50: 638–691.

Caldwell, D. M., A. E. Ades, and J. P. T. Higgins. 2005. Simultaneous comparison of multiple treatments: Combining direct and indirect evidence. *British Medical Journal* 331: 897–900.

Casper, C., and O. A. Perez. 2014. ldbounds: Lan–DeMets method for group sequential boundaries. R package version 1.1-1. http://CRAN.R-project.org/package=ldbounds.

Chaimani, A., and G. Salanti. Forthcoming. Visualizing assumptions and results in network meta-analysis: The network graphs package. *Stata Journal* (Reprinted in this collection on pp. 355–400.)

Crowther, M. J., K. R. Abrams, and P. C. Lambert. 2012. Flexible parametric joint modelling of longitudinal and survival data. *Statistics in Medicine* 31: 4456–4471.

Crowther, M. J., S. R. Hinchliffe, A. Donald, and A. J. Sutton. 2013. Simulation-based sample-size calculation for designing new clinical trials and diagnostic test accuracy studies to update an existing meta-analysis. *Stata Journal* 13: 451–473. (Reprinted in this collection on pp. 476–498.)

Fisher, D. J. 2015. Two-stage individual participant data meta-analysis and generalized forest plots. *Stata Journal* 15: 369–396. (Reprinted in this collection on pp. 280–307.)

Harris, R. J., M. J. Bradburn, J. J. Deeks, R. M. Harbord, D. G. Altman, and J. A. C. Sterne. 2008. metan: Fixed- and random-effects meta-analysis. *Stata Journal* 8: 3–28. (Reprinted in this collection on pp. 29–54.)

Kontopantelis, E., and D. Reeves. 2010. metaan: Random-effects meta-analysis. *Stata Journal* 10: 395–407. (Reprinted in this collection on pp. 55–67.)

————. 2013. A short guide and a forest plot command (ipdforest) for one-stage meta-analysis. *Stata Journal* 13: 574–587. (Reprinted in this collection on pp. 321–354.)

Miladinovic, B., A. Chaimani, I. Hozo, and B. Djulbegovic. 2014. Indirect treatment comparison. *Stata Journal* 14: 76–86. (Reprinted in this collection on pp. 311–320.)

Miladinovic, B., I. Hozo, and B. Djulbegovic. 2013. Trial sequential boundaries for cumulative meta-analyses. *Stata Journal* 13: 77–91. (Reprinted in this collection on pp. 462–475.)

Palmer, T. M., and J. A. C. Sterne. Forthcoming. Fitting fixed- and random-effects meta-analysis models using structural equation modeling with the sem and gsem commands. *Stata Journal* (Reprinted in this collection on pp. 435–461.)

Salanti, G. 2012. Indirect and mixed-treatment comparison, network, or multiple-treatments meta-analysis: Many names, many benefits, many concerns for the next generation evidence synthesis tool. *Research Synthesis Methods* 3: 80–97.

Salanti, G., J. P. T. Higgins, A. E. Ades, and J. P. A. Ioannidis. 2008. Evaluation of networks of randomized trials. *Statistical Methods in Medical Research* 17: 279–301.

Sterne, J. A. C. 1998. sbe22: Cumulative meta analysis. *Stata Technical Bulletin* 42: 13–16. Reprinted in *Stata Technical Bulletin Reprints*, vol. 7, pp. 143–147. College Station, TX: Stata Press. (Reprinted in this collection on pp. 68–77.)

Sutton, A. J., N. J. Cooper, D. R. Jones, P. C. Lambert, J. R. Thompson, and K. R. Abrams. 2007. Evidence-based sample size calculations based upon updated meta-analysis. *Statistics in Medicine* 26: 2479–2500.

White, I. R. 2011. Multivariate random-effects meta-regression: Updates to mvmeta. *Stata Journal* 11: 255–270. (Reprinted in this collection on pp. 249–264.)

————. Forthcoming. Network meta-analysis. *Stata Journal* (Reprinted in this collection on pp. 321–354.)

Introduction to the first edition

This first collection of articles from the *Stata Technical Bulletin* and the *Stata Journal* brings together updated user-written commands for meta-analysis, which has been defined as a statistical analysis that combines or integrates the results of several independent studies considered by the analyst to be combinable (Huque 1988). The statistician Karl Pearson is commonly credited with performing the first meta-analysis more than a century ago (Pearson 1904)—the term "meta-analysis" was first used by Glass (1976). The rapid increase over the last three decades in the number of meta-analyses reported in the social and medical literature has been accompanied by extensive research on the underlying statistical methods. It is therefore surprising that the major statistical software packages have been slow to provide meta-analytic routines (Sterne, Egger, and Sutton 2001).

During the mid-1990s, Stata users recognized that the ease with which new commands could be written and distributed, and the availability of improved graphics programming facilities, provided an opportunity to make meta-analysis software widely available. The first command, `meta`, was published in 1997 (Sharp and Sterne 1997), while the `metan` command—now the main Stata meta-analysis command—was published shortly afterward (Bradburn, Deeks, and Altman 1998). A major motivation for writing `metan` was to provide independent validation of the routines programmed into the specialist software written for the Cochrane Collaboration, an international organization dedicated to improving health care decision-making globally, through systematic reviews of the effects of health care interventions, published in The Cochrane Library (see www.cochrane.org). The groups responsible for the `meta` and `metan` commands combined to produce a major update to `metan` that was published in 2008 (Harris et al. 2008). This update uses the most recent Stata graphics routines to provide flexible displays combining text and figures. Further articles describe commands for cumulative meta-analysis (Sterne 1998) and for meta-analysis of p-values (Tobias 1999), which can be traced back to Fisher (1932). Between-study heterogeneity in results, which can cause major difficulties in interpretation, can be investigated using meta-regression (Berkey et al. 1995). The `metareg` command (Sharp 1998) remains one of the few implementations of meta-regression and has been updated to take account of improvements in Stata estimation facilities and recent methodological developments (Harbord and Higgins 2008).

Enthusiasm for meta-analysis has been tempered by a realization that flaws in the conduct of studies (Schulz et al. 1995), and the tendency for the publication process to favor studies with statistically significant results (Begg and Berlin 1988; Dickersin, Min, and Meinert 1992), can lead to the results of meta-analyses mirroring overoptimistic results from the original studies (Egger et al. 1997). A set of Stata commands—`metafunnel`, `confunnel`, `metabias`, and `metatrim`—address these issues both graphically (via routines to draw standard funnel plots and "contour-enhanced" funnel plots) and statistically, by providing tests for funnel plot asymmetry, which can be used to diagnose publication bias and other small-study effects (Sterne, Gavaghan, and Egger 2000; Sterne, Egger, and Moher 2008).

This collection also contains advanced routines that exploit Stata's range of estimation procedures. Meta-analysis of studies that estimate the accuracy of diagnostic tests, implemented in the `metandi` command, is inherently bivariate, because of the trade-off between sensitivity and specificity (Rutter and Gatsonis 2001; Reitsma et al. 2005). Meta-analyses of observational studies will often need to combine dose–response relationships, but reports of such studies often report comparisons between three or more categories. The method of Greenland and Longnecker (1992), implemented in the `glst` command, converts categorical to dose–response comparisons and can thus be used to derive the data needed for dose–response meta-analyses. White and colleagues (White and Higgins 2009; White 2009) have recently provided general routines to deal with missing data in meta-analysis, and for multivariate random-effects meta-analysis.

Finally, the appendix lists user-written meta-analysis commands that have not, so far, been accepted for publication in the *Stata Journal*. For the most up-to-date information on meta-analysis commands in Stata, readers are encouraged to check the Stata frequently asked question on meta-analysis:

http://www.stata.com/support/faqs/stat/meta.html

Those involved in developing Stata meta-analysis commands have been delighted by their widespread worldwide use. However, a by-product of the large number of commands and updates to these commands now available has been that users find it increasingly difficult to identify the most recent version of commands, the commands most relevant to a particular purpose, and the related documentation. This collection aims to provide a comprehensive description of the facilities for meta-analysis now available in Stata and has also stimulated the production and documentation of a number of updates to existing commands, some of which were long overdue. I hope that this collection will be useful to the large number of Stata users already conducting meta-analyses, as well as facilitate interest in and use of the commands by new users.

Jonathan A. C. Sterne
February 2009

2 References

Begg, C. B., and J. A. Berlin. 1988. Publication bias: A problem in interpreting medical data. *Journal of the Royal Statistical Society, Series A* 151: 419–463.

Berkey, C. S., D. C. Hoaglin, F. Mosteller, and G. A. Colditz. 1995. A random-effects regression model for meta-analysis. *Statistics in Medicine* 14: 395–411.

Bradburn, M. J., J. J. Deeks, and D. G. Altman. 1998. sbe24: metan—an alternative meta-analysis command. *Stata Technical Bulletin* 44: 4–15. Reprinted in *Stata Technical Bulletin Reprints*, vol. 8, pp. 86–100. College Station, TX: Stata Press. (Reprinted in this collection on pp. 3–28.)

Dickersin, K., Y. I. Min, and C. L. Meinert. 1992. Factors influencing publication of research results: Follow-up of applications submitted to two institutional review boards. *Journal of the American Medical Association* 267: 374–378.

Egger, M., G. Davey Smith, M. Schneider, and C. Minder. 1997. Bias in meta-analysis detected by a simple, graphical test. *British Medical Journal* 315: 629–634.

Fisher, R. A. 1932. *Statistical Methods for Research Workers.* 4th ed. Edinburgh: Oliver & Boyd.

Glass, G. V. 1976. Primary, secondary, and meta-analysis of research. *Educational Researcher* 10: 3–8.

Greenland, S., and M. P. Longnecker. 1992. Methods for trend estimation from summarized dose–reponse data, with applications to meta-analysis. *American Journal of Epidemiology* 135: 1301–1309.

Harbord, R. M., and J. P. T. Higgins. 2008. Meta-regression in Stata. *Stata Journal* 8: 493–519. (Reprinted in this collection on pp. 85–111.)

Harris, R. J., M. J. Bradburn, J. J. Deeks, R. M. Harbord, D. G. Altman, and J. A. C. Sterne. 2008. metan: Fixed- and random-effects meta-analysis. *Stata Journal* 8: 3–28. (Reprinted in this collection on pp. 29–54.)

Huque, M. F. 1988. Experiences with meta-analysis in NDA submissions. *Proceedings of the Biopharmaceutical Section of the American Statistical Association* 2: 28–33.

Pearson, K. 1904. Report on certain enteric fever inoculation statistics. *British Medical Journal* 2: 1243–1246.

Reitsma, J. B., A. S. Glas, A. W. S. Rutjes, R. J. P. M. Scholten, P. M. Bossuyt, and A. H. Zwinderman. 2005. Bivariate analysis of sensitivity and specificity produces informative summary measures in diagnostic reviews. *Journal of Clinical Epidemiology* 58: 982–990.

Rutter, C. M., and C. A. Gatsonis. 2001. A hierarchical regression approach to meta-analysis of diagnostic test accuracy evaluations. *Statistics in Medicine* 20: 2865–2884.

Schulz, K. F., I. Chalmers, R. J. Hayes, and D. G. Altman. 1995. Empirical evidence of bias. Dimensions of methodological quality associated with estimates of treatment effects in controlled trials. *Journal of the American Medical Association* 273: 408–412.

Sharp, S. 1998. sbe23: Meta-analysis regression. *Stata Technical Bulletin* 42: 16–22. Reprinted in *Stata Technical Bulletin Reprints*, vol. 7, pp. 148–155. College Station, TX: Stata Press. (Reprinted in this collection on pp. 112–120.)

Sharp, S., and J. A. C. Sterne. 1997. sbe16: Meta-analysis. *Stata Technical Bulletin* 38: 9–14. Reprinted in *Stata Technical Bulletin Reprints*, vol. 7, pp. 100–106. College Station, TX: Stata Press.

Sterne, J. A. C. 1998. sbe22: Cumulative meta analysis. *Stata Technical Bulletin* 42: 13–16. Reprinted in *Stata Technical Bulletin Reprints*, vol. 7, pp. 143–147. College Station, TX: Stata Press. (Reprinted in this collection on pp. 68–77.)

Sterne, J. A. C., M. Egger, and D. Moher. 2008. Addressing reporting biases. In *Cochrane Handbook for Systematic Reviews of Interventions*, ed. J. P. T. Higgins and S. Green, 297–334. Chichester, UK: Wiley.

Sterne, J. A. C., M. Egger, and A. J. Sutton. 2001. Meta-analysis software. In *Systematic Reviews in Health Care: Meta-Analysis in Context*, 2nd edition, ed. M. Egger, G. Davey Smith, and D. G. Altman, 336–346. London: BMJ Books.

Sterne, J. A. C., D. Gavaghan, and M. Egger. 2000. Publication and related bias in meta-analysis: Power of statistical tests and prevalence in the literature. *Journal of Clinical Epidemiology* 53: 1119–1129.

Tobias, A. 1999. sbe28: Meta-analysis of p-values. *Stata Technical Bulletin* 49: 15–17. Reprinted in *Stata Technical Bulletin Reprints*, vol. 9, pp. 138–140. College Station, TX: Stata Press.

White, I. R. 2009. Multivariate random-effects meta-analysis. *Stata Journal* 9: 40–56. (Reprinted in this collection on pp. 232–248.)

White, I. R., and J. P. T. Higgins. 2009. Meta-analysis with missing data. *Stata Journal* 9: 57–69. (Reprinted in this collection on pp. 422–434.)

Install the software

The most recent version of the majority of the user-written commands that are described in this collection are available from the Statistical Software Components archive. To install commands and other ancillary files from the archive for use with your own research, type `ssc install` followed by the user-written command name.

For example, to install all files in the `metan` package, type

```
. ssc install metan
```

To keep your installed packages up to date, type `adoupdate` regularly.

In rare cases, authors have chosen not to make their software available on the Statistical Software Components archive. For these commands, the software is available from the *Stata Journal* website.

For *Stata Journal* articles, you can type `net sj` followed by the volume, number, and package ID, all of which appear in the upper-left or upper-right corner of the first page of the article, depending on whether the first page of the article is on an even or odd page. For *Stata Technical Bulletin* articles, you can type `net stb` followed by the volume and package ID, which appear in the upper-left or upper-right corner of the first page of the article, depending on whether the first page of the article is on an even or odd page. After either command, you will need to type `net install` followed by the package ID.

For example, to install the `indirect` command and associated files that were affiliated with the article published in the *Stata Journal*, Volume 14, Number 1, with package ID st0325, you would type

```
. net sj 14-1 st0325
. net install st0325
```

To install the `metap` command and associated files that were affiliated with the article published in the *Stata Technical Bulletin*, Volume 49, with package ID sbe28, you would type

```
. net stb 49 sbe28
. net install sbe28
```

   Some readers may also wish to duplicate the results displayed in the articles. To
do this, you will need to install the version of the command available at the time the
article was published. These archival versions are available from the `sj` or `stb` net site
and may be installed using `net install` as shown above.

Part 1

Meta-analysis in Stata: metan, metaan, metacum, and metap

The `metan` command is the main Stata meta-analysis command. In its latest version, it provides highly flexible facilities for doing meta-analyses and graphing their results. Its worldwide use testifies to the dedication and skills of Michael Bradburn, who did most of the original programming and then added a range of facilities in response to user requests, and Ross Harris, who redesigned the graphics and updated them to Stata 9 and added further options.

`metan` is described in two articles. The first—by Bradburn, Deeks, and Altman—was published in the *Stata Technical Bulletin* in 1998. For the first edition of this collection, it was updated to describe and use the most recent `metan` syntax, with the graphics also having been updated. Editorial notes explain where other commands originally published and distributed with `metan` have now been superseded. The additional facilities made available since the publication of the original `metan` command are described in the 2008 *Stata Journal* article by Harris et al.

`metan` contains several different estimators for fitting fixed- and random-effects meta-analysis models.

Kontopantelis and Reeves (2010) introduced the `metaan` command, which includes additional estimators for random-effects meta-analysis, such as a random-effects model with bootstrapped heterogeneity parameters, profile likelihood estimation for the random-effects model, and a permutation-based random-effects model. `metaan` can also produce an alternative style of forest plot. The `metaan` command has been updated to include a sensitivity analysis that varies the degree of heterogeneity in the meta-analysis (Kontopantelis, Springate, and Reeves 2013).

The evolution of evidence over time can be described and displayed using cumulative meta-analysis. `metacum`—a command for cumulative analysis—was described by Sterne in the *Stata Technical Bulletin* in 1998. For the first edition of this collection, the `metacum` command was updated by Ross Harris (2008) to use version 9 graphics and the same syntax as `metan`, and the original article has been updated to reflect this.

In some circumstances, only the p-value from each study is available. Meta-analysis of p-values, implemented in the `metap` command (Tobias 1999), can be traced back to Fisher (1932). However, users should be aware that such analyses ignore the direction of the effect in individual studies and so are best seen as providing an overall test of the null hypothesis of no effect.

1   References

Bradburn, M. J., J. J. Deeks, and D. G. Altman. 1998. sbe24: metan—an alternative meta-analysis command. *Stata Technical Bulletin* 44: 4–15. Reprinted in *Stata Technical Bulletin Reprints*, vol. 8, pp. 86–100. College Station, TX: Stata Press. (Reprinted in this collection on pp. 3–28.)

Fisher, R. A. 1932. *Statistical Methods for Research Workers*. 4th ed. Edinburgh: Oliver & Boyd.

Harris, R. J., M. J. Bradburn, J. J. Deeks, R. M. Harbord, D. G. Altman, and J. A. C. Sterne. 2008. metan: Fixed- and random-effects meta-analysis. *Stata Journal* 8: 3–28. (Reprinted in this collection on pp. 29–54.)

Kontopantelis, E., and D. Reeves. 2010. metaan: Random-effects meta-analysis. *Stata Journal* 10: 395–407. (Reprinted in this collection on pp. 55–67.)

Kontopantelis, E., D. A. Springate, and D. Reeves. 2013. A re-analysis of the Cochrane Library data: The dangers of unobserved heterogeneity in meta-analyses. *PLOS ONE* 8: e69930.

Sterne, J. A. C. 1998. sbe22: Cumulative meta analysis. *Stata Technical Bulletin* 42: 13–16. Reprinted in *Stata Technical Bulletin Reprints*, vol. 7, pp. 143–147. College Station, TX: Stata Press. (Reprinted in this collection on pp. 68–77.)

Tobias, A. 1999. sbe28: Meta-analysis of p-values. *Stata Technical Bulletin* 49: 15–17. Reprinted in *Stata Technical Bulletin Reprints*, vol. 9, pp. 138–140. College Station, TX: Stata Press.

The Stata Technical Bulletin (1998)
STB-44, sbe24, pp. 4–15

metan—a command for meta-analysis in Stata[1]

Michael J. Bradburn
Clinical Trials Research Unit
ScHARR
Sheffield, UK
m.bradburn@sheffield.ac.uk

Jonathan J. Deeks
Unit of Public Health, Epidemiology, and Biostatistics
University of Birmingham
Birmingham, UK
j.deeks@bham.ac.uk

Douglas G. Altman
Centre for Statistics in Medicine
University of Oxford
Oxford, UK

1 Background

When several studies are of a similar design, it often makes sense to try to combine the information from them all to gain precision and to investigate consistencies and discrepancies between their results. In recent years, there has been a considerable growth of this type of analysis in several fields, and in medical research in particular. In medicine, such studies usually relate to controlled trials of therapy, but the same principles apply in any scientific area; for example in epidemiology, psychology, and educational research. The essence of meta-analysis is to obtain a single estimate of the effect of interest (effect size) from some statistic observed in each of several similar studies. All methods of meta-analysis estimate the overall effect by computing a weighted average of the studies' individual estimates of effect.

metan provides methods for the meta-analysis of studies with two groups. With binary data, the effect measure can be the difference between proportions (sometimes called the risk difference or absolute risk reduction), the ratio of two proportions (risk ratio or relative risk), or the odds ratio. With continuous data, both observed differences in means or standardized differences in means (effect sizes) can be used. For both binary and continuous data, either fixed-effects or random-effects models can be fitted (Fleiss 1993). There are also other approaches, including empirical and fully Bayesian methods. Meta-analysis can be extended to other types of data and study designs, but these are not considered here.

---

1. The original title was *metan—an alternative meta-analysis command*. The updated syntax is by Ross Harris, Centre for Infections, Health Protection Agency, London.—Ed.

As well as the primary pooling analysis, there are secondary analyses that are often performed. One common additional analysis is to test whether there is excess heterogeneity in effects across the studies. There are also several graphs that can be used to supplement the main analysis.

2   Data structure

Consider a meta-analysis of k studies. When the studies have a binary outcome, the results of each study can be presented in a 2×2 table (table 1) giving the numbers of subjects who do or do not experience the event in each of the two groups (here called intervention and control).

Table 1. Binary data

Study i; $1 \leq i \leq k$	Event	No event
Intervention	a_i	b_i
Control	c_i	d_i

If the outcome is a continuous measure, the number of subjects in each of the two groups, their mean response, and the standard deviation of their responses are required to perform meta-analysis (table 2).

Table 2. Continuous data

Study i; ($1 \leq i \leq k$)	Group size	Mean response	Standard deviation
Intervention	n_{1i}	m_{1i}	sd_{1i}
Control	n_{2i}	m_{2i}	sd_{2i}

3   Analysis of binary data using fixed-effects models

There are two alternative fixed-effects analyses. The inverse variance method (sometimes referred to as Woolf's method) computes an average effect by weighting each study's log odds-ratio, log relative-risk, or risk difference according to the inverse of their sampling variance, such that studies with higher precision (lower variance) are given higher weights. This method uses large sample asymptotic sampling variances, so it may perform poorly for studies with very low or very high event rates or small sample sizes. In other situations, the inverse variance method gives a minimum variance unbiased estimate.

The Mantel–Haenszel method uses an alternative weighting scheme originally derived for analyzing stratified case–control studies. The method was first described for the odds ratio by Mantel and Haenszel (1959) and extended to the relative risk and risk difference by Greenland and Robins (1985). The estimate of the variance of the overall odds ratio was described by Robins, Greenland, and Breslow (1986). These methods are preferable to the inverse variance method as they have been shown to be robust when data are sparse, and give similar estimates to the inverse variance method in other situations. They are the default in the `metan` command. Alternative formulations of the Mantel–Haenszel methods more suited to analyzing stratified case–control studies are available in the `epitab` commands.

Peto proposed an assumption free method for estimating an overall odds ratio from the results of several large clinical trials (Yusuf et al. 1985). The method sums across all studies the difference between the observed $\{O(a_i)\}$ and expected $\{E(a_i)\}$ numbers of events in the intervention group (the expected number of events being estimated under the null hypothesis of no treatment effect). The expected value of the sum of $O - E$ under the null hypothesis is zero. The overall log odds-ratio is estimated from the ratio of the sum of the $O - E$ and the sum of the hypergeometric variances from individual trials. This method gives valid estimates when combining large balanced trials with small treatment effects, but has been shown to give biased estimates in other situations (Greenland and Salvan 1990).

If a study's 2×2 table contains one or more zero cells, then computational difficulties may be encountered in both the inverse variance and the Mantel–Haenszel methods. These can be overcome by adding a standard correction of 0.5 to all cells in the 2×2 table, and this is the approach adopted here. However, when there are no events in one whole column of the 2×2 table (i.e., all subjects have the same outcome regardless of group), the odds ratio and the relative risk cannot be estimated, and the study is given zero weight in the meta-analysis. Such trials are included in the risk difference methods as they are informative that the difference in risk is small.

4   Analysis of continuous data using fixed-effects models

The weighted mean difference meta-analysis combines the differences between the means of intervention and control groups $(m_{1i} - m_{2i})$ to estimate the overall mean difference (Sinclair and Bracken 1992, chap. 2). A prerequisite of this method is that the response is measured in the same units using comparable devices in all studies. Studies are weighted using the inverse of the variance of the differences in means. Normality within trial arms is assumed, and between trial variations in standard deviations are attributed to differences in precision, and are assumed equal in both study arms.

An alternative approach is to pool standardized differences in means, calculated as the ratio of the observed difference in means to an estimate of the standard deviation of the response. This approach is especially appropriate when studies measure the same concept (e.g., pain or depression) but use a variety of continuous scales. By standardization, the study results are transformed to a common scale (standard deviation units) that facilitates pooling. There are various methods for computing the standardized study results: Glass's method (Glass, McGaw, and Smith 1981) divides the differences in means by the control group standard deviation, whereas Cohen's and Hedges' methods use the same basic approach, but divide by an estimate of the standard deviation obtained from pooling the standard deviations from both experimental and control groups (Rosenthal 1994). Hedges' method incorporates a small sample bias correction factor (Hedges and Olkin 1985, chap. 5). An inverse variance weighting method is used in all the formulations. Normality within trial arms is assumed, and all differences in standard deviations between trials are attributed to variations in the scale of measurement.

5   Test for heterogeneity

For all the above methods, the consistency or homogeneity of the study results can be assessed by considering an appropriately weighted sum of the differences between the k individual study results and the overall estimate. The test statistic has a χ^2 distribution with $k - 1$ degrees of freedom (DerSimonian and Laird 1986).

6   Analysis of binary or continuous data using random-effects models

An approach developed by DerSimonian and Laird (1986) can be used to perform random-effects meta-analysis for all the effect measures discussed above (except the Peto method). Such models assume that the treatment effects observed in the trials are a random sample from a distribution of treatment effects with a variance τ^2. This is in contrast to the fixed-effects models which assume that the observed treatment effects are all estimates of a single treatment effect. The DerSimonian and Laird methods incorporate an estimate of the between-study variation τ^2 into both the study weights (which are the inverse of the sum of the individual sampling variance and the between studies variance τ^2) and the standard error of the estimate of the common effect. Where there are computational problems for binary data due to zero cells the same approach is used as for fixed-effects models.

Where there is excess variability (heterogeneity) between study results, random-effects models typically produce more conservative estimates of the significance of the treatment effect (i.e., a wider confidence interval) than fixed-effects models. As they give proportionately higher weights to smaller studies and lower weights to larger studies than fixed-effects analyses, there may also be differences between fixed and random models in the estimate of the treatment effect.

7   Tests of overall effect

For all analyses, the significance of the overall effect is calculated by computing a z score as the ratio of the overall effect to its standard error and comparing it with the standard normal distribution. Alternatively, for the Mantel–Haenszel odds-ratio and Peto odds-ratio method, χ^2 tests of overall effect are available (Breslow and Day 1993).

8   Graphical analyses

Three plots are available in these programs. The most common graphical display to accompany a meta-analysis shows horizontal lines for each study, depicting estimates and confidence intervals, commonly called a forest plot. The size of the plotting symbol for the point estimate in each study is proportional to the weight that each trial contributes in the meta-analysis. The overall estimate and confidence interval are marked by a diamond. For binary data, a L'Abbé plot (L'Abbé, Detsky, and O'Rourke 1987) plots the event rates in control and experimental groups by study. For all data types a funnel plot shows the relation between the effect size and precision of the estimate. It can be used to examine whether there is asymmetry suggesting possible publication bias (Egger et al. 1997), which usually occurs where studies with negative results are less likely to be published than studies with positive results.

Each trial i should be allocated one row in the dataset. There are three commands for invoking the routines; `metan`, `funnel`, and `labbe`, which are detailed below.

9   Syntax for metan

`metan` *varlist* $\left[\,if\,\right]$ $\left[\,in\,\right]$ $\left[\,,\,\left[\,binary\_data\_options\,|\,continuous\_data\_options\,|\right.\right.$
    $precalculated\_effect\_estimates\_options\,\right]$ *measure_and_model_options output_options*
    $forest\_plot\_options\,\right]$

binary_data_options

    or rr rd fixed random fixedi randomi peto cornfield chi2
    breslow <u>noint</u>eger cc($\#$)

continuous_data_options

    cohen hedges glass nostandard fixed random <u>noint</u>eger

precalculated_effect_estimates_options

    fixed random

measure_and_model_options

    wgt(*wgtvar*) second(*model | estimates_and_description*)
    first(*estimates_and_description*)

output_options

> by(*byvar*) nosubgroup sgweight log eform efficacy ilevel(*#*)
> olevel(*#*) sortby(*varlist*)
> label([namevar = *namevar*], [yearvar = *yearvar*]) nokeep notable
> nograph nosecsub

forest_plot_options

> xlabel(*#*, ...) xtick(*#*, ...) boxsca(*#*) textsize(*#*) nobox
> nooverall nowt nostats counts group1(*string*) group2(*string*)
> effect(*string*) force lcols(*varlist*) rcols(*varlist*) astext(*#*) double
> nohet summaryonly rfdist rflevel(*#*) null(*#*) nulloff
> favours(*string # string*) firststats(*string*) secondstats(*string*)
> boxopt(*marker_options*) diamopt(*line_options*)
> pointopt(*marker_options* | *marker_label_options*) ciopt(*line_options*)
> olineopt(*line_options*) classic nowarning *graph_options*

10   Options for metan

10.1   binary_data_options

or pools ORs.

rr pools RRs; this is the default.

rd pools risk differences.

fixed specifies a fixed-effects model using the Mantel–Haenszel method; this is the default.

random specifies a random-effects model using the DerSimonian and Laird method, with the estimate of heterogeneity being taken.

fixedi specifies a fixed-effects model using the inverse-variance method.

randomi specifies a random-effects model using the DerSimonian and Laird method, with the estimate of heterogeneity being taken from the inverse-variance fixed-effects model.

peto specifies that the Peto method is used to pool ORs.

cornfield computes confidence intervals for ORs using Cornfield's method, rather than the (default) Woolf method.

chi2 displays a chi-squared statistic (instead of z) for the test of significance of the pooled effect size. This option is available only for ORs pooled using the Peto or Mantel–Haenszel methods.

breslow produces a Breslow–Day test for homogeneity of ORs.

nointeger allows the cell counts to be nonintegers. This option may be useful when a variable continuity correction is sought for studies containing zero cells but also may be used in other circumstances, such as where a cluster-randomized trial is to be incorporated and the "effective sample size" is less than the total number of observations.

cc(#) defines a fixed-continuity correction to add where a study contains a zero cell. By default, metan8 adds 0.5 to each cell of a trial where a zero is encountered when using inverse-variance, DerSimonian and Laird, or Mantel–Haenszel weighting to enable finite variance estimators to be derived. However, the cc() option allows the use of other constants (including none). See also the nointeger option.

10.2 continuous_data_options

cohen pools standardized mean differences by the Cohen method; this is the default.

hedges pools standardized mean differences by the Hedges method.

glass pools standardized mean differences by the Glass method.

nostandard pools unstandardized mean differences.

fixed specifies a fixed-effects model using the Mantel–Haenszel method; this is the default.

random specifies a random-effects model using the DerSimonian and Laird method, with the estimate of heterogeneity being taken.

nointeger denotes that the number of observations in each arm does not need to be an integer. By default, the first and fourth variables specified (containing N_intervention and N_control, respectively) may occasionally be noninteger (see nointeger in section 10.1).

10.3 precalculated_effect_estimates_options

fixed specifies a fixed-effects model using the Mantel–Haenszel method; this is the default.

random specifies a random-effects model using the DerSimonian and Laird method, with the estimate of heterogeneity being taken.

10.4 measure_and_model_options

wgt(*wgtvar*) specifies alternative weighting for any data type. The effect size is to be computed by assigning a weight of *wgtvar* to the studies. When RRs or ORs are declared, their logarithms are weighted. This option should be used only if you are satisfied that the weights are meaningful.

second(*model* | *estimates_and_description*) specifies that a second analysis may be per-
formed using another method: fixed, random, or peto. Users may also define their
own estimate and 95% CI based on calculations performed externally to metan, along
with a description of their method, in the format *es lci uci description*. The results
of this analysis are then displayed in the table and forest plot. If by() is used,
subestimates from the second method are not displayed with user-defined estimates
for obvious reasons.

first(*estimates_and_description*) completely changes the way metan operates, as results are no longer based on any standard methods. Users define their own estimate,
95% CI, and description, as in the above option, and must supply their own weight-
ings using wgt(*wgtvar*) to control the display of box sizes. Data must be supplied
in the 2 or 3 variable syntax (*theta se_theta* or *es lci uci*) and by() may not be used
for obvious reasons.

10.5   output_options

by(*byvar*) specifies that the meta-analysis is to be stratified according to the variable
declared.

nosubgroup specifies that no within-group results be presented. By default, metan pools
trials both within and across all studies.

sgweight specifies that the display is to present the percentage weights within each
subgroup separately. By default, metan presents weights as a percentage of the
overall total.

log reports the results on the log scale (valid only for ORs and RRs analyses from raw
data counts).

eform exponentiates all effect sizes and confidence intervals (valid only when the input
variables are log ORs or log hazard-ratios with standard error or confidence intervals).

efficacy specifies results as the vaccine efficacy (the proportion of cases that would have
been prevented in the placebo group had they received the vaccination). Available
only with ORs or RRs.

ilevel(#) specifies the coverage (e.g., 90%, 95%, 99%) for the individual trial confidence intervals; the default is $S_level. ilevel() and olevel() need not be the
same. See [U] **20.7 Specifying the width of confidence intervals**.

olevel(#) specifies the coverage (e.g., 90%, 95%, 99%) for the overall (pooled) trial
confidence intervals; the default is $S_level. ilevel() and olevel() need not be
the same. See [U] **20.7 Specifying the width of confidence intervals**.

sortby(*varlist*) sorts by variable(s) in *varlist*.

label([namevar=*namevar*], [yearvar=*yearvar*]) labels the data by its name, year,
or both. Either or both variable lists may be left blank. For the table display, the

overall length of the label is restricted to 20 characters. If the `lcols()` option is also specified, it will override the `label()` option.

`nokeep` prevents the retention of study parameters in permanent variables (see *Variables generated* below).

`notable` prevents the display of a results table.

`nograph` prevents the display of a graph.

`nosecsub` prevents the display of subestimates using the second method if `second()` is used. This option is invoked automatically with user-defined estimates.

10.6   forest_plot_options

`xlabel(#,...)` defines x-axis labels. This option has been modified so that any number of points may be defined. Also, checks are no longer made as to whether these points are sensible, so the user may define anything if the `force` option is used. Points must be comma separated.

`xtick(#,...)` adds tick marks to the x axis. Points must be comma separated.

`boxsca(#)` controls box scaling. This option has been modified so that the default is `boxsca(100)` (as in 100%) and the percentage may be increased or decreased (e.g., 80 or 120 for 20% smaller or larger, respectively).

`textsize(#)` specifies the font size for the text display on the graph. This option has been modified so that the default is `textsize(100)` (as in 100%) and the percentage may be increased or decreased (e.g., 80 or 120 for 20% smaller or larger, respectively).

`nobox` prevents a "weighted box" from being drawn for each study; only markers for point estimates are shown.

`nooverall` specifies that the overall estimate not be displayed, for example, when it is inappropriate to meta-analyze across groups. (This option automatically enforces the `nowt` option.)

`nowt` prevents the display of study weight on the graph.

`nostats` prevents the display of study statistics on the graph.

`counts` displays data counts (n/N) for each group when using binary data or the sample size, mean, and standard deviation for each group if mean differences are used (the latter is a new feature).

`group1(string)` and `group2(string)` may be used with the `counts` option, and the text should contain the names of the two groups.

`effect(string)` allows the graph to name the summary statistic used when the effect size and its standard error are declared.

`force` forces the x-axis scale to be in the range specified by `xlabel()`.

lcols(*varlist*) and rcols(*varlist*) define columns of additional data to the left or right
    of the graph. The first two columns on the right are automatically set to effect size
    and weight, unless suppressed by using the options nostats and nowt. If counts
    is used, this will be set as the third column. textsize() can be used to fine-tune
    the size of the text to achieve a satisfactory appearance. The columns are labeled
    with the variable label or the variable name if this is not defined. The first variable
    specified in lcols() is assumed to be the study identifier and this is used in the
    table output.

astext(#) specifies the percentage of the graph to be taken up by text. The default
    is 50%, and the percentage must be in the range 10–90.

double allows variables specified in lcols() and rcols() to run over two lines in the
    plot. This option may be of use if long strings are used.

nohet prevents the display of heterogeneity statistics in the graph.

summaryonly shows only summary estimates in the graph. This option may be of use
    for multiple subgroup analyses.

rfdist displays the confidence interval of the approximate predictive distribution of a
    future trial, based on the extent of heterogeneity. This option incorporates uncer-
    tainty in the location and spread of the random-effects distribution using the formula
    $t(\mathrm{df}) \times \sqrt{\mathrm{se2} + \mathrm{tau2}}$, where t is the t distribution with $k-2$ degrees of freedom, se2
    is the squared standard error, and tau2 the heterogeneity statistic. The confidence
    interval is then displayed with lines extending from the diamond. With more than
    3 studies, the distribution is inestimable and effectively infinite. It is thus displayed
    with dotted lines. Where heterogeneity is zero, there is still a slight extension as
    the t statistic is always greater than the corresponding normal deviate. For further
    information, see Higgins and Thompson (2006).

rflevel(#) specifies the coverage (e.g., 90%, 95%, 99%) for the confidence interval of
    the predictive distribution. The default is $S_level. See [U] **20.7 Specifying the
    width of confidence intervals**.

null(#) displays the null line at a user-defined value rather than at 0 or 1.

nulloff removes the null hypothesis line from the graph.

favours(*string* # *string*) applies a label saying something about the treatment effect
    to either side of the graph (strings are separated by the # symbol). This option
    replaces the feature available in b1title in the previous version of metan.

firststats(*string*) and secondstats(*string*) label overall user-defined estimates when
    these have been specified. Labels are displayed in the position usually given to the
    heterogeneity statistics.

boxopt(*marker_options*), diamopt(*line_options*),
    pointopt(*marker_options* | *marker_label_options*) ciopt(*line_options*), and
    olineopt(*line_options*) specify options for the graph routines within the program,
    allowing the user to alter the appearance of the graph. Any options associated with

a particular graph command may be used, except some that would cause incorrect graph appearance. For example, diamonds are plotted using the `twoway pcspike` command, so options for line styles are available (see [G] *line_options*); however, altering the $x - -y$ orientation with the option `horizontal` or `vertical` is not allowed. So, `diamopt(lcolor(green) lwidth(thick))` feeds into a command such as `pcspike(y1 x1 y2 x2, lcolor(green) lwidth(thick))`.

`boxopt`(*marker_options*) controls the boxes and uses options for a weighted marker (e.g., shape and color, but not size). See [G] *marker_options*.

`diamopt`(*line_options*) controls the diamonds and uses options for `twoway pcspike` (not horizontal/vertical). See [G] *line_options*.

`pointopt`(*marker_options* | *marker_label_options*) controls the point estimate by using marker options. See [G] *marker_options* and [G] *marker_label_options*.

`ciopt`(*line_options*) controls the confidence intervals for studies by using options for `twoway pcspike` (not horizontal/vertical). See [G] *line_options*.

`olineopt`(*line_options*) controls the overall effect line with options for another line (not position). See [G] *line_options*.

`classic` specifies that solid black boxes without point estimate markers are used, as in the previous version of `metan`.

`nowarning` switches off the default display of a note warning that studies are weighted from random-effects analyses.

graph_options are any of the options documented in [G] *twoway_options*. These allow the addition of titles, subtitles, captions, etc.; control of margins, plot regions, graph size, and aspect ratio; and the use of schemes. Because titles may be added with *graph_options*, previous options such as `b2title` are no longer necessary.

11   Saved results from metan (macros)

As with many Stata commands, macros are left behind containing the results of the analysis. These include the pooled-effect size and its standard error; or, as described above regarding generated variables, the standard error of the log-effect size for odds and risk ratios. If two methods are specified, by using the option `second()`, some of these are repeated; for example, `r(ES)` and `r(ES_2)` give the pooled-effects estimates for each method. Subgroup statistics when using the `by()` option are not saved; if these are required for storage, it is recommended that a program be written that analyzes subgroups separately (perhaps using the `nograph` and `notable` options).

Name	Second	Description
r(ES)	r(ES_2)	pooled-effect size (if the log option is specified with or or rr, this is the pooled log OR or log RR)
r(seES)	r(seES_2)	standard error of pooled-effect size with symmetrical CI, i.e., mean differences, risk difference, log OR, and log RR using log option
r(selogES)	r(selogES_2)	standard error of log OR or log RR when ORs or RRs are combined without the log option
r(ci_low)	r(ci_low_2)	lower CI of pooled-effect size
r(ci_upp)	r(ci_upp_2)	upper CI of pooled-effect size
r(z)		z-value of effect size
r(p_z)		p-value for significance of effect size
r(het)		chi-squared test for heterogeneity
r(df)		degrees of freedom (number of informative studies minus 1)
r(p_het)		p-value for significance of test for heterogeneity
r(i_sq)		the I^2 statistic
r(tau2)		estimated between-study variance (random-effects analyses only)
r(chi2)		chi-squared test for significance of odds ratio (fixed-effects OR only)
r(p_chi2)		p-value for the above test
r(rger)		overall event rate, group 1 (if binary data are combined)
r(cger)		overall event rate, group 2 (see above)
r(measure)		effect measure (e.g., RR, SMD)
r(method_1)	r(method_2)	analysis method (e.g., MH, DL)

Also, the following variables are added to the dataset by default (to override this use the `nokeep` option):

Variable name	Definition
_ES	Effect size (ES)
_seES	Standard error of ES
_LCI	Lower confidence limit for ES
_UCI	Upper confidence limit for ES
_WT	Study weight
_SS	Study sample size

12 Syntax for funnel

Editorial note: The `funnel` command is no longer distributed with the `metan` package. More recent commands, `metafunnel` and `confunnel`, which use up-to-date Stata graphics and are described later in this collection, are recommended for funnel plots.

funnel [*precision_var effect_size*] [*if*] [*in*] [, *options*]

If the `funnel` command is invoked following `metan` with no parameters specified it will produce a standard funnel plot of precision (1/SE) against treatment effect. Addition of the `noinvert` option will produce a plot of standard error against treatment effect. The alternative sample size version of the funnel plot can be obtained by using the `sample` option (this automatically selects the `noinvert` option). Alternative plots can be created by specifying *precision_var* and *effect_size*. If the effect size is a relative risk or odds ratio, then the `xlog` graph option should be used to create a symmetrical plot.

13 Options for funnel

All options for `graph` are valid. Additionally, the following may be specified:

<u>sample</u> denotes that the y axis is the sample size and not a standard error.

<u>noinvert</u> prevents the values of the precision variable from being inverted.

<u>ysqrt</u> represents the y axis on a square-root scale.

<u>overall</u>(x) draws a dashed vertical line at the overall effect size given by x.

14   Syntax for labbe

labbe *varlist* [*if*] [*in*] [*weight*] [, nowt <u>percent</u> or(#) rd(#) rr(#)
    rrn(#) null logit wgt(*weightvar*) symbol(*symbolstyle*) nolegend id(*idvar*)
    textsize(#) clockvar(*clockvar*) gap(#) *graph_options*]

15   Options for labbe

nowt declares that the plotted data points are to be the same size.

percent displays the event rates as percentages rather than proportions.

or(#) draws a line corresponding to a fixed odds ratio of #.

rd(#) draws a line corresponding to a fixed risk difference of #.

rr(#) draws a line corresponding to a fixed risk ratio of #. See also the rrn() option.

rrn(#) draws a line corresponding to a fixed risk ratio (for the nonevent) of #. The
    rr() and rrn() options may require explanation. Whereas the OR and RD are
    invariant to the definition of which of the binary outcomes is the "event" and which
    is the "nonevent", the RR is not. That is, while the command metan a b c d,
    or gives the same result as metan b a d c, or (with direction changed), an RR
    analysis does not. The L'Abbe plot allows the display of either or both to be
    superimposed risk difference.

null draws a line corresponding to a null effect (i.e., $p1 = p2$).

logit is for use with the or() option; it displays the probabilities on the logit scale,
    i.e., $\log(p/1-p)$. On the logit scale, the odds ratio is a linear effect, making it easier
    to assess the "fit" of the line.

wgt(*weightvar*) specifies alternative weighting by the specified variable; the default is
    sample size.

symbol(*symbolstyle*) allows the symbol to be changed (see [G] *symbolstyle*); the default
    being hollow circles (or points if weights are not used).

nolegend suppresses a legend from being displayed (the default if more than one line
    corresponding to effect measures are specified).

id(*idvar*) displays marker labels with the specified ID variable *idvar*. clockvar() and
    gap() may be used to fine-tune the display, which may become unreadable if studies
    are clustered together in the graph.

textsize(#) increases or decreases the text size of the ID label by specifying # to
    be more or less than unity. The default is usually satisfactory but may need to be
    adjusted.

clockvar(*clockvar*) specifies the position of *idvar* around the study point, as if it were a clock face (values must be integers; see [G] ***clockposstyle***). This option may be used to organize labels where studies are clustered together. By default, labels are positioned to the left (9 o'clock) if above the null and to the right (3 o'clock) if below. Missing values in *clockvar* will be assigned the default position, so this need not be specified for all observations.

gap(*#*) increases or decreases the gap between the study marker and the ID label by specifying *#* to be more or less than unity. The default is usually satisfactory but may need to be adjusted.

graph_options specifies overall graph options that would appear at the end of a twoway graph command. This allows the addition of titles, subtitles, captions, etc., control of margins, plot regions, graph size, aspect ratio, and the use of schemes. See [G] ***twoway_options*** for a list of available options.

16 Example 1: Interventions in smoking cessation

Silagy and Ketteridge (1997) reported a systematic review of randomized controlled trials investigating the effects of physician advice on smoking cessation. In their review, they considered a meta-analysis of trials which have randomized individuals to receive either a minimal smoking cessation intervention from their family doctor or no intervention. An intervention was considered to be "minimal" if it consisted of advice provided by a physician during a single consultation lasting less than 20 minutes (possibly in combination with an information leaflet) with at most one follow-up visit. The outcome of interest was cessation of smoking. The data are presented below:

```
. use example1

. describe

Contains data from example1.dta
  obs:              16
  vars:              6                              17 Jul 2008 13:28
  size:            320 (99.9% of memory free)
```

variable name	storage type	display format	value label	variable label
name	str8	%9s		
year	int	%8.0g		
a	byte	%8.0g		
r1	int	%8.0g		
c	byte	%8.0g		
r2	int	%8.0g		

```
Sorted by:

. list
```

	name	year	a	r1	c	r2
1.	Slama	1990	1	104	1	106
2.	Porter	1972	5	101	4	90
3.	Demers	1990	15	292	5	292
4.	Stewart	1982	11	504	4	187
5.	Page	1986	8	114	5	68
6.	Slama	1995	42	2199	5	929
7.	Haug	1994	20	154	7	109
8.	Russell	1979	34	1031	8	1107
9.	Wilson	1982	21	106	11	105
10.	McDowell	1985	12	85	11	78
11.	Janz	1987	28	144	12	106
12.	Wilson	1990	43	577	17	532
13.	Vetter	1990	34	237	20	234
14.	Higashi	1995	53	468	35	489
15.	Russell	1983	43	761	35	659
16.	Jamrozik	1984	77	512	58	549

We start by producing the data in the format of table 1, and pooling risk ratios by the Mantel–Haenszel fixed-effects method.

```
. generate b = r1-a

. generate d = r2-c

. label var name "Study"

. label var year "Year of publication"
```

```
. metan a b c d, rr lcols(name year) xlabel(0.1,0.2,0.5,1,2,5,10)
> title(Impact of physician advice in smoking cessation) boxsca(50)
          Study       |    RR     [95% Conf. Interval]    % Weight
----------------------+-----------------------------------------------
Slama                 |   1.019    0.065    16.081          0.40
Porter                |   1.114    0.309     4.021          1.71
Demers                |   3.000    1.105     8.147          2.02
Stewart               |   1.020    0.329     3.165          2.36
Page                  |   0.954    0.325     2.800          2.53
Slama                 |   3.549    1.409     8.941          2.84
Haug                  |   2.022    0.886     4.615          3.32
Russell               |   4.563    2.122     9.811          3.12
Wilson                |   1.891    0.960     3.724          4.47
McDowell              |   1.001    0.469     2.137          4.64
Janz                  |   1.718    0.917     3.219          5.59
Wilson                |   2.332    1.347     4.038          7.16
Vetter                |   1.678    0.996     2.829          8.14
Higashi               |   1.582    1.053     2.379         13.85
Russell               |   1.064    0.689     1.642         15.18
Jamrozik              |   1.424    1.035     1.958         22.65
----------------------+-----------------------------------------------
M-H pooled RR         |   1.676    1.440     1.951        100.00
----------------------+-----------------------------------------------

    Heterogeneity chi-squared =   21.51 (d.f. = 15) p = 0.121
    I-squared (variation in RR attributable to heterogeneity) = 30.3%

    Test of RR=1 : z=   6.66 p = 0.000
```

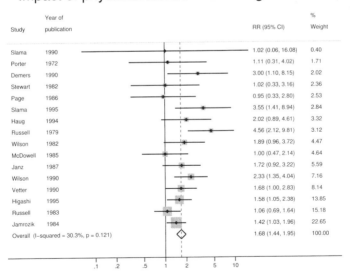

Figure 1. Forest plot for example 1

It appears that there is a significant benefit of such minimal intervention. The non-significance of the test for heterogeneity suggests that the differences between the studies are explicable by random variation, although this test has low statistical power. The L'Abbé plot provides an alternative way of displaying the data which allows inspection of the variability in experimental and control group event rates.

```
. labbe a b c d, xlabel(0,0.1,0.2,0.3) ylabel(0,0.1,0.2,0.3) psize(50)
> t1(Impact of physician advice in smoking cessation:)
> t2(Proportion of patients ceasing to smoke)
> l1(Physician intervention group patients) b2(Control group patients)
(See figure 2 below)
```

A funnel plot can be used to investigate the possibility that the studies which were included in the review were a biased selection. The alternative command `metabias` (Steichen 1998) additionally gives a formal test for nonrandom inclusion of studies in the review.

```
. funnel, xlog ylabel(0,2,4,6) xlabel(0.5,1,2,5) xli(1) overall(1.68)
> b2(Risk Ratio)
(See figure 3 below)
```

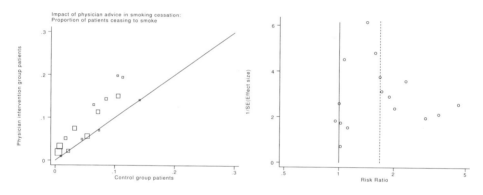

Figure 2. L'Abbé plot for example 1   Figure 3. Funnel plot for example 1

Interpretation of funnel plots can be difficult, as a certain degree of asymmetry is to be expected by chance.

17   Example 2

D'Agostino and Weintraub (1995) reported a meta-analysis of the effects of antihistamines in common cold preparations on the severity of sneezing and runny nose. They combined data from nine randomized trials in which participants with new colds were randomly assigned to an active antihistamine treatment or placebo. The effect of the treatment was measured as the change in severity of runny nose following one day's treatment. The trials used a variety of scales for measuring severity. Due to this, standardized mean differences are used in the analysis. We choose to use Cohen's method (the default) to compute the standardized mean difference.

```
. use example2, clear
. list
```

	n1	mean1	sd1	n2	mean2	sd2
1.	11	.273	.786	16	-.188	.834
2.	128	.932	.593	136	.81	.556
3.	63	.73	.745	64	.578	.773
4.	22	.35	1.139	22	.339	.744
5.	16	.422	2.209	15	-.17	1.374
6.	39	.256	1.666	41	.537	1.614
7.	21	2.831	1.753	21	1.396	1.285
8.	13	2.687	1.607	8	1.625	2.089
9.	194	.49	.895	193	.264	.828

```
. metan n1 mean1 sd1 n2 mean2 sd2, xlabel(-1.5,-1,-0.5,0,0.5,1,1.5)
> title("Effect of antihistamines on cold severity")
> xtitle("Standardized mean difference") textsize(125)
```

Study		SMD	[95% Conf. Interval]		% Weight
1		0.566	-0.218	1.349	2.48
2		0.212	-0.030	0.455	26.01
3		0.200	-0.149	0.549	12.53
4		0.011	-0.580	0.602	4.36
5		0.319	-0.390	1.029	3.03
6		-0.171	-0.611	0.268	7.90
7		0.934	0.295	1.572	3.74
8		0.590	-0.310	1.491	1.88
9		0.262	0.062	0.462	38.06
I-V pooled SMD		0.237	0.113	0.360	100.00

```
  Heterogeneity chi-squared =    9.91 (d.f. = 8) p = 0.271
  I-squared (variation in SMD attributable to heterogeneity) =   19.3%

  Test of SMD=0 : z=   3.76 p = 0.000
```

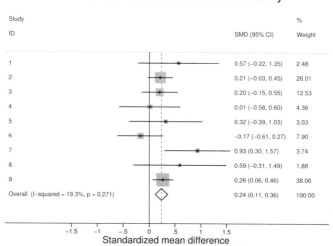

Figure 4. Forest plot for example 2

The patients given antihistamines appear to have a greater reduction in severity of cold symptoms in the first 24 hours of treatment. Again the between-study differences are explicable by random variation.

18   Formulas

19   Individual study responses: binary outcomes

For study i denote the cell counts as in table 1, and let $n_{1i} = a_i + b_i$, $n_{2i} = c_i + d_i$ (the number of participants in the treatment and control groups respectively) and $N_i = n_{1i} + n_{2i}$ (the number in the study). For the Peto method, the individual odds ratios are given by

$$\widehat{\mathrm{OR}}_i = \exp\left[\{a_i - E\left(a_i\right)\}/v_i\right]$$

with its logarithm having standard error

$$\mathrm{se}\{\ln(\widehat{\mathrm{OR}}_i)\} = \sqrt{1/v_i}$$

where $E(a_i) = n_{1i}(a_i + c_i)/N_i$ (the expected number of events in the exposure group) and

$$v_i = \{n_{1i}n_{2i}(a_i + c_i)(b_i + d_i)\}/\{N_i^2(N_i - 1)\} \text{ (the hypergeometric variance of } a_i).$$

For other methods of combining trials, the odds ratio for each study is given by

$$\widehat{\mathrm{OR}}_i = a_i d_i / b_i c_i$$

the standard error of the log odds-ratio being

$$\mathrm{se}\{\ln(\widehat{\mathrm{OR}}_i)\} = \sqrt{1/a_i + 1/b_i + 1/c_i + 1/d_i}$$

The risk ratio for each study is given by

$$\widehat{\mathrm{RR}}_i = (a_i/n_{1i})/(c_i/n_{2i})$$

the standard error of the log risk-ratio being

$$\mathrm{se}\{\ln(\widehat{\mathrm{RR}}_i)\} = \sqrt{1/a_i + 1/c_i - 1/n_{1i} - 1/n_{2i}}$$

The risk difference for each study is given by

$$\widehat{\mathrm{RD}}_i = (a_i/n_{1i}) - (c_i/n_{2i}) \text{ with standard error } \mathrm{se}(\widehat{\mathrm{RD}}_i) = \sqrt{a_i b_i/n_{1i}^3 + c_i d_i/n_{2i}^3}$$

where zero cells cause problems with computation of the standard errors, 0.5 is added to all cells (a_i, b_i, c_i, d_i) for that study.

20 Individual study responses: continuous outcomes

Denote the number of subjects, mean, and standard deviation as in table 1, and let

$$N_i = n_{1i} + n_{2i}$$

and

$$s_i = \sqrt{\{(n_{1i} - 1)sd_{1i}^2 + (n_{2i} - 1)sd_{2i}^2\}/(N_i - 2)}$$

be the pooled standard deviation of the two groups. The weighted mean difference is given by

$$\widehat{\mathrm{WMD}}_i = m_{1i} - m_{2i} \text{ with standard error } \mathrm{se}(\widehat{\mathrm{WMD}}_i) = \sqrt{sd_{1i}^2/n_{1i} + sd_{2i}^2/n_{2i}}$$

There are three formulations of the standardized mean difference. The default is the measure suggested by Cohen (Cohen's d), which is the ratio of the mean difference to the pooled standard deviation s_i; i.e.,

$$\widehat{d}_i = (m_{1i} - m_{2i})/s_i \text{ with standard error } \mathrm{se}(\widehat{d}_i) = \sqrt{N_i/(n_{1i}n_{2i}) + \widehat{d}_i^2/2(N_i - 2)}$$

Hedges suggested a small-sample adjustment to the mean difference (Hedges adjusted g), to give

$$\widehat{g}_i = \{(m_{1i} - m_{2i})/s_i\}\{1 - 3/(4N_i - 9)\} \text{ with standard error }$$
$$\mathrm{se}(\widehat{g}_i) = \sqrt{N_i/(n_{1i}n_{2i}) + \widehat{g}_i^2/2(N_i - 3.94)}$$

Glass suggested using the control group standard deviation as the best estimate of the scaling factor to give the summary measure (Glass's $\widehat{\Delta}$), where

$$\widehat{\Delta}_i = (m_{1i} - m_{2i})/sd_{2i}, \text{ with standard error } \mathrm{se}(\Delta_i) = \sqrt{N_i/(n_{1i}n_{2i}) + \widehat{\Delta}_i^2/2(n_{2i} - 1)}$$

21   Mantel–Haenszel methods for combining trials

For each study, the effect size from each trial $\widehat{\theta}_i$ is given weight w_i in the analysis. The overall estimate of the pooled effect, $\widehat{\theta}_{\text{MH}}$ is given by

$$\widehat{\theta}_{\text{MH}} = (\textstyle\sum w_i \widehat{\theta}_i)/(\sum w_i)$$

For combining odds ratios, each study's OR is given weight

$$w_i = b_i c_i / N_i$$

and the logarithm of $\widehat{\text{OR}}_{\text{MH}}$ has standard error given by

$$\text{se}\{\ln(\widehat{\text{OR}}_{\text{MH}})\} = \sqrt{(PR)/2R^2 + \{(PS+QR)/2(R \times S)\} + (QS)/2S^2}$$

where

$$R = \textstyle\sum a_i d_i / N_i \qquad S = \sum b_i c_i / N_i$$
$$PR = \textstyle\sum (a_i + d_i) a_i d_i / N_i^2 \qquad PS = \sum (a_i + d_i) b_i c_i / N_i^2$$
$$QR = \textstyle\sum (b_i + c_i) a_i d_i / N_i^2 \qquad QS = \sum (b_i + c_i) b_i c_i / N_i^2$$

For combining risk ratios, each study's RR is given weight

$$w_i = (n_{1i} c_i)/N_i$$

and the logarithm of $\widehat{\text{RR}}_{\text{MH}}$ has standard error given by

$$\text{se}\{\ln(\widehat{\text{RR}}_{\text{MH}})\} = \sqrt{P/(R \times S)}$$

where

$$P = \textstyle\sum (n_{1i} n_{2i}(a_i + c_i) - a_i c_i N_i)/N_i^2 \qquad R = \sum a_i n_{2i}/N_i \qquad S = \sum c_i n_{1i}/N_i$$

For risk differences, each study's RD has the weight

$$w_i = n_{1i} n_{2i}/N_i$$

and $\widehat{\text{RD}}_{\text{MH}}$ has standard error given by

$$\text{se}\{\widehat{\text{RD}}_{\text{MH}}\} = \sqrt{(P/Q^2)}$$

where

$$P = \textstyle\sum (a_i b_i n_{2i}^3 + c_i d_i n_{1i}^3)/n_{1i} n_{2i} N_i^2; \qquad Q = \sum n_{1i} n_{2i}/N_i$$

The heterogeneity statistic is given by

$$Q = \textstyle\sum w_i (\widehat{\theta}_i - \widehat{\theta}_{\text{MH}})^2$$

where θ is the log odds-ratio, log relative-risk, or risk difference. Under the null hypothesis that there are no differences in treatment effect between trials, this follows a χ^2 distribution on $k - 1$ degrees of freedom.

22   Inverse variance methods for combining trials

Here, when considering odds ratios or risk ratios, we define the effect size θ_i to be the natural logarithm of the trial's OR or RR; otherwise, we consider the summary statistic (RD, SMD, or WMD) itself. The individual effect sizes are weighted according to the reciprocal of their variance (calculated as the square of the standard errors given in the individual study section above) giving

$$w_i = 1/\text{se}(\widehat{\theta}_i)^2$$

These are combined to give a pooled estimate

$$\widehat{\theta}_{\text{IV}} = (\textstyle\sum w_i \widehat{\theta}_i)/(\sum w_i)$$

with

$$\text{se}\{\widehat{\theta}_{\text{IV}}\} = 1/\sqrt{\textstyle\sum w_i}$$

The heterogeneity statistic is given by a similar formula as for the Mantel–Haenszel method, using the inverse variance form of the weights, w_i

$$Q = \textstyle\sum w_i(\widehat{\theta}_i - \widehat{\theta}_{\text{IV}})^2$$

23   Peto's assumption free method for combining trials

Here, the overall odds ratio is given by

$$\widehat{\text{OR}}_{\text{Peto}} = \exp\{\textstyle\sum w_i \ln(\widehat{\text{OR}}_i)/\sum w_i\}$$

where the odds ratio $\widehat{\text{OR}}_i$ is calculated using the approximate method described in the individual trial section, and the weights, w_i are equal to the hypergeometric variances, v_i.

The logarithm of the odds ratio has standard error

$$\text{se}\{\ln(\widehat{\text{OR}}_{\text{Peto}})\} = 1/\sqrt{\textstyle\sum w_i}$$

The heterogeneity statistic is given by

$$Q = \textstyle\sum w_i\{(\ln \widehat{\text{OR}}_i)^2 - (\ln \widehat{\text{OR}}_{\text{Peto}})^2\}$$

24   DerSimonian and Laird random-effects models

Under the random-effects model, the assumption of a common treatment effect is relaxed, and the effect sizes are assumed to have a distribution

$$\theta_i \sim N(\theta, \tau^2)$$

The estimate of τ^2 is given by

$$\widehat{\tau}^2 = \max[\{Q - (k-1)\}/\{\textstyle\sum w_i - (\sum(w_i^2)/\sum w_i)\}, 0]$$

The estimate of the combined effect for heterogeneity may be taken as either the Mantel–Haenszel or the inverse variance estimate. Again, for odds ratios and risk ratios, the effect size is taken as the natural logarithm of the OR and RR. Each study's effect size is given weight

$$w_i = 1/\{\mathrm{se}(\widehat{\theta}_i)^2 + \widehat{\tau}^2\}$$

The pooled effect size is given by

$$\widehat{\theta}_{\mathrm{DL}} = (\sum w_i \widehat{\theta}_i)/(\sum w_i)$$

and

$$\mathrm{se}\{\widehat{\theta}_{\mathrm{DL}}\} = 1/\sqrt{\sum w_i}$$

Note that in the case where the heterogeneity statistic Q is less than or equal to its degrees of freedom $(k-1)$, the estimate of the between trial variation, $\widehat{\tau}^2$, is zero, and the weights reduce to those given by the inverse variance method.

25   Confidence intervals

The $100(1-\alpha)\%$ confidence interval for $\widehat{\theta}$ is given by

$$\widehat{\theta} - \mathrm{se}(\widehat{\theta})\Phi(1-\alpha/2), \quad \text{to} \quad \widehat{\theta} + \mathrm{se}(\widehat{\theta})\Phi(1-\alpha/2)$$

where $\widehat{\theta}$ is the log odds-ratio, log relative-risk, risk difference, mean difference, or standardized mean difference, and Φ is the standard normal distribution function. The Cornfield confidence intervals for odds ratios are calculated as explained in the Stata manual for the `epitab` command.

26   Test statistics

In all cases, the test statistic is given by

$$z = \widehat{\theta}/\mathrm{se}(\widehat{\theta})$$

where the odds ratio or risk ratio is again considered on the log scale.

For odds ratios pooled by method of Mantel and Haenszel or Peto, an alternative test statistic is available, which is the χ^2 test of the observed and expected events rate in the exposure group. The expectation and the variance of a_i are as given earlier in the Peto odds-ratio section. The test statistic is

$$\chi^2 = [\sum\{a_i - E(a_i)\}]^2/\sum \mathrm{Var}(a_i)$$

on one degree of freedom. Note that in the case of odds ratios pooled by method of Peto, the two test statistics are identical; the χ^2 test statistic is simply the square of the z score.

27 Acknowledgments

The statistical methods programmed in `metan` utilize several of the algorithms used by the MetaView software (part of the Cochrane Library), which was developed by Gordon Dooley of Update Software, Oxford and Jonathan Deeks of the Statistical Methods Working Group of the Cochrane Collaboration. We have also used a subroutine written by Patrick Royston of the Royal Postgraduate Medical School, London.

28 References

Breslow, N. E., and N. E. Day. 1993. *Statistical Methods in Cancer Research: Volume I—The Analysis of Case–Control Studies.* Lyon: International Agency for Research on Cancer.

D'Agostino, R. B., and M. Weintraub. 1995. Meta-analysis: A method for synthesizing research. *Clinical Pharmacology and Therapeutics* 58: 605–616.

DerSimonian, R., and N. Laird. 1986. Meta-analysis in clinical trials. *Controlled Clinical Trials* 7: 177–188.

Egger, M., G. Davey Smith, M. Schneider, and C. Minder. 1997. Bias in meta-analysis detected by a simple, graphical test. *British Medical Journal* 315: 629–634.

Fleiss, J. L. 1993. The statistical basis of meta-analysis. *Statistical Methods in Medical Research* 2: 121–145.

Glass, G. V., B. McGaw, and M. L. Smith. 1981. *Meta-Analysis in Social Research.* Beverly Hills, CA: Sage.

Greenland, S., and J. Robins. 1985. Estimation of a common effect parameter from sparse follow-up data. *Biometrics* 41: 55–68.

Greenland, S., and A. Salvan. 1990. Bias in the one-step method for pooling study results. *Statistics in Medicine* 9: 247–252.

Hedges, L. V., and I. Olkin. 1985. *Statistical Methods for Meta-analysis.* San Diego: Academic Press.

L'Abbé, K. A., A. S. Detsky, and K. O'Rourke. 1987. Meta-analysis in clinical research. *Annals of Internal Medicine* 107: 224–233.

Mantel, N., and W. Haenszel. 1959. Statistical aspects of the analysis of data from retrospective studies of disease. *Journal of the National Cancer Institute* 22: 719–748.

Robins, J., S. Greenland, and N. E. Breslow. 1986. A general estimator for the variance of the Mantel–Haenszel odds ratio. *American Journal of Epidemiology* 124: 719–723.

Rosenthal, R. 1994. Parametric measures of effect size. In *The Handbook of Research Synthesis*, ed. H. Cooper and L. V. Hedges. New York: Russell Sage Foundation.

Silagy, C., and S. Ketteridge. 1997. Physician advice for smoking cessation. In *Tobacco Addiction Module of the Cochrane Database of Systematic Reviews*, ed. T. Lancaster, C. Silagy, and D. Fullerton. Oxford: The Cochrane Collaboration. Available in the Cochrane Library (subscription database and CDROM), issue 4.

Sinclair, J. C., and M. B. Bracken. 1992. *Effective Care of the Newborn Infant*. Oxford: Oxford University Press.

Steichen, T. J. 1998. sbe19: Tests for publication bias in meta-analysis. *Stata Technical Bulletin* 41: 9–15. Reprinted in *Stata Technical Bulletin Reprints*, vol. 7, pp. 125–133. College Station, TX: Stata Press. (Reprinted in this collection on pp. 166–176.)

Yusuf, S., R. Peto, J. Lewis, R. Collins, and P. Sleight. 1985. Beta blockade during and after myocardial infarction: An overview of the randomized trials. *Progress in Cardiovascular Diseases* 27: 335–371.

The Stata Journal (2008)
8, Number 1, sbe24_2, pp. 3–28

metan: fixed- and random-effects meta-analysis

Ross J. Harris
Centre for Infections
Health Protection Agency
London, UK
ross.harris@hpa.org.uk

Michael J. Bradburn
Clinical Trials Research Unit
ScHARR
Sheffield, UK
m.bradburn@sheffield.ac.uk

Jonathan J. Deeks
Unit of Public Health, Epidemiology, and Biostatistics
University of Birmingham
Birmingham, UK
j.deeks@bham.ac.uk

Roger M. Harbord
Department of Social Medicine
University of Bristol
Bristol, UK

Douglas G. Altman
Centre for Statistics in Medicine
University of Oxford
Oxford, UK

Jonathan A. C. Sterne
Department of Social Medicine
University of Bristol
Bristol, UK

Abstract. This article describes updates of the meta-analysis command `metan` and options that have been added since the command's original publication (Bradburn, Deeks, and Altman, metan – an alternative meta-analysis command, *Stata Technical Bulletin Reprints*, vol. 8, pp. 86–100). These include version 9 graphics with flexible display options, the ability to meta-analyze precalculated effect estimates, and the ability to analyze subgroups by using the `by()` option. Changes to the output, saved variables, and saved results are also described.

Keywords: sbe24_2, metan, meta-analysis, forest plot

1   Introduction

Meta-analysis is a two-stage process involving the estimation of an appropriate summary statistic for each of a set of studies followed by the calculation of a weighted average of these statistics across the studies (Deeks, Altman, and Bradburn 2001). Odds ratios, risk ratios, and risk differences may be calculated from binary data, or a difference in means obtained from continuous data. Alternatively, precalculated effect estimates and their standard errors from each study may be pooled, for example, adjusted log odds-ratios from observational studies. The summary statistics from each study can be combined by using a variety of meta-analytic methods, which are classified as fixed-effects models in which studies are weighted according to the amount of information they contain; or random-effects models, which incorporate an estimate of between-study variation (heterogeneity) in the weighting. A meta-analysis will customarily include a forest plot, in which results from each study are displayed as a square and a horizontal line, representing the intervention effect estimate together with its confidence interval. The area of the square reflects the weight that the study contributes to the meta-analysis. The combined-effect estimate and its confidence interval are represented by a diamond.

Here we present updates to the `metan` command and other previously undocumented additions that have been made since its original publication (Bradburn, Deeks, and Altman 1998). New features include

- Version 9 graphics

- Flexible display of tabular data in the forest plot

- Results from a second type of meta-analysis displayed in the same forest plot

- `by()` group processing

- Analysis of precalculated effect estimates

- Prediction intervals for the intervention effect in a new study from random-effects analyses

There are a substantial number of options for the `metan` command because of the variety of meta-analytic techniques and the need for flexible graphical displays. We recommend that new users not try to learn everything at once but to learn the basics and build from there as required. Clickable examples of `metan` are available in the help file, and the dialog box may also be a good way to start using `metan`.

2   Example data

The dataset used in subsequent examples is taken from the meta-analysis published as table 1 in Colditz et al. (1994, 699). The aim of the analysis was to quantify the efficacy of BCG vaccine against tuberculosis, and data from 11 trials are included here. There was considerable between-trial heterogeneity in the effect of the vaccine; it has been suggested that this might be explained by the latitude of the region in which the trial was conducted (Fine 1995).

▷ **Example**

Details of the dataset are shown below by using `describe` and `list` commands.

```
. use bcgtrial
(BCG and tuberculosis)

. describe

Contains data from bcgtrial.dta
  obs:             11                          BCG and tuberculosis
  vars:            12                          31 May 2007 17:11
  size:           693 (99.9% of memory free)   (_dta has notes)
```

variable name	storage type	display format	value label	variable label
trial	byte	%8.0g		Trial number
trialnam	str14	%14s		Trial name
authors	str20	%20s		Authors of trial
startyr	int	%8.0g		Year trial started
latitude	byte	%8.0g		Latitude of trial area
alloc	byte	%33.0g	alloc	Allocation method
tcases	int	%8.0g		BCG vaccinated cases
tnoncases	float	%9.0g		BCG vaccinated noncases
ccases	int	%8.0g		Unvaccinated cases
cnoncases	float	%9.0g		Unvaccinated noncases
ttotal	long	%12.0g		BCG vaccinated population
ctotal	long	%12.0g		Unvaccinated population

```
Sorted by:  startyr  authors

. list trialnam startyr tcases tnoncases ccases cnoncases, clean noobs
> abbreviate(10)
```

trialnam	startyr	tcases	tnoncases	ccases	cnoncases
Canada	1933	6	300	29	274
Northern USA	1935	4	119	11	128
Chicago	1941	17	1699	65	1600
Georgia (Sch)	1947	5	2493	3	2338
Puerto Rico	1949	186	50448	141	27197
Georgia (Comm)	1950	27	16886	29	17825
Madanapalle	1950	33	5036	47	5761
UK	1950	62	13536	248	12619
South Africa	1965	29	7470	45	7232
Haiti	1965	8	2537	10	619
Madras	1968	505	87886	499	87892

Trial name and number identify each study, and we have information on the authors and the year the trial started. There are also two variables relating to study characteristics: the latitude of the area in which the trial was carried out, and the method of allocating patients to the vaccine and control groups—either at random or in some systematic way. The variables tcases, tnoncases, ccases, and cnoncases contain the data from the 2×2 table from each study (the number of cases and noncases in the vaccination group and nonvaccination group). The variables ttotal and ctotal are the total number of individuals (the sum of the cases and noncases) in the vaccine and control groups. Displayed below is the 2×2 table for the first study (Canada, 1933):

	cases	noncases	total
treated	6	300	306
control	29	274	303

The risk ratio (RR), log risk-ratio (log RR), standard error of log RR (SE log RR), 95% confidence interval (CI) for log RR, and 95% CI for RR may be calculated as follows (see, for example, Kirkwood and Sterne 2003).

$$\text{Risk in treated population} = \frac{\texttt{tcases}}{\texttt{ttotal}} = \frac{6}{306} = 0.0196$$

$$\text{Risk in control population} = \frac{\texttt{ccases}}{\texttt{ctotal}} = \frac{29}{303} = 0.0957$$

$$\text{RR} = \frac{\text{Risk in treated population}}{\text{Risk in control population}} = \frac{0.0196}{0.0957} = 0.2049$$

$$\log \text{RR} = \log(\text{RR}) = -1.585$$

$$\text{SE}(\log \text{RR}) = \sqrt{\frac{1}{\texttt{tcases}} + \frac{1}{\texttt{ccases}} - \frac{1}{\texttt{ttotal}} - \frac{1}{\texttt{ctotal}}}$$

$$= \sqrt{\frac{1}{6} + \frac{1}{29} - \frac{1}{306} - \frac{1}{303}} = 0.441$$

$$95\% \text{ CI for } \log \text{RR} = \log \text{RR} \pm 1.96 \times \text{SE}(\log \text{RR}) = -2.450 \text{ to } -0.720$$

$$95\% \text{ CI for RR} = \exp(-2.450) \text{ to } \exp(-0.720) = 0.086 \text{ to } 0.486$$

◁

3   Syntax

metan *varlist* $\big[$ *if* $\big]$ $\big[$ *in* $\big]$ $\big[$, $\big[$ *binary_data_options* | *continuous_data_options* |
   precalculated_effect_estimates_options $\big]$ *measure_and_model_options output_options*
   forest_plot_options $\big]$

binary_data_options

   or rr rd fixed random fixedi randomi peto cornfield chi2
   breslow <u>noint</u>eger cc(#)

continuous_data_options

   cohen hedges glass nostandard fixed random <u>noint</u>eger

precalculated_effect_estimates_options

    fixed random

measure_and_model_options

    wgt(*wgtvar*) second(*model | estimates_and_description*)
    first(*estimates_and_description*)

output_options

    by(*byvar*) nosubgroup sgweight log eform efficacy ilevel(*#*)
    olevel(*#*) sortby(*varlist*)
    label([namevar = *namevar*], [yearvar = *yearvar*]) nokeep notable
    nograph nosecsub

forest_plot_options

    xlabel(*#*, ...) xtick(*#*, ...) boxsca(*#*) textsize(*#*) nobox
    nooverall nowt nostats counts group1(*string*) group2(*string*)
    effect(*string*) force lcols(*varlist*) rcols(*varlist*) astext(*#*) double
    nohet summaryonly rfdist rflevel(*#*) null(*#*) nulloff
    favours(*string # string*) firststats(*string*) secondstats(*string*)
    boxopt(*marker_options*) diamopt(*line_options*)
    pointopt(*marker_options | marker_label_options*) ciopt(*line_options*)
    olineopt(*line_options*) classic nowarning *graph_options*

For a full description of the syntax, see Bradburn, Deeks, and Altman (1998). We will focus on the new options, most of which come under *forest_plot_options*; previously undocumented options such as by() (and related options), breslow, cc(), nointeger; and changes to the output such as the display of the I^2 statistic. Syntax will be explained in the appropriate sections.

4 Basic use

4.1 2 × 2 data

For binary data, the input variables required by metan should contain the cells of the 2×2 table; i.e., the number of individuals who did and did not experience the outcome event in the treatment and control groups for each study. When analyzing 2×2 data a range of methods are available. The default is the Mantel–Haenszel method (fixed). The inverse-variance fixed-effects method (fixedi) or the Peto method for estimating summary odds ratios (peto) may also be chosen. The DerSimonian and Laird random-effects method may be specified with random. See Deeks, Altman, and Bradburn (2001) for a discussion of these methods.

4.2   Display options

Previous versions of the `metan` command used the syntax `label(namevar` = *namevar*,
`yearvar` = *yearvar*) to specify study information in the table and forest plot. This
syntax still functions but has been superseded by the more flexible `lcols(`*varlist*`)` and
`rcols(`*varlist*`)` options. The use of these options is described in more detail in section 5.
The option `favours(`*string* # *string*`)` allows the user to display text information about
the direction of the treatment effect, which appears under the graph (e.g., exposure
good, exposure bad). `favours()` replaces the option `b2title()`. The # is required to
split the two strings, which appear to either side of the null line.

▷ **Example**

Here we use `metan` to derive an inverse-variance weighted (fixed effect) meta-analysis
of the BCG trial data. Risk ratios are specified as the summary statistic, and the trial
name and the year the trial started are displayed in the forest plot using `lcols()` (see
section 5).

```
. metan tcases tnoncases ccases cnoncases, rr fixedi lcols(trialnam startyr)
> xlabel(0.1, 10) favours(BCG reduces risk of TB # BCG increases risk of TB)
```

Study	RR	[95% Conf.	Interval]	% Weight
Canada	0.205	0.086	0.486	1.11
Northern USA	0.411	0.134	1.257	0.66
Chicago	0.254	0.149	0.431	2.96
Georgia (Sch)	1.562	0.374	6.528	0.41
Puerto Rico	0.712	0.573	0.886	17.42
Georgia (Comm)	0.983	0.582	1.659	3.03
Madanapalle	0.804	0.516	1.254	4.22
UK	0.237	0.179	0.312	10.81
South Africa	0.625	0.393	0.996	3.83
Haiti	0.198	0.078	0.499	0.97
Madras	1.012	0.895	1.145	54.58
I-V pooled RR	0.730	0.667	0.800	100.00

```
Heterogeneity chi-squared = 125.63 (d.f. = 10) p = 0.000
I-squared (variation in RR attributable to heterogeneity) =  92.0%

Test of RR=1 : z=   6.75 p = 0.000
```

The output table contains effect estimates (here RRs), CIs, and weights for each study,
followed by the overall (combined) effect estimate. The results for the Canada study are
identical to those derived in section 2. Heterogeneity statistics relating to the extent that
RRs vary between studies are displayed, including the I^2 statistic, which is a previously
undocumented addition. The I^2 statistic (see section 9.1) is the percentage of between-
study heterogeneity that is attributable to variability in the true treatment effect, rather
than sampling variation (Higgins and Thompson 2004, Higgins et al. 2003). Here there
is substantial between-study heterogeneity. Finally, a test of the null hypothesis that
the vaccine has no effect (RR=1) is displayed. There is strong evidence against the null
hypothesis, but the presence of between-study heterogeneity means that the fixed-effects

assumption (that the true treatment effect is the same in each study) is incorrect. The forest plot displayed by the command is shown in figure 1.

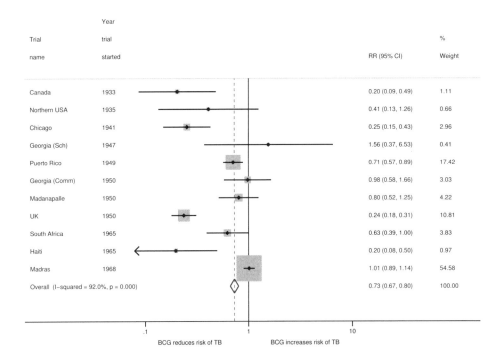

Figure 1. Forest plot displaying an inverse-variance weighted fixed-effects meta-analysis of the effect of BCG vaccine on incidence of tuberculosis.

◁

4.3 Precalculated effect estimates

The `metan` command may also be used to meta-analyze precalculated effect estimates, such as log odds-ratios and their standard errors or 95% CI, using syntax similar to the alternative Stata meta-analysis command `meta` (Sharp and Sterne 1997, 1998). Here only the inverse-variance fixed-effects and DerSimonian and Laird random-effects methods are available, because other methods require the 2×2 cell counts or the means and standard deviations in each group. The `fixed` option produces an inverse-variance weighted analysis when precalculated effect estimates are analyzed.

When analyzing ratio measures (RRs or odds ratios), the log ratio with its standard error or 95% CI should be used as inputs to the command. The `eform` option can then be used to display the output on the ratio scale (as for the `meta` command).

▷ **Example**

We will illustrate this feature by generating the log RR and its standard error in each study from the 2×2 data, and then by meta-analyzing these variables.

```
. gen logRR = ln( (tcases/ttotal) / (ccases/ctotal) )
. gen selogRR = sqrt( 1/tcases +1/ccases -1/ttotal -1/ctotal )
. metan logRR selogRR, fixed eform nograph
```

Study	ES	[95% Conf. Interval]	% Weight
(table of study results omitted)			
I-V pooled ES	0.730	0.667 0.800	100.00

```
Heterogeneity chi-squared = 125.63 (d.f. = 10) p = 0.000
I-squared (variation in ES attributable to heterogeneity) =  92.0%
Test of ES=1 : z=   6.75 p = 0.000
```

The results are identical to those derived directly from the 2×2 data in section 4.1; we would have observed minor differences if the default Mantel–Haenszel method had been used previously. When analyzing precalculated estimates, `metan` does not know what these measures are, so the summary estimate is named "ES" (effect size) in the output.

◁

4.4   Specifying two analyses

`metan` now allows the display of a second meta-analytic estimate in the same output table and forest plot. A typical use is to compare fixed-effects and random-effects analyses, which can reveal the presence of small-study effects. These may result from publication or other biases (Sterne, Gavaghan, and Egger 2000). See Poole and Greenland (1999) for a discussion of the ways in which fixed-effects and random-effects analyses may differ. The syntax is to specify the method for the second meta-analytic estimate as `second(`*method*`)`, where *method* is any of the standard `metan` options.

▷ **Example**

Here we use `metan` to analyze 2×2 data as in section 4.1, specifying an inverse-variance weighted (fixed effect) model for the first method and a DerSimonian and Laird (random effects) model for the second method:

```
. metan tcases tnoncases ccases cnoncases, rr fixedi second(random)
> lcols(trialnam startyr) nograph
```

Study	RR	[95% Conf. Interval]	% Weight
(table of study results omitted)			
I-V pooled RR	0.730	0.667 0.800	100.00
D+L pooled RR	0.508	0.336 0.769	100.00

```
Heterogeneity chi-squared = 125.63 (d.f. = 10) p = 0.000
I-squared (variation in RR attributable to heterogeneity) =  92.0%

Test of RR=1 : z=   6.75 p = 0.000
```

The results of the second analysis are displayed in the table: a forest plot using the second() option is derived in the next section and displayed in figure 2. The protective effect of BCG against tuberculosis appears greater in the random-effects analysis than in the fixed-effects analysis, although CI is wider. This reflects the greater uncertainty in the random-effects analysis, which allows for the true effect of the vaccine to vary between studies. Random-effects analyses give relatively greater weight to smaller studies than fixed-effects analyses, and so these results suggest that the estimated effect of BCG was greater in the smaller studies. It is also possible to supply a precalculated pooled-effect estimate with second(); see section 7.2 for details.

◁

5   Displaying data columns in graphs

The options lcols(*varlist*) and rcols(*varlist*) produce columns to the left or right of the forest plot. String (character) or numeric variables can be displayed. If numeric variables have value labels, these will be displayed in the graph. If the variable itself is labeled, this will be used as the column header, allowing meaningful names to be used. Up to four lines are used for the heading, so names can be long without taking up too much graph width.

The first variable in lcols() is used to identify studies in the table output, and summary statistics and study weight are always the first columns on the right of the forest plot. These can be switched off by using the options nostats and nowt, but the order cannot be changed.

If lengthy string variables are to be displayed, the double option may be used to allow output to spread over two lines per study in the forest plot. The percentage of the forest plot given to text may be adjusted using astext(#), which can be between 10 and 90 (the default is 50).

A previously undocumented option that affects columns is counts. When this option is specified, more columns will appear on the right of the graph displaying the raw data; either the 2×2 table for binary data or the sample size, mean, and standard deviation in each group if the data are continuous. The groups may be labeled by using group1(*string*) and group2(*string*), although the defaults *Treatment* and *Control* will often be acceptable for the analysis of randomized controlled trials (RCTs).

▷ **Example**

We now present an example command that uses these features, as well as the `second()` option. The resulting forest plot is displayed in figure 2:

```
. metan tcases tnoncases ccases cnoncases, rr fixedi second(random)
> lcols(trialnam authors startyr alloc latitude) counts astext(70)
> textsize(200) boxsca(80) xlabel(0.1,10) notable xsize(10) ysize(6)
```

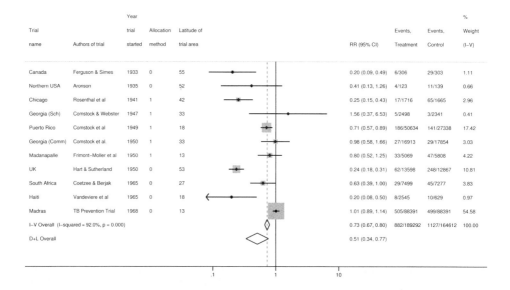

Figure 2. Forest plot displaying an inverse-variance weighted fixed-effects meta-analysis of the effect of BCG vaccine on incidence of tuberculosis. Columns of data are displayed in the plot.

Note the specification of x-axis labels and text and box sizes. The graph is also reshaped by using the standard Stata graph options `xsize()` and `ysize()`; see section 10.2 for more details. Box and text sizes are expressed as a percentage of standard size with the default as 100, such that 50 will halve the size and 200 will double it.

◁

6   by() processing

A major addition to `metan` is the ability to perform stratified or subgroup analyses. These may be used to investigate the possibility that treatment effects vary between subgroups; however, formal comparisons between subgroups are best performed by using meta-regression; see Harbord and Higgins (2008) or Higgins and Thompson (2004). We may also want to display results for different groups of studies in the same plot, even though it is inappropriate to meta-analyze across these groups.

6.1 Syntax and options for by()

nooverall specifies that the overall estimate not be displayed, for example, when it is inappropriate to meta-analyze across groups.

sgweight requests that weights be displayed such that they sum to 100% within each subgroup. This option is invoked automatically with nooverall.

nosubgroup specifies that studies be arranged by the subgroup specified, but estimates for each subgroup not be displayed.

nosecsub specifies that subestimates using the method defined by second() not be displayed.

summaryonly specifies that individual study estimates not be displayed, for example, to produce a summary of different groups in a compact graph.

▷ **Example**

Fine (1995) suggested that there is a relationship between the effect of BCG and the latitude of the area in which the trial was conducted. Here we may want to use meta-regression to further investigate this tendency (see Harbord and Higgins 2008). To illustrate the by() option, we will classify the studies into three groups defined by latitude. We define these groups as tropical (≤ 23.5 degrees), midlatitude (between 23.5 and 40 degrees), and northern (≥ 40 degrees).

```
. gen lat_cat = ""
(11 missing values generated)
. replace lat_cat = "Tropical, < 23.5 latitude" if latitude <= 23.5
lat_cat was str1 now str27
(4 real changes made)
. replace lat_cat = "23.5-40 latitude" if latitude > 23.5 & latitude < 40
(3 real changes made)
. replace lat_cat = "Northern, > 40 latitude" if latitude >= 40 & latitude < .
(4 real changes made)
. assert lat_cat != ""
. label var lat_cat "Latitude region"
```

```
. metan tcases tnoncases ccases cnoncases, rr fixedi second(random) nosecsub
> lcols(trialnam startyr latitude) astext(60) by(lat_cat) xlabel(0.1,10)
> xsize(10) ysize(8)
```

Study	RR	[95% Conf. Interval]		% Weight
Northern, > 40 lat				
Canada	0.205	0.086	0.486	1.11
Northern USA	0.411	0.134	1.257	0.66
Chicago	0.254	0.149	0.431	2.96
UK	0.237	0.179	0.312	10.81
Sub-total				
I-V pooled RR	0.243	0.193	0.306	15.54
23.5-40 latitude				
Georgia (Sch)	1.562	0.374	6.528	0.41
Georgia (Comm)	0.983	0.582	1.659	3.03
South Africa	0.625	0.393	0.996	3.83
Sub-total				
I-V pooled RR	0.795	0.567	1.114	7.27
Tropical, < 23.5 l				
Puerto Rico	0.712	0.573	0.886	17.42
Madanapalle	0.804	0.516	1.254	4.22
Haiti	0.198	0.078	0.499	0.97
Madras	1.012	0.895	1.145	54.58
Sub-total				
I-V pooled RR	0.904	0.815	1.003	77.19
Overall				
I-V pooled RR	0.730	0.667	0.800	100.00
D+L pooled RR	0.508	0.336	0.769	

```
Test(s) of heterogeneity:
                Heterogeneity   degrees of
                  statistic      freedom      P      I-squared**
Northern, > 40 lat   1.06           3        0.787      0.0%
23.5-40 latitude     2.51           2        0.285     20.2%
Tropical, < 23.5 l  18.42           3        0.000     83.7%
Overall            125.63          10        0.000     92.0%
Overall Test for heterogeneity between sub-groups:
                   103.64           2        0.000
```

** I-squared: the variation in RR attributable to heterogeneity)

Considerable heterogeneity observed (up to 83.7%) in one or more sub-groups,
Test for heterogeneity between sub-groups likely to be invalid

Significance test(s) of RR=1

```
Northern, > 40 lat   z= 12.00     p = 0.000
23.5-40 latitude     z=  1.33     p = 0.183
Tropical, < 23.5 l   z=  1.90     p = 0.058
Overall              z=  6.75     p = 0.000
```

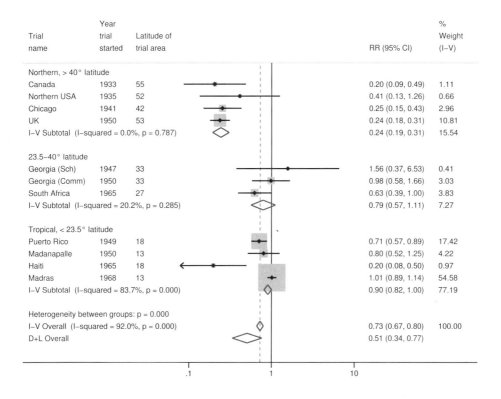

Figure 3. Forest plot displaying an inverse-variance weighted fixed-effects meta-analysis of the effect of BCG vaccine on incidence of tuberculosis. Results are stratified by latitude region, and the overall random-effects estimate is also displayed.

The output table is now stratified by latitude group, and pooled estimates for each group are displayed. Tests of heterogeneity and the null hypothesis are displayed for each group and overall. With the inverse-variance method, a test of heterogeneity between groups is also displayed; note the warning in the output that the test may be invalid because of within-subgroup heterogeneity. Output is similar in the forest plot, displayed in figure 3. Examining each subgroup in turn, it appears that much of the heterogeneity is accounted for by latitude: for two of the groups there is little or no evidence of heterogeneity. The only group to show a strong treatment effect is the ≥40 degree group.

◁

The test for between-group heterogeneity is an issue of current debate, as it is strictly valid only when using the fixed-effects inverse-variance method, and p-values will be too small if there is heterogeneity within any of the subgroups. Therefore, the test is performed only with the inverse-variance method (`fixedi`), and warnings will appear if there is evidence of within-group heterogeneity. Despite these caveats, this method is better than other, seriously flawed, methods such as testing the significance of a treatment effect in each group rather than testing for differences between the groups. As explained at the start of this section, meta-regression is the best way to examine and test for between-group differences.

7   User-defined analyses

7.1   Study weights

The `wgt(`*wgtvar*`)` option allows the studies to be combined by using specific weights that are defined by the variable *wgtvar*. The user must ensure that the weights chosen are meaningful. Typical uses are when analyzing precalculated effect estimates that require weights that are not based on standard error or to assess the robustness of conclusions by assigning alternative weights.

7.2   Pooled estimates

Pooled estimates may be derived by using another package and presented in a forest plot by using the `first()` option to supply these to the `metan` command. Here `wgt(`*wgtvar*`)` is used merely to specify box sizes in the forest plot, no heterogeneity statistics are produced, and no values are returned. When using this feature, stratified analyses are not allowed.

An alternative method is to provide the user-supplied meta-analytic estimate by using the `second()` option. Data are analyzed by using standard methods, and the resulting pooled estimate is displayed together with the user-defined estimate (which need not be derived by using `metan`), allowing a comparison. When using this feature, the option `nosecsub` is invoked, as stratification using the user-defined method is not possible.

When these options are specified, the user must supply the pooled estimate with its standard error or CI and a method label. The user may also supply text to be displayed at the bottom of the forest plot, in the position normally given to heterogeneity statistics, using `firststats(`*string*`)` and `secondstats(`*string*`)`.

▷ **Example**

The BCG data were analyzed by using a fully Bayesian random-effects model with WinBUGS software (Lunn et al. 2000). This analysis used the methods described by Warn, Thompson, and Spiegelhalter (2002) to deal with RRs. The chosen model incorporated a noninformative prior (mean 0, precision 0.001). The resulting RR of 0.518 (95% CI: 0.300, 0.824) is similar to that derived from a DerSimonian and Laird random-effects analysis. However, the CI from the Bayesian analysis is wider, because it allows for the uncertainty in estimating the between-study variance. The following syntax supplies the summary estimates in `second()` and compares this result with the random-effects analysis. The resulting forest plot is displayed in figure 4.

```
. metan logRR selogRR, random second(-.6587 -1.205 -.1937 Bayes)
> secondstats(Noninformative prior: d~dnorm(0.0, 0.001)) eform
> notable astext(60) textsize(130) lcols(trialnam startyr latitude)
> xlabel(0.1,10)
```

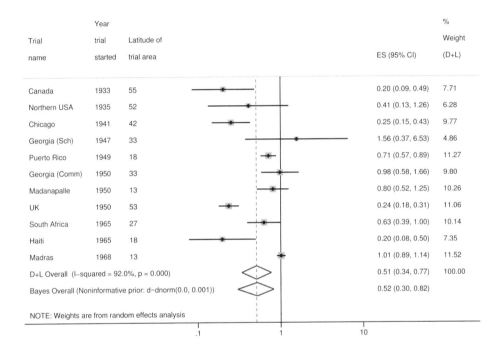

Figure 4. Forest plot displaying a fully Bayesian meta-analysis of the effect of BCG vaccine on incidence of tuberculosis. A noninformative prior has been specified, resulting in a pooled-effect estimate similar to the random-effects analysis.

◁

8   New analysis options

Here we discuss previously undocumented options added to `metan` since its original publication.

8.1 Dealing with zero cells

The cc(#) option allows the user to choose what value (if any) is to be added to the cells of the 2×2 table for a study in which one or more of the cell counts equals zero. Here the default is to add 0.5 to all cells of the 2×2 table for the study (except for the Peto method, which does not require a correction). This approach has been criticized, and other approaches (including making no correction) may be preferable (see Sweeting, Sutton, and Lambert [2004] for a discussion). The number declared in cc(#) must be between zero and one and will be added to each cell. When no events are recorded and RRs or odds ratios are to be combined the study is omitted, although for risk differences the effect is still calculable and the study is included. If no adjustment is made in the presence of zero cells, odds ratios and their standard errors cannot be calculated. Risk ratios and their standard errors cannot be calculated when the number of events in either the treatment or control group is zero.

8.2 Noninteger sample size

The nointeger option allows the number of observations in each arm (cell counts for binary data or the number of observations for continuous data) to be noninteger. By default, the sample size is assumed to be a whole number for both binary and continuous data. However, it may make sense for this not to be so, for example, to use a more flexible continuity correction with a different number added to each cell or when the meta-analysis incorporates cluster randomized trials and the effective-sample size is less than the total number of observations.

8.3 Breslow and Day test for heterogeneity

The breslow option can be used to perform the Breslow–Day test for heterogeneity of the odds ratio (Breslow and Day 1993). A review article by Reis, Hirji, and Afifi (1999) compared several different tests of heterogeneity and found this test to perform well in comparison to other asymptotic tests.

9 New output

9.1 The I^2 statistic

metan now displays the I^2 statistic as well as Cochran's Q to quantify heterogeneity, based on the work by Higgins and Thompson (2004) and Higgins et al. (2003). Briefly, I^2 is the percentage of variation attributable to heterogeneity and is easily interpretable. Cochran's Q can suffer from low power when the number of studies is low or excessive power when the number of studies is large. I^2 is calculated from the results of the meta-analysis by

$$I^2 = 100\% \times \frac{(Q - \mathrm{df})}{Q}$$

where Q is Cochran's heterogeneity statistic and df is the degrees of freedom. Negative values of I^2 are set to zero so that I^2 lies between 0% and 100%. A value of 0% indicates no observed heterogeneity, and larger values show increasing heterogeneity. Although there can be no absolute rule for when heterogeneity becomes important, Higgins et al. (2003) tentatively suggest adjectives of low for I^2 values between 25%–50%, moderate for 50%–75%, and high for $\geq 75\%$.

9.2   Prediction interval for the random-effects distribution

The presentation of summary random-effects estimates may sometimes be misleading, as the CI refers to the average true treatment effect, but this is assumed under the random-effects model to vary between studies. A CI derived from a larger number of studies exhibiting a high degree of heterogeneity could be of similar width to a CI derived from a smaller number of more homogeneous studies, but in the first situation, we will be much less sure of the range within which the treatment effect in a new study will lie (Higgins and Thompson 2001). The prediction interval for the treatment effect in a new trial may be approximated by using the formula

$$\mathrm{mean} \pm t_{\mathrm{df}} \times \sqrt{(\mathrm{se}^2 + \tau^2)}$$

where t is the appropriate centile point (e.g., 95%) of the t distribution with $k-2$ degrees of freedom, se^2 is the squared standard error, and τ^2 the between-study variance. This incorporates uncertainty in the location and spread of the random-effects distribution. The approximate prediction interval can be displayed in the forest plot, with lines extending from the summary diamond, by using the option rfdist. With ≤ 2 studies, the distribution is inestimable and effectively infinite; thus the interval is displayed with dotted lines. When heterogeneity is estimated to be zero, the prediction interval is still slightly wider than the summary diamond as the t statistic is always greater than the corresponding normal deviate. The coverage (e.g., 90%, 95%, or 99%) for the interval may be set by using the command rflevel(#).

▷ Example

Here we display the prediction intervals corresponding to the stratified analyses derived in section 6.1. The resulting forest plot is displayed in figure 5.

```
. metan tcases tnoncases ccases cnoncases, rr random rfdist
> lcols(trialnam startyr latitude) astext(60) by(lat_cat) xlabel(0.1,10)
> xsize(10) ysize(8) notable
```

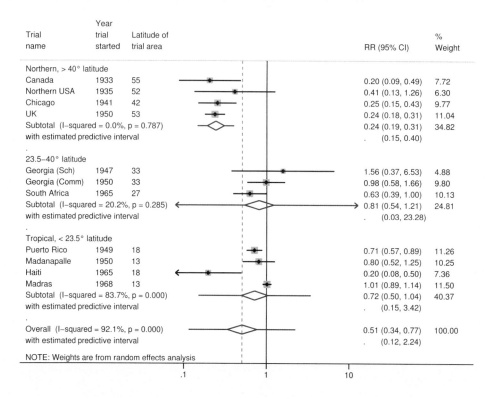

Figure 5. Forest plot displaying a random-effects meta-analysis of the effect of BCG vaccine on incidence of tuberculosis. Results are stratified by latitude region and the prediction interval for a future trial is displayed for each and overall.

◁

9.3   Vaccine efficacy

Results from the analysis of 2×2 data from vaccine trials may be reexpressed as the *vaccine efficacy* (also known as the *relative-risk reduction*); defined as the proportion of cases that would have been prevented in the placebo group had they received the vaccination (Kirkwood and Sterne 2003). The formula is

$$\text{Vaccine efficacy (VE)} = 100\% \times \left(1 - \frac{\text{risk of disease in vaccinated}}{\text{risk of disease in unvaccinated}} \right)$$

$$= 100\% \times (1 - \text{RR})$$

In `metan`, data are entered in the same way as any other analysis of 2×2 data and the option `efficacy` added. Results are displayed as odds ratios or RRs in the table and forest plot, but another column is added to the plot showing the results reexpressed as vaccine efficacy.

▷ **Example**

The BCG data are reanalyzed here, with results also displayed in terms of vaccine efficacy. The resulting forest plot is displayed in figure 6.

```
. metan tcases tnoncases ccases cnoncases, rr random efficacy
> lcols(trialnam startyr) textsize(150) notable xlabel(0.1, 10)
```

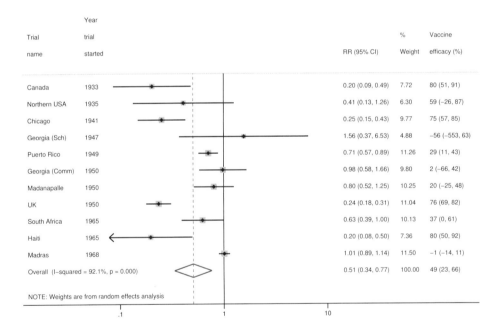

Figure 6. Forest plot displaying a random-effects meta-analysis of the effect of BCG vaccine on incidence of tuberculosis. Results are also displayed in terms of vaccine efficacy; estimates with a RR of greater than 1 produce a negative vaccine efficacy.

10   More graph options

10.1   metan graph options

Previous users of `metan` may find that they do not like the new box style and prefer a solid black box without the point estimate marker. The option `classic` changes back to this style. There are also options available to change the boxes, diamonds, and other lines. This is achieved by using options that change the standard graph commands that `metan` uses. For instance, the vertical line representing the overall effect may be changed using `olineopt()`, which can take standard Stata *line_options* such as `lwidth()`, `lcolor()`, and `lpattern()`. Boxes are weighted markers and not much can be changed, although shape and color may be modified by using *marker_options* in the `boxopt()` option, such as `msymbol()` and `mcolor()`, or we can dispense with the boxes entirely by using the option `nobox`. The point estimate markers have more flexibility and may also be modified by using *marker_options* in the `pointopt()` option; for instance, labels may by attached to them by using `mlabel()`. The CIs and diamonds may be changed by using *line_options* in the options `ciopt()` and `diamopt()`. For more details, see the `metan` help file and the Stata *Graphics Reference Manual* ([G] **graph**).

▷ **Example**

   Here many aspects of the graph are changed and a raw data variable is defined (as in `counts`) and attached to the point estimates in the graph. The resulting graph is not shown here, but a similar application is shown in section 10.3.

```
. gen counts = string(tcases) + "/" + string(tcases+tnoncases) + "," +
> string(ccases) + "/" + string(ccases+cnoncases)

. metan tcases tnoncases ccases cnoncases, rr fixedi second(random) nosecsub
> notable olineopt(lwidth(thick) lcolor(navy) lpattern(dot))
> boxopt(msymbol(triangle) mcolor(dkgreen))
> pointopt(mlabel(counts) mlabsize(tiny) mlabposition(5))
```

                                                                              ◁

10.2   Overall graph options

Any graph options that come under the *overall*, *note*, and *caption* sections of Stata's `graph twoway` command may be added to a `metan` command, and the x axis (and y axis if required) may have a title added. The options `aspect()` or `xsize()` and `ysize()` may be used to specify different aspect ratios (e.g., portrait). The default aspect ratio of a Stata graph is around 0.7 (height/width), and `metan` tries to stick to this shape; although graphs that are more naturally displayed as long or wide will be reshaped to some degree. Use of the above options will control this more precisely.

   Finally, the use of schemes is also supported. As colors of boxes and so on are defined within `metan`, these will not always give the desired result but may produce some interesting effects. Try, for example, using the scheme `economist`. More on schemes can be found in [G] **schemes intro**.

10.3 Notes on graph building

It can be useful to declare local or global macros that contain portions of code that are frequently used. For example, if the forest plot always has triangular "boxes" in forest green, contains the same columns of data, and so on, global macros may be declared for these bits of code. These can then be reused for a series of meta-analyses to specify the look and contents of the graphs. These could also be declared in an ado-file so that they are ready to use in every Stata session. This idea is similar to using Stata graph schemes.

▷ **Example**

Macros are defined to control various aspects of the graph and then used in the `metan` command. The resulting forest plot is displayed in figure 7.

```
. global metamethod rr fixedi second(random) nosecsub
. global metacolumns lcols(trialnam startyr latitude) astext(60)
. global metastyle boxopt(mcolor(forest_green) msymbol(triangle))
> pointopt(msymbol(smtriangle) mcolor(gold) msize(tiny))
> mlabel(counts) mlabsize(tiny) mlabposition(2) mlabcolor(brown))
> diamopt(lcolor(black) lwidth(medthick)) graphregion(fcolor(gs10)) boxsca(80)
. global metaopts favours(decreases TB # increases TB)
> xlabel(0.1, 0.2, 0.5, 2, 5, 10) notable
. metan tcases tnoncases ccases cnoncases,
> $metamethod $metacolumns $metastyle $metaopts by(lat_cat) xsize(10) ysize(8)
```

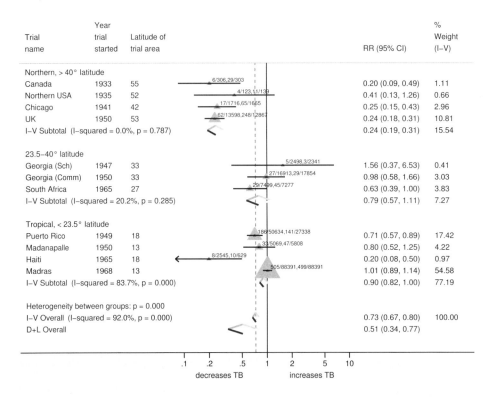

Figure 7. Forest plot displaying an inverse-variance weighted fixed-effects meta-analysis of the effect of BCG vaccine on incidence of tuberculosis. Results are stratified by latitude region, and the overall random-effects estimate is also displayed. Various options have been used to change the display of the graph.

◁

11   Variables and results produced by metan

11.1   Variables generated

When odds ratios (OR) or RRs are combined from 2×2 data and the `log` option is not used, the SE log OR or log RR is saved in a variable named `_selogES`, to make clear that it is the SE log OR or RR and not on the same scale. If the `log` option is used, the standard error is named `_seES`, as it is on the same scale as the estimate itself. In both cases, the estimate is called `_ES`.

It is possible to calculate the standard error of ORs and RRs by the delta method; this is what Stata does, for example, with the results reported by the logistic command.

However, the distribution of ratios is in general highly skewed, and for this reason, `metan` does not attempt to record the standard error of either the OR or RR.

Absolute measures (risk differences or mean differences) are symmetric and may be assumed to be normally distributed via the central limit theorem. Here `metan` stores these quantities in _ES and their standard errors in _seES. The derived variables incorporate the correction for zero cells (see section 8.1).

_ES	Effect size (ES)
_seES	Standard error of ES
_selogES	Standard error of log ES
_LCI	Lower confidence limit for ES
_UCI	Upper confidence limit for ES
_WT	Study percentage weight
_SS	Study sample size

11.2   Saved results (macros)

As with many Stata commands, macros are left behind containing the results of the analysis. If two methods are specified by using the option `second()`, some of these are repeated; for example, `r(ES)` and `r(ES_2)` give the pooled-effects estimates for each method. Subgroup statistics when using the `by()` option are not saved; if these are required for storage, it is recommended that a program be written that analyzes subgroups separately (perhaps using the `nograph` and `notable` options).

Name	Second	Description
r(ES)	r(ES_2)	pooled-effect size (if the log option is specified with or or rr, this is the pooled log OR or log RR)
r(seES)	r(seES_2)	standard error of pooled-effect size with symmetrical CI, i.e., mean differences, risk difference, log OR, and log RR using log option
r(selogES)	r(selogES_2)	standard error of log OR or log RR when ORs or RRs are combined without the log option
r(ci_low)	r(ci_low_2)	lower CI of pooled-effect size
r(ci_upp)	r(ci_upp_2)	upper CI of pooled-effect size
r(z)		z-value of effect size
r(p_z)		p-value for significance of effect size
r(het)		chi-squared test for heterogeneity
r(df)		degrees of freedom (number of informative studies minus 1)
r(p_het)		p-value for significance of test for heterogeneity
r(i_sq)		the I^2 statistic
r(tau2)		estimated between-study variance (random-effects analyses only)
r(chi2)		chi-squared test for significance of odds ratio (fixed-effects OR only)
r(p_chi2)		p-value for the above test
r(rger)		overall event rate, group 1 (if binary data are combined)
r(cger)		overall event rate, group 2 (see above)
r(measure)		effect measure (e.g., RR, SMD)
r(method_1)	r(method_2)	analysis method (e.g., M-H, D+L)

12   References

Bradburn, M. J., J. J. Deeks, and D. G. Altman. 1998. sbe24: metan—an alternative meta-analysis command. *Stata Technical Bulletin* 44: 4–15. Reprinted in *Stata Technical Bulletin Reprints*, vol. 8, pp. 86–100. College Station, TX: Stata Press. (Reprinted in this collection on pp. 3–28.)

Breslow, N. E., and N. E. Day. 1993. *Statistical Methods in Cancer Research: Volume I—The Analysis of Case–Control Studies.* Lyon: International Agency for Research on Cancer.

Colditz, G. A., T. F. Brewer, C. S. Berkey, M. E. Wilson, E. Burdick, H. V. Fineberg, and F. Mosteller. 1994. Efficacy of BCG vaccine in the prevention of tuberculosis.

Meta-analysis of the published literature. *Journal of the American Medical Association* 271: 698–702.

Deeks, J. J., D. G. Altman, and M. J. Bradburn. 2001. Statistical methods for examining heterogeneity and combining results from several studies in meta-analysis. In *Systematic Reviews in Health Care: Meta-analysis in Context*, ed. M. Egger, G. Davey Smith, and D. G. Altman, 2nd ed., 285–312. London: BMJ Books.

Fine, P. E. M. 1995. Variation in protection by BCG: Implications of and for heterologous immunity. *Lancet* 346: 1339–1345.

Harbord, R. M., and J. P. T. Higgins. 2008. Meta-regression in Stata. *Stata Journal* 8: 493–519. (Reprinted in this collection on pp. 85–111.)

Higgins, J. P. T., and S. G. Thompson. 2001. Presenting random effects meta-analyses: Where are we going wrong? In *9th International Cochrane Colloquium*. Lyon, France.

———. 2004. Controlling the risk of spurious findings from meta-regression. *Statistics in Medicine* 23: 1663–1682.

Higgins, J. P. T., S. G. Thompson, J. J. Deeks, and D. G. Altman. 2003. Measuring inconsistency in meta-analyses. *British Medical Journal* 327: 557–560.

Kirkwood, B. R., and J. A. C. Sterne. 2003. *Essential Medical Statistics.* 2nd ed. Oxford: Blackwell.

Lunn, D. J., A. Thomas, N. Best, and D. J. Spiegelhalter. 2000. WinBUGS – A Bayesian modelling framework: Concepts, structure and extensibility. *Statistics and Computing* 10: 325–337.

Poole, C., and S. Greenland. 1999. Random-effects meta-analyses are not always conservative. *American Journal of Epidemiology* 150: 469–475.

Reis, I. M., K. F. Hirji, and A. A. Afifi. 1999. Exact and asymptotic tests for homogeneity in several 2×2 tables. *Statistics in Medicine* 18: 893–906.

Sharp, S., and J. A. C. Sterne. 1997. sbe16: Meta-analysis. *Stata Technical Bulletin* 38: 9–14. Reprinted in *Stata Technical Bulletin Reprints*, vol. 7, pp. 100–106. College Station, TX: Stata Press.

———. 1998. sbe16.1: New syntax and output for the meta-analysis command. *Stata Technical Bulletin* 42: 6–8. Reprinted in *Stata Technical Bulletin Reprints*, vol. 7, pp. 106–108. College Station, TX: Stata Press.

Sterne, J. A. C., D. Gavaghan, and M. Egger. 2000. Publication and related bias in meta-analysis: Power of statistical tests and prevalence in the literature. *Journal of Clinical Epidemiology* 53: 1119–1129.

Sweeting, M. J., A. J. Sutton, and P. C. Lambert. 2004. What to add to nothing? Use and avoidance of continuity corrections in meta-analysis of sparse data. *Statistics in Medicine* 23: 1351–1375.

Warn, D. E., S. G. Thompson, and D. J. Spiegelhalter. 2002. Bayesian random effects meta-analysis of trials with binary outcomes: Methods for absolute risk difference and relative risk scales. *Statistics in Medicine* 21: 1601–1623.

The Stata Journal (2010)
10, Number 3, st0201, pp. 395–407

metaan: Random-effects meta-analysis

Evangelos Kontopantelis
National Primary Care
Research & Development Centre
University of Manchester
Manchester, UK
e.kontopantelis@manchester.ac.uk

David Reeves
Health Sciences Primary Care
Research Group
University of Manchester
Manchester, UK
david.reeves@manchester.ac.uk

Abstract. This article describes the new meta-analysis command `metaan`, which can be used to perform fixed- or random-effects meta-analysis. Besides the standard DerSimonian and Laird approach, `metaan` offers a wide choice of available models: maximum likelihood, profile likelihood, restricted maximum likelihood, and a permutation model. The command reports a variety of heterogeneity measures, including Cochran's Q, I^2, H_M^2, and the between-studies variance estimate $\hat{\tau}^2$. A forest plot and a graph of the maximum likelihood function can also be generated.

Keywords: st0201, metaan, meta-analysis, random effect, effect size, maximum likelihood, profile likelihood, restricted maximum likelihood, REML, permutation model, forest plot

1 Introduction

Meta-analysis is a statistical methodology that integrates the results of several independent clinical trials in general that are considered by the analyst to be "combinable" (Huque 1988). Usually, this is a two-stage process: in the first stage, the appropriate summary statistic for each study is estimated; then in the second stage, these statistics are combined into a weighted average. Individual patient data (IPD) methods exist for combining and meta-analyzing data across studies at the individual patient level. An IPD analysis provides advantages such as standardization (of marker values, outcome definitions, etc.), follow-up information updating, detailed data-checking, subgroup analyses, and the ability to include participant-level covariates (Stewart 1995; Lambert et al. 2002). However, individual observations are rarely available; additionally, if the main interest is in mean effects, then the two-stage and the IPD approaches can provide equivalent results (Olkin and Sampson 1998).

This article concerns itself with the second stage of the two-stage approach to meta-analysis. At this stage, researchers can select between two main approaches—the fixed-effects (FE) or the random-effects model—in their efforts to combine the study-level summary estimates and calculate an overall average effect. The FE model is simpler and assumes the true effect to be the same (homogeneous) across studies. However, homogeneity has been found to be the exception rather than the rule, and some degree of true effect variability between studies is to be expected (Thompson and Pocock 1991). Two sorts of between-studies heterogeneity exist: clinical heterogeneity stems from differences in populations, interventions, outcomes, or follow-up times, and methodological heterogeneity stems from differences in trial design and quality (Higgins and Green 2009; Thompson 1994). The most common approach to modeling the between-studies variance is the model proposed by DerSimonian and Laird (1986), which is widely used in generic and specialist meta-analysis statistical packages alike. In Stata, the DerSimonian–Laird (DL) model is used in the most popular meta-analysis commands—the recently updated `metan` and the older but still useful `meta` (Harris et al. 2008). However, the between-studies variance component can be estimated using more-advanced (and computationally expensive) iterative techniques: maximum likelihood (ML), profile likelihood (PL), and restricted maximum likelihood (REML) (Hardy and Thompson 1996; Thompson and Sharp 1999). Alternatively, the estimate can be obtained using nonparametric approaches, such as the permutations (PE) model proposed by Follmann and Proschan (1999).

We have implemented these models in `metaan`, which performs the second stage of a two-stage meta-analysis and offers alternatives to the DL random-effects model. The command requires the studies' effect estimates and standard errors as input. We have also created `metaeff`, a command that provides support in the first stage of the two-stage process and complements `metaan`. The `metaeff` command calculates for each study the effect size (standardized mean difference) and its standard error from the input parameters supplied by the user, using one of the models described in the *Cochrane Handbook for Systematic Reviews of Interventions* (Higgins and Green 2006). For more details, type `ssc describe metaeff` in Stata or see Kontopantelis and Reeves (2009).

The `metaan` command does not offer the plethora of options `metan` does for inputting various types of binary or continuous data. Other useful features in `metan` (unavailable in `metaan`) include stratified meta-analysis, user-input study weights, vaccine efficacy calculations, the Mantel–Haenszel FE method, L'Abbe plots, and funnel plots. The REML model, assumed to be the best model for fitting a random-effects meta-analysis model even though this assumption has not been thoroughly investigated (Thompson and Sharp 1999), has recently been coded in the updated metaregression command `metareg` (Harbord and Higgins 2008) and the new multivariate random-effects meta-analysis command `mvmeta` (White 2009). However, the output and options provided by `metaan` can be more useful in the univariate meta-analysis context.

2   The metaan command

2.1   Syntax

`metaan` *varname1 varname2* [*if*] [*in*], {fe|dl|ml|reml|pl|pe} [varc
    label(*varname*) forest forestw(*#*) plplot(*string*)]

where

> *varname1* is the study effect size.
>
> *varname2* is the study effect variation, with standard error used as the default.

2.2 Options

fe fits an FE model that assumes there is no heterogeneity between the studies. The model assumes that within-study variances may differ, but that there is homogeneity of effect size across studies. Often the homogeneity assumption is unlikely, and variation in the true effect across studies is to be expected. Therefore, caution is required when using this model. Reported heterogeneity measures are estimated using the dl option. You must specify one of fe, dl, ml, reml, pl, or pe.

dl fits a DL random-effects model, which is the most commonly used model. The model assumes heterogeneity between the studies; that is, it assumes that the true effect can be different for each study. The model assumes that the individual-study true effects are distributed with a variance τ^2 around an overall true effect, but the model makes no assumptions about the form of the distribution of either the within-study or the between-studies effects. Reported heterogeneity measures are estimated using the dl option. You must specify one of fe, dl, ml, reml, pl, or pe.

ml fits an ML random-effects model. This model makes the additional assumption (necessary to derive the log-likelihood function, and also true for reml and pl, below) that both the within-study and the between-studies effects have normal distributions. It solves the log-likelihood function iteratively to produce an estimate of the between-studies variance. However, the model does not always converge; in some cases, the between-studies variance estimate is negative and set to zero, in which case the model is reduced to an fe specification. Estimates are reported as missing in the event of nonconvergence. Reported heterogeneity measures are estimated using the ml option. You must specify one of fe, dl, ml, reml, pl, or pe.

reml fits an REML random-effects model. This model is similar to ml and uses the same assumptions. The log-likelihood function is maximized iteratively to provide estimates, as in ml. However, under reml, only the part of the likelihood function that is location invariant is maximized (that is, maximizing the portion of the likelihood that does not involve μ if estimating τ^2, and vice versa). The model does not always converge; in some cases, the between-studies variance estimate is negative and set to zero, in which case the model is reduced to an fe specification. Estimates are reported as missing in the event of nonconvergence. Reported heterogeneity measures are estimated using the reml option. You must specify one of fe, dl, ml, reml, pl, or pe.

pl fits a PL random-effects model. This model uses the same likelihood function as ml but takes into account the uncertainty associated with the between-studies variance estimate when calculating an overall effect, which is done by using nested iterations to converge to a maximum. The confidence intervals (CIs) provided by the model are asymmetric, and hence so is the diamond in the forest plot. However, the model does not always converge. Values that were not computed are reported as missing. Reported heterogeneity measures are estimated using the `ml` option because $\widehat{\mu}$ and $\widehat{\tau}^2$, the effect and between-studies variance estimates, are the same. Only their CIs are reestimated. The model also provides a CI for the between-studies variance estimate. You must specify one of `fe`, `dl`, `ml`, `reml`, `pl`, or `pe`.

`pe` fits a PE random-effects model. This model can be described in three steps. First, in line with a null hypothesis that all true study effects are zero and observed effects are due to random variation, a dataset of all possible combinations of observed study outcomes is created by permuting the sign of each observed effect. Then, the `dl` model is used to compute an overall effect for each combination. Finally, the resulting distribution of overall effect sizes is used to derive a CI for the observed overall effect. The CI provided by the model is asymmetric, and hence so is the diamond in the forest plot. Reported heterogeneity measures are estimated using the `dl` option. You must specify one of `fe`, `dl`, `ml`, `reml`, `pl`, or `pe`.

`varc` specifies that the study-effect variation variable, *varname2*, holds variance values. If this option is omitted, `metaan` assumes that the variable contains standard-error values (the default).

`label(`*varname*`)` selects labels for the studies. One or two variables can be selected and converted to strings. If two variables are selected, they will be separated by a comma. Usually, the author names and the year of study are selected as labels. The final string is truncated to 20 characters.

`forest` requests a forest plot. The weights from the specified analysis are used for plotting symbol sizes (`pe` uses `dl` weights). Only one graph output is allowed in each execution.

`forestw(`*#*`)` requests a forest plot with adjusted weight ratios for better display. The value can be in the $[1, 50]$ range. For example, if the largest to smallest weight ratio is 60 and the graph looks awkward, the user can use this command to improve the appearance by requesting that the weight be rescaled to a largest/smallest weight ratio of 30. Only the weight squares in the plot are affected, not the model. The CIs in the plot are unaffected. Only one graph output is allowed in each execution.

`plplot(`*string*`)` requests a plot of the likelihood function for the average effect or between-studies variance estimate of the `ml`, `pl`, or `reml` model. The `plplot(mu)` option fixes the average effect parameter to its model estimate in the likelihood function and creates a two-way plot of τ^2 versus the likelihood function. The `plplot(tsq)` option fixes the between-studies variance to its model estimate in the likelihood function and creates a two-way plot of μ versus the likelihood function. Only one graph output is allowed in each execution.

2.3   Saved results

metaan saves the following in r() (some varying by selected model):

Scalars
r(Hsq)	heterogeneity measure H^2_M
r(Q)	Cochran's Q value
r(df)	degrees of freedom
r(effvar)	effect variance
r(efflo)	effect size, lower 95% CI
r(Isq)	heterogeneity measure I^2
r(Qpval)	p-value for Cochran's Q
r(eff)	effect size
r(effup)	effect size, upper 95% CI

In addition to the standard results, metaan, fe and metaan, dl save the following in r():

Scalars
r(tausq_dl)	$\hat{\tau}^2$, from the DL model

In addition to the standard results, metaan, ml saves the following in r():

Scalars
r(tausq_dl)	$\hat{\tau}^2$, from the DL model
r(conv_ml)	ML convergence information
r(tausq_ml)	$\hat{\tau}^2$, from the ML model

In addition to the standard results, metaan, reml saves the following in r():

Scalars
r(tausq_dl)	$\hat{\tau}^2$, from the DL model
r(conv_reml)	REML convergence information
r(tausq_reml)	$\hat{\tau}^2$, from the REML model

In addition to the standard results, metaan, pl saves the following in r():

Scalars
r(tausq_dl)	$\hat{\tau}^2$, from the DL model
r(tausqlo_pl)	$\hat{\tau}^2$ (PL), lower 95% CI
r(cloeff_pl)	convergence information, PL effect size (lower CI)
r(ctausqlo_pl)	convergence information, PL $\hat{\tau}^2$ (lower CI)
r(conv_ml)	ML convergence information
r(tausq_pl)	$\hat{\tau}^2$, from the PL model
r(tausqup_pl)	$\hat{\tau}^2$ (PL), upper 95% CI
r(cupeff_pl)	convergence information, PL effect size (upper CI)
r(ctausqup_pl)	convergence information, PL $\hat{\tau}^2$ (upper CI)

In addition to the standard results, metaan, pe saves the following in r():

Scalars
r(tausq_dl)	$\hat{\tau}^2$, from the DL model
r(exec_pe)	information on PE execution

In each case, heterogeneity measures H_M^2 and I^2 are computed using the returned between-variances estimate $\widehat{\tau}^2$. Convergence and PE execution information is returned as 1 if successful and as 0 otherwise. `r(effvar)` cannot be computed for PE. `r(effvar)` is the same for ML and PL, but for PL the CIs are "amended" to take into account the $\widehat{\tau}^2$ uncertainty.

3 Methods

The `metaan` command offers six meta-analysis models for calculating a mean effect estimate and its CIs: FE model, random-effects DL method, ML random-effects model, REML random-effects model, PL random-effects model, and PE method using a DL random-effects model. Models of the random-effects family take into account the identified between-studies variation, estimate it, and usually produce wider CIs for the overall effect than would an FE analysis. Brief descriptions of the models have been provided in section 2.2. In this section, we will provide a few more details and practical advice in selecting among the models. Their complexity prohibits complete descriptions in this article, and users wishing to look into model details are encouraged to refer to the original articles that described them (DerSimonian and Laird 1986; Hardy and Thompson 1996; Follmann and Proschan 1999; Brockwell and Gordon 2001).

The three ML models are iterative and usually computationally expensive. ML and PL derive the μ (overall effect) and τ^2 estimates by maximizing the log-likelihood function in (1) under different conditions. REML estimates τ^2 and μ by maximizing the restricted log-likelihood function in (2).

$$\log L(\mu, \tau^2) = -\frac{1}{2}\left[\sum_{i=1}^{k}\log\left\{2\pi\left(\widehat{\sigma}_i^2 + \tau^2\right)\right\} + \sum_{i=1}^{k}\frac{(\widehat{y}_i - \mu)^2}{\widehat{\sigma}_i^2 + \tau^2}\right], \quad \mu \in \Re \; \& \; \tau^2 \geq 0 \quad (1)$$

$$\log L'(\mu, \tau^2) = -\frac{1}{2}\left[\sum_{i=1}^{k}\log\left\{2\pi\left(\widehat{\sigma}_i^2 + \tau^2\right)\right\} + \sum_{i=1}^{k}\frac{(\widehat{y}_i - \widehat{\mu})^2}{\widehat{\sigma}_i^2 + \tau^2}\right]$$
$$-\frac{1}{2}\log\sum_{i=1}^{k}\frac{1}{\widehat{\sigma}_i^2 + \tau^2}, \quad \widehat{\mu} \in \Re \; \& \; \tau^2 \geq 0 \quad (2)$$

where k is the number of studies to be meta-analyzed, $\widehat{y}_i$ and $\widehat{\sigma}_i^2$ are the effect and variance estimates for study i, and $\widehat{\mu}$ is the overall effect estimate.

ML follows the simplest approach, maximizing (1) in a single iteration loop. A criticism of ML is that it takes no account of the loss in degrees of freedom that results from estimating the overall effect. REML derives the likelihood function in a way that adjusts for this and removes downward bias in the between-studies variance estimator. A useful description for REML, in the meta-analysis context, has been provided by Normand (1999). PL uses the same likelihood function as ML, but uses nested iterations to take into account the uncertainty associated with the between-studies variance estimate when calculating an overall effect. By incorporating this extra factor of uncertainty, PL yields

CIs that are usually wider than for DL and also are asymmetric. PL has been shown to outperform DL in various scenarios (Brockwell and Gordon 2001).

The PE model (Follmann and Proschan 1999) can be described as follows: First, in line with a null hypothesis that all true study effects are zero and observed effects are due to random variation, a dataset of all possible combinations of observed study outcomes is created by permuting the sign of each observed effect. Next the `dl` model is used to compute an overall effect for each combination. Finally, the resulting distribution of overall effect sizes is used to derive a CI for the observed overall effect.

Method performance is known to be affected by three factors: the number of studies in the meta-analysis, the degree of heterogeneity in true effects, and—provided there is heterogeneity present—the distribution of the true effects (Brockwell and Gordon 2001). Heterogeneity, which is attributed to clinical or methodological diversity (Higgins and Green 2006), is a major problem researchers have to face when combining study results in a meta-analysis. The variability that arises from different interventions, populations, outcomes, or follow-up times is described by clinical heterogeneity, while differences in trial design and quality are accounted for by methodological heterogeneity (Thompson 1994). Traditionally, heterogeneity is tested with Cochran's Q, which provides a p-value for the test of homogeneity, when compared with a χ^2_{k-1} distribution where k is the number of studies (Brockwell and Gordon 2001). However, the test is known to be poor at detecting heterogeneity because its power is low when the number of studies is small (Hardy and Thompson 1998). An alternative measure is I^2, which is thought to be more informative in assessing inconsistency between studies. I^2 values of 25%, 50%, and 75% correspond to low, moderate, and high heterogeneity, respectively (Higgins et al. 2003). Another measure is H^2_M, the measure least affected by the value of k. It takes values in the $[0, +\infty)$ range, with 0 indicating perfect homogeneity (Mittlböck and Heinzl 2006). Obviously, the between-studies variance estimate $\hat{\tau}^2$ can also be informative about the presence or absence of heterogeneity.

The test for heterogeneity is often used as the basis for applying an FE or a random-effects model. However, the often low power of the Q test makes it unwise to base a decision on the result of the test alone. Research studies, even on the same topic, can vary on a large number of factors; hence, homogeneity is often an unlikely assumption and some degree of variability between studies is to be expected (Thompson and Pocock 1991). Some authors recommend the adoption of a random-effects model unless there are compelling reasons for doing otherwise, irrespective of the outcome of the test for heterogeneity (Brockwell and Gordon 2001).

However, even though random-effects methods model heterogeneity, the performance of the ML models (ML, REML, and PL) in situations where the true effects violate the assumptions of a normal distribution may not be optimal (Brockwell and Gordon 2001; Hardy and Thompson 1998; Böhning et al. 2002; Sidik and Jonkman 2007). The number of studies in the analysis is also an issue, because most meta-analysis models (including DL, ML, REML, and PL—but not PE) are only asymptotically correct; that is, they provide the theoretical 95% coverage only as the number of studies increases (approaches infinity). Method performance is therefore affected when the number of studies is small, but the extent depends on the model (some are more susceptible), along with the degree of heterogeneity and the distribution of the true effects (Brockwell and Gordon 2001).

4   Example

As an example, we apply the `metaan` command to health-risk outcome data from seven studies. The information was collected for an unpublished meta-analysis, and the data are available from the authors. Using the `describe` and `list` commands, we provide details of the dataset and proceed to perform a univariate meta-analysis with `metaan`.

```
. use metaan_example
. describe
Contains data from metaan_example.dta
  obs:             7
  vars:            4                               19 Apr 2010 12:19
  size:          560 (99.9% of memory free)
```

variable name	storage type	display format	value label	variable label
study	str16	%16s		First author and year
outcome	str48	%35s		Outcome description
effsize	float	%9.0g		effect sizes
se	float	%9.0g		SE of the effect sizes

```
Sorted by:  study  outcome
. list study outcome effsize se, noobs clean
```

study	outcome	effsize	se
Bakx A, 1985	Serum cholesterol (mmol/L)	-.3041526	.0958199
Campbell A, 1998	Diet	.2124063	.0812414
Cupples, 1994	BMI	.0444239	.090661
Eckerlund	SBP	-.3991309	.12079
Moher, 2001	Cholesterol (mmol/l)	-.9374746	.0691572
Woolard A, 1995	Alcohol intake (g/week)	-.3098185	.206331
Woolard B, 1995	Alcohol intake (g/week)	-.4898825	.2001602

```
. metaan effsize se, pl label(study) forest

Profile Likelihood method selected
```

Study	Effect	[95% Conf. Interval]		% Weight
Bakx A, 1985	-0.304	-0.492	-0.116	15.09
Campbell A, 1998	0.212	0.053	0.372	15.40
Cupples, 1994	0.044	-0.133	0.222	15.20
Eckerlund	-0.399	-0.636	-0.162	14.49
Moher, 2001	-0.937	-1.073	-0.802	15.62
Woolard A, 1995	-0.310	-0.714	0.095	12.01
Woolard B, 1995	-0.490	-0.882	-0.098	12.19
Overall effect (pl)	-0.308	-0.622	0.004	100.00

```
ML method succesfully converged
PL method succesfully converged for both upper and lower CI limits

Heterogeneity Measures
```

	value	df	p-value
Cochrane Q	139.81	6	0.000
I^2 (%)	91.96		
H^2	11.44		

	value	[95% Conf. Interval]	
tau^2 est	0.121	0.000	0.449

```
Estimate obtained with Maximum likelihood - Profile likelihood provides the CI
PL method succesfully converged for both upper and lower CI limits of the tau^2
> estimate
```

The PL model used in the example converged successfully, as did ML, whose convergence is a prerequisite. The overall effect is not found to be significant at the 95% level, and there is considerable heterogeneity across studies, according to the measures. The model also displays a 95% CI for the between-studies variance estimate $\hat{\tau}^2$ (provided that convergence is achieved, as is the case in this example). The forest plot created by the command is displayed in figure 3.

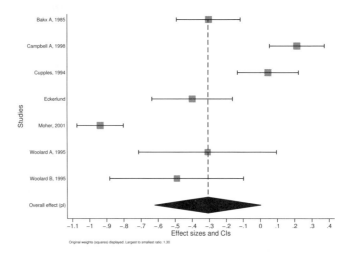

Figure 1. Forest plot displaying PL meta-analysis

When we reexecute the analysis with the `plplot(mu)` and `plplot(tsq)` options, we obtain the log-likelihood function plots shown in figures 2 and 3.

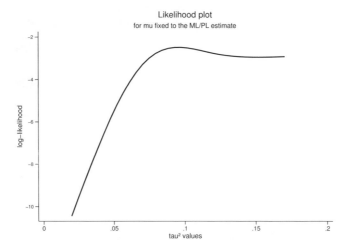

Figure 2. Log-likelihood function plot for μ fixed to the model estimate

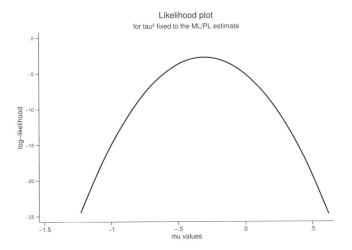

Figure 3. Log-likelihood function plot for τ^2 fixed to the model estimate

5 Discussion

The `metaan` command can be a useful meta-analysis tool that includes newer and, in certain circumstances, better-performing models than the standard DL random-effects model. Unpublished results exploring model performance in various scenarios are available from the authors. Future work will involve implementing more models in the `metaan` command and embellishing the forest plot.

6 Acknowledgments

We would like to thank the authors of `meta` and `metan` for all their work and the anonymous reviewer whose useful comments improved the article considerably.

7 References

Böhning, D., U. Malzahn, E. Dietz, P. Schlattmann, C. Viwatwongkasem, and A. Biggeri. 2002. Some general points in estimating heterogeneity variance with the DerSimonian–Laird estimator. *Biostatistics* 3: 445–457.

Brockwell, S. E., and I. R. Gordon. 2001. A comparison of statistical methods for meta-analysis. *Statistics in Medicine* 20: 825–840.

DerSimonian, R., and N. Laird. 1986. Meta-analysis in clinical trials. *Controlled Clinical Trials* 7: 177–188.

Follmann, D. A., and M. A. Proschan. 1999. Valid inference in random effects meta-analysis. *Biometrics* 55: 732–737.

Harbord, R. M., and J. P. T. Higgins. 2008. Meta-regression in Stata. *Stata Journal* 8: 493–519. (Reprinted in this collection on pp. 85–111.)

Hardy, R. J., and S. G. Thompson. 1996. A likelihood approach to meta-analysis with random effects. *Statistics in Medicine* 15: 619–629.

———. 1998. Detecting and describing heterogeneity in meta-analysis. *Statistics in Medicine* 17: 841–856.

Harris, R. J., M. J. Bradburn, J. J. Deeks, R. M. Harbord, D. G. Altman, and J. A. C. Sterne. 2008. metan: Fixed- and random-effects meta-analysis. *Stata Journal* 8: 3–28. (Reprinted in this collection on pp. 29–54.)

Higgins, J. P. T., and S. Green. 2006. *Cochrane Handbook for Systematic Reviews of Interventions Version 4.2.6.* http://www2.cochrane.org/resources/handbook/Handbook4.2.6Sep2006.pdf.

———. 2009. *Cochrane Handbook for Systematic Reviews of Interventions Version 5.0.2.* http://www.cochrane-handbook.org/.

Higgins, J. P. T., S. G. Thompson, J. J. Deeks, and D. G. Altman. 2003. Measuring inconsistency in meta-analyses. *British Medical Journal* 327: 557–560.

Huque, M. F. 1988. Experiences with meta-analysis in NDA submissions. *Proceedings of the Biopharmaceutical Section of the American Statistical Association* 2: 28–33.

Kontopantelis, E., and D. Reeves. 2009. MetaEasy: A meta-analysis add-in for Microsoft Excel. *Journal of Statistical Software* 30: 1–25.

Lambert, P. C., A. J. Sutton, K. R. Abrams, and D. R. Jones. 2002. A comparison of summary patient-level covariates in meta-regression with individual patient data meta-analysis. *Journal of Clinical Epidemiology* 55: 86–94.

Mittlböck, M., and H. Heinzl. 2006. A simulation study comparing properties of heterogeneity measures in meta-analyses. *Statistics in Medicine* 25: 4321–4333.

Normand, S.-L. T. 1999. Meta-analysis: Formulating, evaluating, combining, and reporting. *Statistics in Medicine* 18: 321–359.

Olkin, I., and A. Sampson. 1998. Comparison of meta-analysis versus analysis of variance of individual patient data. *Biometrics* 54: 317–322.

Sidik, K., and J. N. Jonkman. 2007. A comparison of heterogeneity variance estimators in combining results of studies. *Statistics in Medicine* 26: 1964–1981.

Stewart, L. A. 1995. Practical methodology of meta-analyses (overviews) using updated individual patient data. *Statistics in Medicine* 14: 2057–2079.

Thompson, S. G. 1994. Systematic review: Why sources of heterogeneity in meta-analysis should be investigated. *British Medical Journal* 309: 1351–1355.

Thompson, S. G., and S. J. Pocock. 1991. Can meta-analyses be trusted? *Lancet* 338: 1127–1130.

Thompson, S. G., and S. Sharp. 1999. Explaining heterogeneity in meta-analysis: A comparison of methods. *Statistics in Medicine* 18: 2693–2708.

White, I. R. 2009. Multivariate random-effects meta-analysis. *Stata Journal* 9: 40–56. (Reprinted in this collection on pp. 232–248.)

The Stata Technical Bulletin (1998)
STB-42, sbe22, pp. 13–16

Cumulative meta-analysis[1]

Jonathan A. C. Sterne
Department of Social Medicine
University of Bristol
Bristol, UK

Meta-analysis is used to combine the results of several studies, and the Stata command `metan` (Bradburn, Deeks, and Altman 1998; Harris et al. 2008)[2] can be used to perform meta-analyses and graph the results. In cumulative meta-analysis (Lau et al. 1992), the pooled estimate of the treatment effect is updated each time the results of a new study are published. This makes it possible to track the accumulation of evidence on the effect of a particular treatment.

The command `metacum` performs cumulative meta-analysis (using fixed- or random-effects models) and, optionally, graphs the results.

1 Syntax

`metacum` *varlist* [*if*] [*in*] [, [*binary_data_options* | *continuous_data_options* |
precalculated_effect_estimates_options] *measure_and_model_option* *output_options*
forest_plot_options]

binary_data_options

    `or rr rd fixed random fixedi randomi peto` <u>noint</u>`eger cc(#)`

continuous_data_options

    `cohen hedges glass nostandard fixed random` <u>noint</u>`eger`

precalculated_effect_estimates_options

    `fixed random`

measure_and_model_option

    `wgt(`*wgtvar*`)`

---

1. The updated syntax is by Ross Harris, Centre for Infections, Health Protection Agency, London.—
Ed.

2. The original command to be installed was `meta`; see Sharp and Sterne (1997, 1998).—Ed.

output_options

> by(*byvar*) log eform <u>il</u>evel(*#*)
> sortby(*varlist*)
> label([namevar = *namevar*], [yearvar = *yearvar*]) notable nograph

forest_plot_options

> <u>xl</u>abel(*#*, ...) <u>xt</u>ick(*#*, ...) <u>t</u>extsize(*#*) nowt nostats counts
> group1(*string*) group2(*string*) effect(*string*) force lcols(*varlist*)
> rcols(*varlist*) astext(*#*) double summaryonly null(*#*) nulloff
> favours(*string # string*)
> pointopt(*marker_options* | *marker_label_options*) ciopt(*line_options*)
> olineopt(*line_options*) classic nowarning *graph_options*

2 Options

2.1 binary_data_options

or pools ORs.

rr pools RRs; this is the default.

rd pools risk differences.

fixed specifies a fixed-effects model using the Mantel–Haenszel method; this is the default.

random specifies a random-effects model using the DerSimonian and Laird method, with the estimate of heterogeneity being taken.

fixedi specifies a fixed-effects model using the inverse-variance method.

randomi specifies a random-effects model using the DerSimonian and Laird method, with the estimate of heterogeneity being taken from the inverse-variance fixed-effects model.

peto specifies that the Peto method is used to pool ORs.

nointeger allows the cell counts to be nonintegers. This option may be useful when a variable continuity correction is sought for studies containing zero cells but also may be used in other circumstances, such as where a cluster-randomized trial is to be incorporated and the "effective sample size" is less than the total number of observations.

cc(*#*) defines a fixed-continuity correction to add where a study contains a zero cell. By default, metan8 adds 0.5 to each cell of a trial where a zero is encountered when using inverse-variance, DerSimonian and Laird, or Mantel–Haenszel weighting to enable finite variance estimators to be derived. However, the cc() option allows the use of other constants (including none). See also the nointeger option.

2.2   continuous_data_options

`cohen` pools standardized mean differences by the Cohen method; this is the default.

`hedges` pools standardized mean differences by the Hedges method.

`glass` pools standardized mean differences by the Glass method.

`nostandard` pools unstandardized mean differences.

`fixed` specifies a fixed-effects model using the Mantel–Haenszel method; this is the default.

`random` specifies a random-effects model using the DerSimonian and Laird method, with the estimate of heterogeneity being taken.

`nointeger` denotes that the number of observations in each arm does not need to be an integer. By default, the first and fourth variables specified (containing N_intervention and N_control, respectively) may occasionally be noninteger (see `nointeger` in section 2.1).

2.3   precalculated_effect_estimates_options

`fixed` specifies a fixed-effects model using the Mantel–Haenszel method; this is the default.

`random` specifies a random-effects model using the DerSimonian and Laird method, with the estimate of heterogeneity being taken.

2.4   measure_and_model_option

`wgt`(*wgtvar*) specifies alternative weighting for any data type. The effect size is to be computed by assigning a weight of *wgtvar* to the studies. When RRs or ORs are declared, their logarithms are weighted. This option should be used only if you are satisfied that the weights are meaningful.

2.5   output_options

`by`(*byvar*) specifies that the meta-analysis is to be stratified according to the variable declared.

`log` reports the results on the log scale (valid only for ORs and RRs analyses from raw data counts).

`eform` exponentiates all effect sizes and confidence intervals (valid only when the input variables are log ORs or log hazard-ratios with standard error or confidence intervals).

`ilevel`(*#*) specifies the coverage (e.g., 90%, 95%, 99%) for the individual trial confidence intervals; the default is $S_level. `ilevel()` and `olevel()` need not be the same. See [U] **20.7 Specifying the width of confidence intervals**.

sortby(*varlist*) sorts by variable(s) in *varlist*.

label([namevar=*namevar*], [yearvar=*yearvar*]) labels the data by its name, year, or both. Either or both variable lists may be left blank. For the table display, the overall length of the label is restricted to 20 characters. If the lcols() option is also specified, it will override the label() option.

notable prevents the display of a results table.

nograph prevents the display of a graph.

2.6 forest_plot_options

xlabel(#,...) defines *x*-axis labels. This option has been modified so that any number of points may be defined. Also, checks are no longer made as to whether these points are sensible, so the user may define anything if the force option is used. Points must be comma separated.

xtick(#,...) adds tick marks to the *x* axis. Points must be comma separated.

textsize(#) specifies the font size for the text display on the graph. This option has been modified so that the default is textsize(100) (as in 100%) and the percentage may be increased or decreased (e.g., 80 or 120 for 20% smaller or larger, respectively).

nowt prevents the display of study weight on the graph.

nostats prevents the display of study statistics on the graph.

counts displays data counts (n/N) for each group when using binary data or the sample size, mean, and standard deviation for each group if mean differences are used (the latter is a new feature).

group1(*string*) and group2(*string*) may be used with the counts option, and the text should contain the names of the two groups.

effect(*string*) allows the graph to name the summary statistic used when the effect size and its standard error are declared.

force forces the x-axis scale to be in the range specified by xlabel().

lcols(*varlist*) and rcols(*varlist*) define columns of additional data to the left or right of the graph. The first two columns on the right are automatically set to effect size and weight, unless suppressed by using the options nostats and nowt. If counts is used, this will be set as the third column. textsize() can be used to fine-tune the size of the text to achieve a satisfactory appearance. The columns are labeled with the variable label or the variable name if this is not defined. The first variable specified in lcols() is assumed to be the study identifier and this is used in the table output.

astext(#) specifies the percentage of the graph to be taken up by text. The default is 50%, and the percentage must be in the range 10–90.

double allows variables specified in `lcols()` and `rcols()` to run over two lines in the plot. This option may be of use if long strings are used.

summaryonly shows only summary estimates in the graph. This option may be of use for multiple subgroup analyses.

null(#) displays the null line at a user-defined value rather than at 0 or 1.

nulloff removes the null hypothesis line from the graph.

favours(*string* **#** *string*) applies a label saying something about the treatment effect to either side of the graph (strings are separated by the **#** symbol). This option replaces the feature available in **b1title** in the previous version of **metan**.

pointopt(*marker_options* | *marker_label_options*), **ciopt**(*line_options*), and **olineopt**(*line_options*) specify options for the graph routines within the program, allowing the user to alter the appearance of the graph. Any options associated with a particular graph command may be used, except some that would cause incorrect graph appearance. For example, diamonds are plotted using the **twoway pcspike** command, so options for line styles are available (see [G] *line_options*); however, altering the $x--y$ orientation with the option **horizontal** or **vertical** is not allowed. So, **ciopt(lcolor(green)**
lwidth(thick)) feeds into a command such as **pcspike(y1 x1 y2 x2,**
lcolor(green) lwidth(thick))

pointopt(*marker_options* | *marker_label_options*), controls the point estimate by using marker options. See [G] *marker_options* and [G] *marker_label_options*.

ciopt(*line_options*) controls the confidence intervals for studies by using options for **twoway pcspike** (not horizontal/vertical). See [G] *line_options*.

olineopt(*line_options*) controls the overall effect line with options for another line (not position). See [G] *line_options*.

classic specifies that solid black boxes without point estimate markers are used, as in the previous version of **metan**.

nowarning switches off the default display of a note warning that studies are weighted from random-effects analyses.

graph_options are any of the options documented in [G] *twoway_options*. These allow the addition of titles, subtitles, captions, etc.; control of margins, plot regions, graph size, aspect ratio; and the use of schemes. Because titles may be added with *graph_options*, previous options such as **b2title** are no longer necessary.

3 Background

The command **metacum** provides an alternative means of presenting the results of a meta-analysis, where instead of the individual study effects and combined estimate, the cumulative evidence up to and including each trial can be printed and/or graphed. The technique was suggested by Lau et al. (1992).

4 Example

The first trial of streptokinase treatment following myocardial infarction was reported in 1959. A further 21 trials were conducted between that time and 1986, when the ISIS-2 multicenter trial (on over 17,000 patients in whom over 1800 deaths were reported) demonstrated conclusively that the treatment reduced the chances of subsequent death.

Lau et al. (1992) pointed out that a meta-analysis of trials performed up to 1977 provided strong evidence that the treatment worked. Despite this, it was another 15 years until the treatment became routinely used.

Dataset `strepto.dta` contains the results of 22 trials of streptokinase conducted between 1959 and 1986.

```
. describe
Contains data from streptok.dta
  obs:            22                          Streptokinase and CHD
 vars:             7                          8 Jun 2005 15:47
 size:           638 (99.9% of memory free)   (_dta has notes)

              storage  display   value
variable name   type   format    label    variable label

trial          byte    %8.0g              trial number
trialnam       str14   %14s               Trial name
year           int     %8.0g              year published
pop1           int     %12.0g             Treated population
deaths1        int     %12.0g             Treated cases
pop0           int     %12.0g             Control population
deaths0        int     %12.0g             Control cases

Sorted by:  trial
. list trialnam year pop1 deaths1 pop0 deaths0, noobs clean
           trialnam   year   pop1   deaths1   pop0   deaths0
           Fletcher   1959     12         1     11         4
              Dewar   1963     21         4     21         7
       1st European   1969     83        20     84        15
        Heikinheimo   1971    219        22    207        17
            Italian   1971    164        19    157        18
       2nd European   1971    373        69    357        94
      2nd Frankfurt   1973    102        13    104        29
      1st Australian  1973    264        26    253        32
         NHLBI SMIT   1974     53         7     54         3
             Valere   1975     49        11     42         9
              Frank   1975     55         6     53         6
          UK Collab   1976    302        48    293        52
              Klein   1976     14         4      9         1
            Austrian  1977    352        37    376        65
           Lasierra   1977     13         1     11         3
           N German   1977    249        63    234        51
            Witchitz  1977     32         5     26         5
      2nd Australian  1977    112        25    118        31
       3rd European   1977    156        25    159        50
               ISAM   1986    859        54    882        63
             GISSI-1  1986   5860       628   5852       758
             ISIS-2   1988   8592       791   8595      1029
```

Before doing our meta-analysis, we calculate the log odds-ratio for each study, and
its corresponding variance. We also create a string variable containing the trial name
and year of publication:

```
. gen logor=log((deaths1/(pop1-deaths1))/((deaths0/(pop0-deaths0))))
. gen selogor=sqrt(1/deaths1+1/(pop1-deaths1)+1/deaths0+1/(pop0-deaths0))
. metan logor selogor, eform fixed label(namevar=trialnam, yearvar=year)
> xlabel(.1,.5,1,2,10) force effect("Odds ratio")
```

Study	ES	[95% Conf. Interval]		% Weight
Fletcher (1959)	0.159	0.015	1.732	0.07
Dewar (1963)	0.471	0.114	1.942	0.21
1st European (1969)	1.460	0.689	3.096	0.75
Heikinheimo (1971)	1.248	0.643	2.423	0.96
Italian (1971)	1.012	0.510	2.008	0.90
2nd European (1971)	0.635	0.447	0.903	3.42
2nd Frankfurt (1973)	0.378	0.183	0.778	0.81
1st Australian (1973	0.754	0.436	1.306	1.41
NHLBI SMIT (1974)	2.587	0.632	10.596	0.21
Valere (1975)	1.061	0.392	2.876	0.43
Frank (1975)	0.959	0.289	3.185	0.29
UK Collab (1976)	0.876	0.570	1.346	2.29
Klein (1976)	3.200	0.296	34.588	0.07
Austrian (1977)	0.562	0.365	0.867	2.26
Lasierra (1977)	0.222	0.019	2.533	0.07
N German (1977)	1.215	0.797	1.853	2.38
Witchitz (1977)	0.778	0.199	3.044	0.23
2nd Australian (1977	0.806	0.440	1.477	1.16
3rd European (1977)	0.416	0.242	0.716	1.44
ISAM (1986)	0.872	0.599	1.270	2.99
GISSI-1 (1986)	0.807	0.721	0.903	33.44
ISIS-2 (1988)	0.746	0.676	0.822	44.19
I-V pooled ES	0.774	0.725	0.826	100.00

```
  Heterogeneity chi-squared =   31.50 (d.f. = 21) p = 0.066
  I-squared (variation in ES attributable to heterogeneity) =  33.3%

  Test of ES=1 : z=   7.71 p = 0.000
```

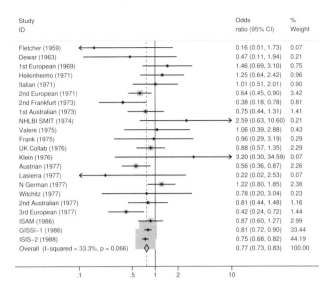

Figure 1. Streptokinase meta-analysis

It can be seen from the fixed-effects weights, and the graphical display, that the results are dominated by the two large trials reported in 1986. We now do a cumulative meta-analysis:

```
. metacum logor selogor, eform fixed label(namevar=trialnam, yearvar=year)
> xlabel(.1,.5,1,2) force effect("Odds ratio")
              Study        |     ES      [95% Conf. Interval]
---------------------------+-------------------------------------------------------
Fletcher (1959)            |   0.159      0.015     1.732
Dewar (1963)               |   0.355      0.105     1.200
1st European (1969)        |   0.989      0.522     1.875
Heikinheimo (1971)         |   1.106      0.698     1.753
Italian (1971)             |   1.076      0.734     1.577
2nd European (1971)        |   0.809      0.624     1.048
2nd Frankfurt (1973)       |   0.742      0.581     0.946
1st Australian (1973       |   0.744      0.595     0.929
NHLBI SMIT (1974)          |   0.767      0.615     0.955
Valere (1975)              |   0.778      0.628     0.965
Frank (1975)               |   0.783      0.634     0.968
UK Collab (1976)           |   0.801      0.662     0.968
Klein (1976)               |   0.808      0.668     0.976
Austrian (1977)            |   0.762      0.641     0.906
Lasierra (1977)            |   0.757      0.637     0.900
N German (1977)            |   0.811      0.691     0.951
Witchitz (1977)            |   0.810      0.691     0.950
2nd Australian (1977       |   0.810      0.695     0.945
3rd European (1977)        |   0.771      0.665     0.894
ISAM (1986)                |   0.784      0.683     0.899
GISSI-1 (1986)             |   0.797      0.731     0.870
ISIS-2 (1988)              |   0.774      0.725     0.826
---------------------------+-------------------------------------------------------
```

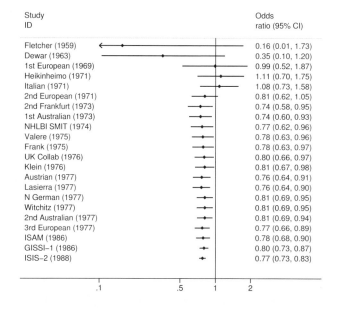

Figure 2. Streptokinase cumulative meta-analysis

By the end of 1977 there was clear evidence that streptokinase treatment prevented death following myocardial infarction. The point estimate of the pooled treatment effect was virtually identical in 1977 (odds ratio = 0.771) and after the results of the large trials in 1986 (odds ratio = 0.774).

5 Note

The command `metan` (Bradburn, Deeks, and Altman 1998; Harris et al. 2008)[3] should be installed before running `metacum`.

6 Acknowledgments

I thank Stephen Sharp for reviewing the command, Matthias Egger for providing the streptokinase data, and Thomas Steichen for providing the alternative forms of command syntax.

7 References

Bradburn, M. J., J. J. Deeks, and D. G. Altman. 1998. sbe24: metan—an alternative meta-analysis command. *Stata Technical Bulletin* 44: 4–15. Reprinted in *Stata Technical Bulletin Reprints*, vol. 8, pp. 86–100. College Station, TX: Stata Press. (Reprinted in this collection on pp. 3–28.)

Harris, R. J., M. J. Bradburn, J. J. Deeks, R. M. Harbord, D. G. Altman, and J. A. C. Sterne. 2008. metan: Fixed- and random-effects meta-analysis. *Stata Journal* 8: 3–28. (Reprinted in this collection on pp. 29–54.)

Lau, J., E. M. Antman, J. Jimenez-Silva, B. Kupelnick, F. Mosteller, and T. C. Chalmers. 1992. Cumulative meta-analysis of therapeutic trials for myocardial infarction. *New England Journal of Medicine* 327: 248–254.

Sharp, S., and J. A. C. Sterne. 1997. sbe16: Meta-analysis. *Stata Technical Bulletin* 38: 9–14. Reprinted in *Stata Technical Bulletin Reprints*, vol. 7, pp. 100–106. College Station, TX: Stata Press.

———. 1998. sbe16.1: New syntax and output for the meta-analysis command. *Stata Technical Bulletin* 42: 6–8. Reprinted in *Stata Technical Bulletin Reprints*, vol. 7, pp. 106–108. College Station, TX: Stata Press.

---

3. The original command to be installed was `meta`; see Sharp and Sterne (1997, 1998).—Ed.

The Stata Technical Bulletin (1999)
STB-49, sbe28, pp. 15–17

Meta-analysis of p-values

Aurelio Tobias
Senior Medical Statistician
National School of Public Health
Instituto de Saluid Carlos III
Madrid, Spain
atobias@isciii.es

Fisher's work on combining p-values (Fisher 1932) has been suggested as the origin of meta-analysis (Jones 1995). However, combination of p-values presents serious disadvantages, relative to combining estimates. For example, when p-values are testing different null hypotheses, they do not consider the direction of the association combining opposing effects, they cannot quantify the magnitude of the association, nor study heterogeneity between studies. Combination of p-values may be the only available option if nonparametric analyses of individual studies have been performed or if little information apart from the p-value is available about the result of a particular study (Jones 1995).

1 Fisher's method

This method (Fisher 1932) combines the probabilities of several hypotheses tests, testing the same null hypothesis

$$U = -2\sum_{j=1}^{k} \ln(p_j)$$

where the p_j are the one-tailed p-values for each study, and k is the number of studies. Then U follows a χ^2 distribution with $2k$ degrees of freedom. This method is not suggested to combine a large number of studies because it tends to reject the null hypothesis routinely (Rosenthal 1984). It also tends to have problems combining studies that are statistically significant, but in opposite directions (Rosenthal 1980).

2 Edgington's methods

The first method (Edgington 1972a) is based on the sum of probabilities

$$p = \left(\sum_{j=1}^{K} p_j\right)^k \bigg/ k!$$

The results obtained are similar to Fisher's method, but it is also restricted for a small number of studies. This method presents problems when the sum of probabilities is higher than one; in this situation the combined probability tends to be conservative (Rosenthal 1980).

An alternative method was also suggested by Edgington (1972b), to combine more than four studies, based on the contrast of the p-value average

$$\bar{p} = \sum_{j=1}^{k} p_j \Big/ k$$

in which case $U = (0.5 - \bar{p})\sqrt{12}$ follows a normal distribution.

3 Syntax

The command `metap` works on a dataset containing the p-values for each study. The syntax is as follows:

`metap` *pvar* $\big[\,if\,\big]$ $\big[\,in\,\big]$ $\big[\,,\,\underline{\text{method}}(\#)\,\big]$

4 Option

`method(#)` combines the p-values using three available methods:

`method(f)`, Fisher's method. This is the default.

`method(ea)`, Edgington's additive method based on the sum of probabilities (Edgington 1972a). This method is suggested to combine a small number of studies, producing similar results as Fisher's method.

`method(en)`, Edgington's normal curve method, based on the contrast of the p-value average (Edgington 1972b). This method is suggested to combine a large number of studies.

5 Example

We consider data from seven placebo-controlled studies on the effect of aspirin in preventing death after myocardial infarction. Fleiss (1993) published an overview of these data. Let us assume that each study included in the meta-analysis is testing the same null hypothesis $H_0\colon \theta \leq 0$ versus the alternative $H_1\colon \theta > 0$. If the estimate of the log odds-ratio and its standard error is available, then one-tailed p-values can easily be generated using the `normprob` function:

```
. generate pvar=normprob(-logrr/logse)

. list studyid logrr logse pvar, noobs
 studyid      logrr       logse        pvar
   MCR-1     0.3289      0.1972    .0476728
     CDP     0.3853      0.2029    .0287845
   MRC-2     0.2192      0.1432    .0629185
    GASP     0.2229      0.2545    .1905599
   PARIS     0.2261      0.1876    .1140584
    AMIS    -0.1249      0.0981    .8985248
  ISIS-2     0.1112      0.0388    .0020786
```

In this situation, all methods to combine *p*-values produce similar results:

```
. metap pvar
Meta-analysis of p_values
----------------------------------------------------------------
  Method              |    chi2         p_value      studies
--------------------+---------------------------------------------
  Fisher              |  38.938235     .00037283     7
----------------------------------------------------------------

. metap pvar, e(a)
Meta-analysis of p_values
----------------------------------------------------------------
  Method              |    .            p_value      studies
--------------------+---------------------------------------------
  Edgington, additive|    .           .00157658     7
----------------------------------------------------------------

. metap pvar, e(n)
Meta-analysis of p_values
----------------------------------------------------------------
  Method              |    Z            p_value      studies
--------------------+---------------------------------------------
  Edgington, Normal   |  2.8220842     .00238563     7
----------------------------------------------------------------
```

These figures agree with the result obtained using the `meta` command introduced in Sharp and Sterne (1997, 1998) on a fixed effects ($z = 3.289$, $p = 0.001$) and random effects ($z = 2.093$, $p = 0.036$) models, respectively. However, the combination of *p*-values presents the serious limitations described previously.

6   Individual or frequency records

As for other meta-analysis commands, `metap` works on data contained in frequency records, one for each study or trial.

7   Saved results

`metap` saves the following results:

S_1	method used to combine the *p*-values
S_2	number of studies
S_3	statistic used to obtain the combined probability
S_4	values of the statistic described in S_3
S_5	combined probability
r(method)	the method used to combine the *p*-values
r(n)	the number of studies
r(stat)	the statistic used to combine the *p*-values
r(z)	the value of the statistic used
r(pvalue)	returns the combined *p*-value

8 References

Edgington, E. S. 1972a. An additive method for combining probability values from independent experiments. *Journal of Psychology* 80: 351–363.

———. 1972b. A normal curve method for combining probability values from independent experiments. *Journal of Psychology* 82: 85–89.

Fisher, R. A. 1932. *Statistical Methods for Research Workers*. 4th ed. Edinburgh: Oliver & Boyd.

Fleiss, J. L. 1993. The statistical basis of meta-analysis. *Statistical Methods in Medical Research* 2: 121–145.

Jones, D. R. 1995. Meta-analysis: Weighing the evidence. *Statistics in Medicine* 14: 137–149.

Rosenthal, R., ed. 1980. *New Directions for Methodology of Social and Behavioral Science*. Volume V. San Francisco: Sage.

Rosenthal, R. 1984. Valid interpretation of quantitative research results. In *New Directions for Methodology of Social and Behavioral Science: Forms of Validity in Research*, ed. D. Brinberg and L. Kidder, 12. San Francisco: Jossey-Bass.

Sharp, S., and J. A. C. Sterne. 1997. sbe16: Meta-analysis. *Stata Technical Bulletin* 38: 9–14. Reprinted in *Stata Technical Bulletin Reprints*, vol. 7, pp. 100–106. College Station, TX: Stata Press.

———. 1998. sbe16.1: New syntax and output for the meta-analysis command. *Stata Technical Bulletin* 42: 6–8. Reprinted in *Stata Technical Bulletin Reprints*, vol. 7, pp. 106–108. College Station, TX: Stata Press.

Part 2

Meta-regression: metareg

Interpretation of the results of meta-analyses is simplest when there is little between-study heterogeneity. It is then appropriate to report a fixed-effect analysis, which assumes that the studies estimate the same underlying effect. Random-effects meta-analyses allow for between-study heterogeneity, but the interpretation of such analyses is more subtle than is commonly realized because rather than estimating a single effect, they estimate a mean effect, while the true effect is assumed to vary between studies around this mean. It is desirable, rather than simply allowing for heterogeneity, to understand the reasons for it. This can be done using meta-regression.

The user-written `metareg` command remains one of the few implementations of meta-regression and has been updated to account for improvements in Stata estimation facilities and recent methodological developments. However, enthusiasm for meta-regression has been tempered over the last decade. These are observational analyses in which the unit of analysis is the study. There are often more potential explanations for heterogeneity than studies in the meta-analysis, and Higgins and Thompson (2004) have shown that, in this context, multiple comparisons can substantially inflate the rate of false-positive findings. The updated `metareg` command includes the permutation test proposed by these authors to deal with the problem of multiple comparisons.

Because Harbord and Higgins (2008) describe a comprehensive update to the command `metareg`, this article appears first; the original syntax continues to work. This syntax and the formulas implemented in the 1998 version of `metareg` are then described in the article by Sharp (1998).

Two articles that appear later in this collection also discuss meta-regression. White (2011) (see part 4) discusses meta-regression for multivariate outcome meta-analysis models using `mvmeta`. Palmer and Sterne (Forthcoming) (see part 7) describe how to fit fixed- and random-effects meta-regression models using the `sem` and `gsem` commands for structural equation modeling.

1   References

Harbord, R. M., and J. P. T. Higgins. 2008. Meta-regression in Stata. *Stata Journal* 8: 493–519. (Reprinted in this collection on pp. 85–111.)

Higgins, J. P. T., and S. G. Thompson. 2004. Controlling the risk of spurious findings from meta-regression. *Statistics in Medicine* 23: 1663–1682.

Palmer, T. M., and J. A. C. Sterne. Forthcoming. Fitting fixed- and random-effects meta-analysis models using structural equation modeling with the sem and gsem commands. *Stata Journal* (Reprinted in this collection on pp. 435–461.)

Sharp, S. 1998. sbe23: Meta-analysis regression. *Stata Technical Bulletin* 42: 16–22. Reprinted in *Stata Technical Bulletin Reprints*, vol. 7, pp. 148–155. College Station, TX: Stata Press. (Reprinted in this collection on pp. 112–120.)

White, I. R. 2011. Multivariate random-effects meta-regression: Updates to mvmeta. *Stata Journal* 11: 255–270. (Reprinted in this collection on pp. 249–264.)

The Stata Journal (2008)
8, Number 4, sbe23_1, pp. 493–519

Meta-regression in Stata

Roger M. Harbord
Department of Social Medicine
University of Bristol
Bristol, UK
roger.harbord@bristol.ac.uk

Julian P. T. Higgins
MRC Biostatistics Unit
Cambridge, UK
julian.higgins@mrc-bsu.cam.ac.uk

Abstract. We present a revised version of the `metareg` command, which performs meta-analysis regression (meta-regression) on study-level summary data. The major revisions involve improvements to the estimation methods and the addition of an option to use a permutation test to estimate p-values, including an adjustment for multiple testing. We have also made additions to the output, added an option to produce a graph, and included support for the `predict` command. Stata 8.0 or above is required.

Keywords: sbe23_1, meta-regression, meta-analysis, permutation test, multiple testing, metareg

1 Introduction

Meta-analysis regression, or meta-regression, is an extension to standard meta-analysis that investigates the extent to which statistical heterogeneity between results of multiple studies can be related to one or more characteristics of the studies (Thompson and Higgins 2002). Like meta-analysis, meta-regression is usually conducted on study-level summary data, because individual observations from all studies (often referred to as individual patient data in medical applications) are frequently not available.

Sharp (1998) introduced the `metareg` command to perform meta-regression on study-level summary data. In this article, we present a substantially updated and largely rewritten version of `metareg`. The planning and interpretation of meta-regression studies raises substantial statistical issues discussed at length elsewhere (Davey Smith, Egger, and Phillips 1997; Higgins et al. 2002; Thompson and Higgins 2002, 2005). In this article, we will concentrate on the rationale for and the implementation and interpretation of the following new features of `metareg`:

- An improved algorithm for the estimation of the between-study variance, τ^2, by residual (restricted) maximum likelihood (REML)

- A modification to the calculation of standard errors, p-values, and confidence intervals for coefficients suggested by Knapp and Hartung (2003)

- Various enhancements to the output

- An option to produce a graph of the fitted model with a single covariate

- An option to calculate permutation-based p-values, including an adjustment for multiple testing based on the work of Higgins and Thompson (2004)

- Support for many of Stata's postestimation commands, including `predict`

We begin with a brief outline in section 2 of the statistical basis of meta-analysis and meta-regression, and we continue with a summary in section 3 of the relationship of `metareg` to other Stata commands. Section 4 introduces two example datasets that we use to illustrate the discussion of new features in section 5, which constitutes the main body of the article and has subsections corresponding to each of the new features listed above. The final two sections are reference material: Section 7 gives the Stata syntax and full list of options for `metareg` and `predict` after `metareg`, and lists the results saved by the command. Finally, section 7 gives details of the methods and formulas used.

2   Basis of meta-regression

In this section, we outline the statistical basis of random- and fixed-effects meta-regression and their relation to random- and fixed-effects meta-analysis. We will use mathematical formulas for brevity and precision. Less mathematically inclined readers or those who are already familiar with the principles of meta-analysis and meta-regression can skip this section.

We assume that study i of a total of n studies provides an estimate, y_i, of the effect of interest, such as a log odds-ratio, log risk-ratio, or difference in means. Each study also provides a standard error for this estimate, σ_i, which we assume is known, as is common in meta-analysis (although in practice, it will have been estimated from the data in that study). Let us start from the simplest model:

- *Fixed-effects meta-analysis* assumes that there is a single true effect size, θ, so that
$$y_i \sim N(\theta, \sigma_i^2)$$
  or equivalently,
$$y_i = \theta + \epsilon_i, \qquad \text{where} \ \ \epsilon_i \sim N(0, \sigma_i^2)$$

- *Random-effects meta-analysis* allows the true effects, θ_i, to vary between studies by assuming that they have a normal distribution around a mean effect, θ:
$$y_i \mid \theta_i \sim N(\theta_i, \sigma_i^2), \qquad \text{where} \ \ \theta_i \sim N(\theta, \tau^2)$$
  So
$$y_i \sim N(\theta, \sigma_i^2 + \tau^2)$$
  or equivalently,
$$y_i = \theta + u_i + \epsilon_i, \qquad \text{where} \ \ u_i \sim N(0, \tau^2) \ \text{and} \ \ \epsilon_i \sim N(0, \sigma_i^2)$$

Here τ^2 is the between-study variance and must be estimated from the data.

- *Fixed-effects meta-regression* extends fixed-effects meta-analysis by replacing the mean, θ, with a linear predictor, $\mathbf{x}_i\boldsymbol{\beta}$:

$$y_i \sim N(\theta_i, \sigma_i^2), \qquad \text{where} \quad \theta_i = \mathbf{x}_i\boldsymbol{\beta}$$

or equivalently,

$$y_i = \mathbf{x}_i\boldsymbol{\beta} + \epsilon_i, \qquad \text{where} \ \epsilon_i \sim N(0, \sigma_i^2)$$

Here $\boldsymbol{\beta}$ is a $k \times 1$ vector of coefficients (including a constant if fitted), and $\mathbf{x}_i$ is a $1 \times k$ vector of covariate values in study i (including a 1 if a constant is fit).

- *Random-effects meta-regression* allows for such residual heterogeneity (between-study variance not explained by the covariates) by assuming that the true effects follow a normal distribution around the linear predictor:

$$y_i \,|\, \theta_i \sim N(\theta_i, \sigma_i^2), \qquad \text{where} \ \theta_i \sim N(\mathbf{x}_i\boldsymbol{\beta}, \tau^2)$$

so

$$y_i \sim N(\mathbf{x}_i\boldsymbol{\beta}, \sigma_i^2 + \tau^2)$$

or equivalently,

$$y_i = \mathbf{x}_i\boldsymbol{\beta} + u_i + \epsilon_i, \qquad \text{where} \ u_i \sim N(0, \tau^2) \text{ and } \epsilon_i \sim N(0, \sigma_i^2)$$

Random-effects meta-regression can be considered either an extension to fixed-effects meta-regression that allows for residual heterogeneity or an extension to random-effects meta-analysis that includes study-level covariates.

Table 1 summarizes the relationships between these models and gives the corresponding Stata commands, which are summarized in the next section.

Table 1. Summary of `metareg` and related Stata commands

	No covariates	With covariate(s)
Fixed-effects model	fixed-effects meta-analysis	fixed-effects meta-regression (not recommended)
	`metan` with `fixedi`, `peto`, or no options	`vwls`
Random-effects model	random-effects meta-analysis	random-effects meta-regression (mixed-effects meta-regression)
	`metan` with `random` or `randomi` options	`metareg`

3   Relation to other Stata commands

Both fixed- and random-effects meta-analysis are available in the user-written package `metan` (Harris et al. 2008). Random-effects meta-analysis can also be performed with `metareg` by not including any covariates (the method-of-moments estimate for between-study variance must be specified to produce identical results to the `metan` command). `metan` can also be used to generate the variables required by `metareg` containing the effect estimate and its standard error for each study from data in various other forms (Harris et al. 2008).

Fixed-effects meta-regression can be fit by weighted least squares by using the official Stata command `vwls` (see [R] **vwls**) with the weights $1/\sigma_i^2$. Fixed-effects meta-regression is not usually recommended, however, because it assumes that all the heterogeneity can be explained by the covariates, and it leads to excessive type I errors when there is residual, or unexplained, heterogeneity (Higgins and Thompson 2004; Thompson and Sharp 1999).

Random-effects meta-regression is closely related to the seldom-used "between-effects" model available in the official Stata command `xtreg` (see [XT] **xtreg**), with studies corresponding to units. Whereas meta-regression assumes that the within-study data have been summarized by an effect estimate, y_i, and its standard error, σ_i, for each study, `xtreg` requires data on individual observations, e.g., individual patient data. Meta-regression is often used on binary outcomes summarized by log odds-ratios or log risk-ratios and their standard errors, whereas `xtreg` is appropriate only for continuous outcomes. `xtreg` also uses different estimators from those available in `metareg`, which are outlined in section 5.1.

4 Background to examples

Our first example is from a meta-analysis of 28 randomized controlled trials of cholesterol-lowering interventions for reducing risk of ischemic heart disease (IHD). The outcome event was death from IHD or nonfatal myocardial infarction. These data are taken from table 1 of Thompson and Sharp (1999). Data from 25 of these trials were also published in Thompson (1993). The measure of effect size is the odds ratio, but statistical analysis is conducted on its natural logarithm, the log odds-ratio, because this has a sampling distribution more closely approximated by a normal distribution. The interventions are varied, with 18 trials of several different drugs, 9 trials of dietary interventions, and 1 trial of a surgical intervention. The eligibility criteria also differed—19 studies recruited only participants without known IHD on entry, 6 recruited only those with IHD, and 3 included those with or without IHD. The reduction in cholesterol varied among trials, as quantified by the difference in mean serum cholesterol concentrations between the treated and control subjects at the end of each trial. Interest focuses on estimating the odds ratio for any given degree of cholesterol reduction (e.g., 1 mmol/L), assuming that any effect on IHD is mediated through the reduction in serum cholesterol. The Stata dataset is named `cholesterol.dta`.

The second example is drawn from a systematic review of 10 randomized controlled trials of exercise as an intervention in the management of depression (Lawlor and Hopker 2001). Here the outcome, severity of depression, was measured on one of two numerical scales, and the measure of effect size was the standardized mean difference. There was considerable between-study heterogeneity in the results of the trials, and the authors considered eight study-level covariates that might explain this heterogeneity. We will focus on the five covariates selected by Higgins and Thompson (2004). The Stata dataset is named `xrcise4deprsn.dta`.

5 New and enhanced features

We now give details of each of the new and enhanced features available in this revision of `metareg`, as listed in section 1. Sections 5.1–5.3 are relevant to all uses of `metareg`. When there is a single continuous covariate, the fitted model can be presented graphically, as shown in section 5.4. Section 5.5 explores a permutation-based approach to calculating p-values, suggested by Higgins and Thompson (2004), who recommended its use when there are few studies and as a way of adjusting for multiple testing when there is more than one covariate of interest. Section 5.6 is intended for more advanced users only; it describes the postestimation facilities available after a `metareg` model has been fit, and it assumes some familiarity with random-effects models, as well as with Stata's graphics commands and postestimation tools.

5.1 Algorithm for REML estimation of τ^2

All algorithms for random-effects meta-regression first estimate the between-study variance, τ^2, and then estimate the coefficients, β, by weighted least squares by using the weights $1/(\sigma_i^2 + \tau^2)$, where σ_i^2 is the standard error of the estimated effect in study i. The default algorithm in `metareg` is REML, as advocated by Thompson and Sharp (1999).

The algorithm for REML estimation has been improved in this update of `metareg`. The original version used an iterative algorithm (Morris 1983) that was not guaranteed to converge and was only an approximation when the within-study standard errors varied. The original version of `metareg` sometimes misleadingly reported an estimate of $\widehat{\tau}^2 = 0$ when the algorithm was in fact diverging (for example, with the `cholesterol` data). This revised version of `metareg` instead directly maximizes the residual (restricted) log likelihood by using Stata's robust and well-tested `ml` command, avoiding the approximations and convergence problems of the previous method.

We decided not to implement the standard maximum likelihood (ML) estimator in this updated version of `metareg`. (To ensure all do-files written for the original version of `metareg` continue to work, however, the code of the original program is included in this package so that a request for the ML estimator can be handled by calling the original code.) Both REML and ML are iterative methods. Unlike REML, however, ML does not account for the degrees of freedom used in estimating the fixed effects. This can make a particular difference in meta-regression because the number of observations (studies) is often small. As a result, the ML estimate of τ^2 is often biased downward, leading to underestimated standard errors and anticonservative inference (Thompson and Sharp 1999; Sidik and Jonkman 2007).

Further details of the methods for the estimation of τ^2 are given in section 7.1.

5.2   Knapp–Hartung variance estimator and associated t test

Knapp and Hartung (2003) introduced a novel estimator for the variances of the effect estimates in meta-regression. Their variance estimator amounts to calculating a quadratic form, q, and multiplying the usual variance estimates by q if $q > 1$. This estimator should be used with a t distribution when calculating p-values and confidence intervals. They found this procedure to have much more appropriate false-positive rates than the standard approach, a finding confirmed by Higgins and Thompson (2004) in more extensive simulations.

We therefore recommend this variance estimator and have made it the default in `metareg`. It is particularly suitable for estimation of standard errors and confidence intervals. However, it can be unreasonably conservative (false-positive rates below the nominal level) when the number of studies is particularly small, further reducing the already limited power. When there are few studies, the permutation test detailed in section 5.5 below has the potential to provide a better, though more computationally intensive, method for calculating p-values.

5.3　Enhancements to the output

The following additions have been made to the output of `metareg` that is displayed above the coefficient table:

- A measure of the percentage of the residual variation that is attributable to between-study heterogeneity (I_{res}^2)

- The proportion of between-study variance explained by the covariates (a type of adjusted R^2 statistic)

- An overall test of all the covariates in the random-effects model

The iteration log is no longer displayed by default.

We will illustrate these additions by using the output of `metareg` in the simplest situation where a single continuous covariate is fit, using the `cholesterol` data as an example:

```
. use cholesterol
(Serum cholesterol reduction & IHD)

. metareg logor cholreduc, wsse(selogor)

Meta-regression                                    Number of obs   =        28
REML estimate of between-study variance            tau2            =     .0097
% residual variation attributable to heterogeneity I-squared_res   =    31.34%
Proportion of between-study variance explained     Adj R-squared   =    69.02%
With Knapp-Hartung modification
```

logor	Coef.	Std. Err.	t	P>|t|	[95% Conf. Interval]	
cholreduc	-.5056849	.1834858	-2.76	0.011	-.8828453	-.1285244
_cons	.1467225	.1374629	1.07	0.296	-.1358367	.4292816

Residual heterogeneity of the fixed-effects model

The residual heterogeneity statistic is the weighted sum of squares of the residuals from the fixed-effects meta-regression model and is a generalization of Cochran's Q from meta-analysis to meta-regression. To distinguish it from the total heterogeneity statistic Q that would be obtained from ordinary meta-analysis, i.e., without fitting any covariates, we will denote it by Q_{res} (Lipsey and Wilson [2001] denote the same statistic by Q_E). A test of the null hypothesis of no residual (unexplained) heterogeneity can be obtained by comparing Q_{res} to a χ^2 distribution with $n - k$ degrees of freedom. However, it is often more useful to quantify heterogeneity than to test for it (Higgins et al. 2003): The proportion of residual between-study variation due to heterogeneity, as opposed to sampling variability, is calculated as $I_{res}^2 = \max[0, \{Q_{res} - (n - k)\}/Q_{res}]$, an obvious extension to the I^2 measure in meta-analysis (Higgins et al. 2003).

From the value of I_{res}^2 in the output above, 31% of the residual variation is due to heterogeneity, with the other 69% attributable to within-study sampling variability.

Adjusted R^2

The proportion of between-study variance explained by the covariates can be calculated by comparing the estimated between-study variance, $\hat{\tau}^2$, with its value when no covariates are fit, $\hat{\tau}_0^2$. Adjusted R^2 is the relative reduction in the between-study variance, $R^2_{\mathrm{adj}} = (\hat{\tau}_0^2 - \hat{\tau}^2)/\hat{\tau}_0^2$. It is possible for this to be negative if the covariates explain less of the heterogeneity than would be expected by chance, but the same is true for adjusted R^2 in ordinary linear regression. It may be more common in meta-regression because the number of studies is often small.

In the above example, 69% of the between-study variance is explained by the covariate `cholreduc`, and the remaining between-study variance appears small at 0.0097. (It is coincidence that the figure of 69% also appears in the preceding subsection.)

Joint test for all covariates

When more than one covariate is fit, `metareg` reports a test of the null hypothesis that the coefficients of the covariates are all zero, obtained from a multiparameter Wald test by using Stata's `test` command (see [R] **test**). The test statistic is compared to the appropriate F distribution if the default Knapp–Hartung adjustment is used. If `metareg`'s z option is used to specify the use of conventional variance estimates and tests for the effect estimates, a χ^2 distribution is used for the joint test. To simplify the output, this test is not displayed when only a single covariate is fit because it would give an identical p-value to the one displayed for the covariate in the regression table.

This gives one way of controlling the risk of false-positive findings when performing meta-regression with multiple covariates: we can use the overall model p-value to assess if there is evidence for an association of *any* of the covariates with the outcome. However, when a small p-value indicates that there is such evidence, it becomes harder to decide which, and how many, of the covariates there is good evidence for. Another method of dealing with this multiplicity issue that may help overcome this problem, though at the expense of longer computation time, is given in section 5.5 below.

▷ **Example**

We illustrate this joint test by using all five covariates available in the data on exercise for depression:

```
. use xrcise4deprsn
(Exercise for depression)

. metareg smd abstract-phd, wsse(sesmd)
Meta-regression                                   Number of obs  =       10
REML estimate of between-study variance           tau2           =        0
% residual variation attributable to heterogeneity  I-squared_res =    0.00%
Proportion of between-study variance explained    Adj R-squared  = 100.00%
Simultaneous test for all covariates              Model F(5,4)   =     6.57
With Knapp-Hartung modification                   Prob > F       =   0.0460
```

smd	Coef.	Std. Err.	t	P>|t|	[95% Conf. Interval]	
abstract	-1.33993	.3892562	-3.44	0.026	-2.420678	-.2591814
duration	.1567629	.0616404	2.54	0.064	-.0143784	.3279041
itt	.4611682	.3883635	1.19	0.301	-.6171018	1.539438
alloc	-.4063866	.3503447	-1.16	0.311	-1.379099	.5663263
phd	-.0138045	.440595	-0.03	0.977	-1.237092	1.209483
_cons	-2.07241	.5683944	-3.65	0.022	-3.650526	-.4942942

Here $\hat{\tau}^2$ is zero, and it follows that $I^2_{\text{res}} = 0\%$ and $R^2_{\text{adj}} = 100\%$. The joint test for all five covariates gives a p-value of 0.046, indicating some evidence for an association of at least one of the covariates with the size of the treatment effect.

◁

5.4 Graph of the fitted model

When a single continuous covariate is fit, one common way to present the fitted model, sometimes referred to as a "bubble plot", is to graph the fitted regression line together with circles representing the estimates from each study, sized according to the precision of each estimate (the inverse of its within-study variance, σ_i^2). The graph option to metareg gives an easy way to produce such a plot, as illustrated in figure 1 for the cholesterol data.

```
. use cholesterol
(Serum cholesterol reduction & IHD)
. metareg logor cholreduc, wsse(selogor) graph
  (output omitted)
```

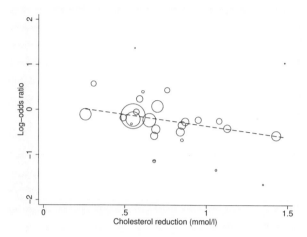

Figure 1. "Bubble plot" with fitted meta-regression line

An additional option, `randomsize`, is provided for those who prefer the size of the circles to depend on the weight of the study in the fitted random-effects meta-regression model (the inverse of its total variance, $\sigma_i^2 + \hat{\tau}^2$). This makes only a slight difference to the example above because the estimated between-study variance, $\hat{\tau}^2$, is small; in general, though, it will give circles that vary less in size.

Those wishing to further customize the plot can use the `predict` command to generate fitted values followed by a `graph twoway` command (see section 5.6).

5.5   Permutation test

Higgins and Thompson (2004) proposed using a permutation test approach to calculating p-values in meta-regression. Permutation tests provide a nonparametric way of simulating data under the null hypothesis (see, e.g., Manly [2007]). Calculation of exact permutation p-values would be feasible when there are few studies by enumeration of all possible permutations, but for simplicity, we have implemented a permutation test based on Monte Carlo simulation, i.e., based on *random* permutations.

The algorithm is similar to other applications of permutation methods, and it is implemented with Stata's `permute` command (see [R] **permute**). The covariates are randomly reallocated to the outcomes many times, and a t statistic is calculated each time. The true p-value for the relationship between a given covariate and the response is computed by counting the number of times these t statistics are greater than or equal to the observed t statistic. When multiple covariates are included in the meta-regression, the covariate values for a given study are kept together to preserve and account for their correlation structure. In meta-regression, unlike other regressions, the outcome consists of both the effect size and its standard error, and these must be kept together. This small complication makes it impossible to use `permute` on `metareg` directly from the command line when there are multiple covariates, so we have written a `permute()` option for `metareg`. This option also implements the following extension, which adjusts *p*-values for multiple tests when there are several covariates.

Multiplicity adjustment

When several covariates are used in meta-regression, either in several separate univariable meta-regressions or in one multiple meta-regression, there is an increased chance of at least one false-positive finding (type I error). The statistics obtained from the random permutations can be used to adjust for such multiple testing by comparing the observed *t* statistic for every covariate with the largest *t* statistic for any covariate in each random permutation. The proportion of times that the former equals or exceeds the latter gives the probability of observing a *t* statistic for any covariate as extreme or more extreme than that observed for a particular covariate, under the complete null hypothesis that all the regression coefficients are zero.

The number of random permutations must be specified—there is deliberately no default. We suggest that a small number (e.g., 100) be specified initially to check that the command is working as expected. The number should then be increased to at least 1,000, but 5,000 or 20,000 permutations may be necessary for sufficient precision (Manly 2007; Westfall and Young 1993). Because the `permute()` option uses Stata's random-number generator, the `set seed` command should be used first if replicability of results is desired. When the `permute()` option is specified, the defaults are to use the method-of-moments estimate of τ^2 for reasons of speed and to not use the Knapp–Hartung modification to the standard errors.

By default, `permute()` performs multivariable meta-regression; i.e., all the covariates are entered into a single model in each permutation.

▷ Example

We illustrate the use of the `permute()` option by using the data on exercise for depression.

```
. use xrcise4deprsn
(Exercise for depression)

. set seed 15160401

. metareg smd abstract-phd, wsse(sesmd) permute(20000)
Monte Carlo permutation test for meta-regression
Moment-based estimate of between-study variance
Without Knapp & Hartung modification to standard errors
P-values unadjusted and adjusted for multiple testing
                        Number of obs =         10
                        Permutations   =      20000

                               P
                smd | Unadjusted   Adjusted
            abstract       0.023      0.089
            duration       0.056      0.201
                 itt       0.311      0.721
               alloc       0.313      0.736
                 phd       0.978      1.000

largest Monte Carlo SE(P) = 0.0033
WARNING:
Monte Carlo methods use random numbers, so results may differ between runs.
Ensure you specify enough permutations to obtain the desired precision.
```

The first column of the results table gives permutation p-values without an adjust-ment for multiplicity. The results are in good agreement with the p-values obtained in section 5.3 without using the permutation option but with the Knapp–Hartung modifi-cation. The second column gives p-values adjusted for multiplicity. We see that all the p-values are increased. After adjusting for multiple testing, there remains some weak evidence that results of studies published as an abstract differ on average from results of studies published as a full article. The adjusted p-value of 0.089 gives the probability under the complete null hypothesis (that all regression coefficients are zero) of a t statis-tic for *any* of the five covariates as extreme or more extreme as that observed for the covariate `abstract`. As Higgins and Thompson (2004) suggest, this can be interpreted as describing the degree of "surprise" one might have about the observed result for this covariate, considering that five covariates are being examined. This is less conservative than the Bonferroni adjusted p-value of $0.0235 \times 5 = 0.1175$.

The output also gives the largest Monte Carlo standard error of the calculated p-values as an indication of the degree of precision obtained by the specified number of random permutations. Standard errors and "exact" confidence intervals for each of the p-values can be obtained by using the `detail` suboption. (These can always be calculated afterward by using the `cii` command if this option was not specified.)

◁

❏ **Technical note**

Higgins and Thompson (2004) originally proposed a slightly different permutation-based multiplicity adjustment: it compared the ith largest t statistic observed (for the "*i*th most significant" covariate) with the *i*th largest *t* statistic in each random permutation. This adjustment was implemented in a revised version of `metareg` released previously on the Statistical Software Components archive. This adjustment has been found to be hard to interpret in practice, however, because for the second most significant covariate it effectively gives a *joint* test of the two covariates with the largest two observed *t* statistics (and similarly for third and subsequent covariates if more than two covariates are supplied). The resulting multiplicity-adjusted *p*-value can turn out to be either larger *or smaller* than the unadjusted *p*-value, which can appear counter-intuitive.

For this release of `metareg`, we have therefore chosen to implement a different permutation-based algorithm for multiplicity adjustment based on the one-step "maxT" method of Westfall and Young (1993). This adjustment compares the *t* statistic for *every* covariate with the largest *t* statistic in each random permutation. The resulting multiplicity-adjusted *p*-values are always as large as or (usually) larger than the unadjusted *p*-values. This procedure ensures weak control of the familywise error rate, defined as the probability that at least one null hypothesis is rejected when *all* the null hypotheses are true (Shaffer 1995). It does not guarantee strong control of the familywise error rate, however; i.e., when one or more null hypotheses are false, it does not guarantee control of the proportion of the remaining true null hypotheses that are incorrectly rejected, though such strong control should be achieved asymptotically as the number of studies increases (Westfall and Young 1993; Shaffer 1995).

The false discovery rate (Benjamini and Hochberg 1995) and related procedures (Newson and The ALSPAC Study Team 2003; Storey, Taylor, and Siegmund 2004; Wacholder et al. 2004) have been suggested as an alternative method of multiplicity adjustment, but we have chosen not to implement such procedures in `metareg`. Such procedures are always either step-up or (more rarely) step-down algorithms. Although stepwise algorithms are suitable for hypothesis testing and often give greater power, the resulting adjusted *p*-values cannot be interpreted as giving the strength of evidence against the null hypothesis, the interpretation increasingly advocated in medicine and epidemiology (Sterne and Davey Smith 2001). In particular, stepwise methods may assign equal adjusted *p*-values to covariates with different unadjusted *p*-values.

❏

Suboptions to permute()

The `permute()` option can also be used to perform a set of single-variable meta-regressions at each permutation by adding the `univariable` suboption. This suboption reports permutation-based *p*-values for fitting a separate model for each covariate rather than including all the covariates in a multiple regression model. With several covariates, the execution time may be considerably longer than for multivariable meta-regression.

▷ **Example**

We add the `univariable` suboption to the previous example but reduce the number of permutations to cut down the computation time:

```
. metareg smd abstract-phd, wsse(sesmd) permute(5000, univariable)
Monte Carlo permutation test for single covariate meta-regressions
Moment-based estimate of between-study variance
Without Knapp & Hartung modification to standard errors
P-values unadjusted and adjusted for multiple testing
                        Number of obs =         10
                        Permutations  =       5000

                                   P
              smd │ Unadjusted  Adjusted
          ────────┼──────────────────────
          abstract │    0.021     0.043
          duration │    0.030     0.115
               itt │    0.384     0.946
             alloc │    0.330     0.861
               phd │    0.715     0.999
          ─────────

largest Monte Carlo SE(P) = 0.0069
WARNING:
Monte Carlo methods use random numbers, so results may differ between runs.
Ensure you specify enough permutations to obtain the desired precision.
```

In these results, unlike those from the previous example, each covariate is fit in a separate model and so is not adjusted for the other covariates. The *p*-values do not differ greatly in this example, however.

◁

There is also a `joint()` suboption that requests a permutation *p*-value for a joint test of the variables specified. This can be particularly useful if a set of indicator variables is used to model a categorical covariate.

A joint test of covariates can be obtained without using a permutation approach by instead using the **test** or **testparm** (see [R] **test**) command after `metareg`.

A *p*-value for the joint test is not included in the multiplicity-adjustment procedure because the two are neither technically nor philosophically compatible.

▷ **Example**

We return to the `cholesterol` data, in which the `ihdentry` variable is a categorical covariate with three categories indicating whether the study included participants with known IHD on entry to the study, without known IHD, or both:

```
. use cholesterol
(Serum cholesterol reduction & IHD)

. tab ihdentry, gen(ihd)
    Ischaemic heart │
    disease on entry │     Freq.     Percent        Cum.
    ────────────────┼──────────────────────────────────────
   Without known IHD │        6       21.43       21.43
            With IHD │       19       67.86       89.29
  With or without IHD │        3       10.71      100.00
    ────────────────┼──────────────────────────────────────
               Total │       28      100.00
```

```
. metareg logor cholreduc ihd2 ihd3, wsse(selogor)
> permute(5000, joint(ihd2 ihd3))
Monte Carlo permutation test for meta-regression

Moment-based estimate of between-study variance
Without Knapp & Hartung modification to standard errors
joint1 : ihd2 ihd3

P-values unadjusted and adjusted for multiple testing
                 Number of obs =       28
                 Permutations  =     5000
```

	P	
logor	Unadjusted	Adjusted
cholreduc	0.009	0.028
ihd2	0.611	0.933
ihd3	0.907	0.999
joint1	0.883	

```
largest Monte Carlo SE(P) = 0.0069
WARNING:
Monte Carlo methods use random numbers, so results may differ between runs.
Ensure you specify enough permutations to obtain the desired precision.
```

The p-value of 0.883 for the joint test of `ihd2` and `ihd3` indicates that there is very little evidence that the log odds-ratio differs among these three categories of studies, after adjusting for the degree of cholesterol reduction achieved in each study.

◁

5.6   Postestimation tools for metareg

`metareg` is programmed as a Stata estimation command and so supports most of Stata's postestimation commands (except when the `permute()` option is used). (One deliberate exception is `lrtest`, which is not appropriate after `metareg` because the REML log likelihood cannot be used to compare models with different fixed effects, while the method of moments does not give a likelihood.)

Several quantities can be obtained by using `predict` after `metareg`, including fitted values and predicted random effects (empirical Bayes estimates). These can be useful for producing graphs of the fitted model and for model checking. Details of the syntax and options are given in sections 6.4 and 6.5, and section 7.4 contains the formulas used.

We now illustrate the use of some of the quantities available from `predict` in a graph. Using the exercise for `depression` data, we conduct a meta-regression of the standardized mean difference on the single covariate duration that describes the duration of follow-up in each study. Figure 2 shows the fitted line and the estimates from the separate studies that would be produced by the `graph` option to `metareg`, and it also includes the empirical Bayes estimates and shaded bands showing both confidence and prediction intervals (we would not recommend including all these features on a single graph in practice). It was produced by the following commands:

```
. use xrcise4deprsn, clear
(Exercise for depression)

. metareg smd duration, wsse(sesmd)
```

Meta-regression	Number of obs	=	10
REML estimate of between-study variance	tau2	=	.2019
% residual variation attributable to heterogeneity	I-squared_res	=	55.83%
Proportion of between-study variance explained	Adj R-squared	=	55.16%
With Knapp-Hartung modification			

smd	Coef.	Std. Err.	t	P>|t|	[95% Conf. Interval]	
duration	.2097633	.0802611	2.61	0.031	.0246808	.3948457
_cons	-2.907511	.7339255	-3.96	0.004	-4.599946	-1.215076

```
. predict fit
(option xb assumed; fitted values)

. predict stdp, stdp

. predict stdf, stdf

. predict xbu, xbu

. local t = invttail(e(df_r)-1, 0.025)

. gen confl = fit - `t´*stdp

. gen confu = fit + `t´*stdp

. gen predl = fit - `t´*stdf

. gen predu = fit + `t´*stdf

. sort duration

. twoway rarea predl predu duration || rarea confl confu duration
> || line fit duration
> || scatter smd duration [aw=1/sesmd^2], msymbol(Oh)
> || scatter xbu duration, msymbol(t)
> ||, legend(label(1 "Prediction interval") label(2 "Confidence interval")
> cols(1))
```

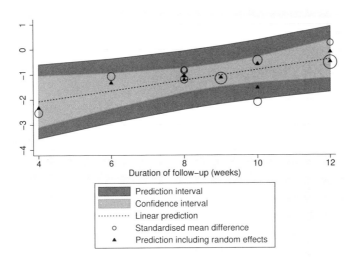

Figure 2. Confidence and prediction intervals and empirical Bayes estimates

The stdp option to predict gives the standard error of the fitted values excluding the random effects, commonly referred to as the standard error of the prediction. This standard error is used to draw a pointwise confidence interval, shown in light gray in figure 2, around the fitted line, illustrating our uncertainty about the position of the line. The stdf option to predict gives the standard deviation of the predicted distribution of the true value of the outcome in a future study with a given value of the covariate(s), commonly referred to as the standard error of the forecast. This standard error is used to draw a prediction interval, shown in dark gray in figure 2, around the fitted line, illustrating our uncertainty about the true effect we would predict in a future study with a known duration of follow-up. The prediction band will be wider than the confidence band unless $\tau^2 = 0$. The use of a t distribution in generating the intervals is an approximation, and opinions differ over the most appropriate degrees of freedom; we use $n - k - 1$ here to be consistent with the $n - 2$ used by Higgins, Thompson, and Spiegelhalter (2009) for confidence and prediction intervals in meta-analysis, where $k = 1$. The xbu option to predict gives the empirical Bayes estimates (predictions including random effects), shown as triangles in figure 2. These are our best estimates of the true effect in each study, assuming the fitted model is correct. If I^2_{res} is small, the empirical Bayes estimates will tend to lie well inside the prediction interval; if $\tau^2 = 0$, implying $I^2_{\text{res}} = 0$, they will all lie on the fitted line.

The statistics available from predict can also be useful for model checking and checking for outliers and influential studies. This checking is best done graphically. One possibility is a normal probability plot of the standardized predicted random effects (equivalently, standardized empirical Bayes residuals, or standardized shrunken residuals; see figure 3). This probability plot can be used to check the assumption of normality of the random effects, although because this assumption has been used in generating the predictions, only gross deviations are likely to be detected. Perhaps more usefully, the probability plot can be used to detect outliers:

```
. use cholesterol, clear
(Serum cholesterol reduction & IHD)
. qui metareg logor cholreduc, wsse(selogor)
. capture drop usta
. predict usta, ustandard
. qnorm usta, mlabel(id)
```

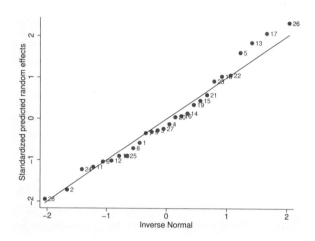

Figure 3. Normal probability plot of standardized shrunken residuals

Figure 3 suggests that the assumption of normal random effects is adequate, and there are no notable outliers because the largest standardized shrunken residual is only slightly over 2.

Other plots useful for model checking and identifying influential points in conventional linear regression may also be useful for meta-regression, for example, leverage–residual (L–R) plots, or plots of residuals versus either fitted values or a predictor; see [R] **regress postestimation** for further details of these and other plots (the various plot commands given there will not work after `metareg`, but it should be fairly straightforward to use `predict` followed by the appropriate `graph twoway` command to produce similar plots).

6   Syntax, options, and saved results

6.1   Syntax

The syntax of `metareg` has been revised somewhat from that of the original version (Sharp 1998). The original syntax should continue to work, but it is not documented here. ML estimation of τ^2 is not supported by the updated `metareg` program, but if the old `bsest(ml)` option is used, the new program simply calls the original version, which is incorporated within the updated `metareg.ado` file.

`metareg` *depvar* $\left[\,indepvars\,\right]$ $\left[\,if\,\right]$ $\left[\,in\,\right]$ `wsse(`*varname*`)` $\big[$, <u>ef</u>orm <u>g</u>raph
    `randomsize` <u>nocons</u>tant `mm reml eb` <u>k</u>napphartung `z` <u>tau2</u>test <u>level</u>(#)
    <u>permute</u>(# $\big[$, <u>u</u>nivariable <u>d</u>etail <u>j</u>oint(*varlist1* $\left[\,|\;\;varlist2\,\ldots\right]$)$\big]$) `log`
    maximize_options $\big]$

    `by` can be used with `metareg`; see [D] **by**.

6.2 Options

`wsse(`*varname*`)` specifies the variable containing σ_i, the standard error of *depvar*, within each study. All values of *varname* must be greater than zero. `wsse()` is required.

`eform` indicates to output the exponentiated form of the coefficients and to suppress reporting of the constant. This option may be useful when *depvar* is the logarithm of a ratio measure, such as a log odds-ratio or a log risk-ratio.

`graph` requests a line graph of fitted values plotted against the first covariate in *indepvars*, together with the estimates from each study represented by circles. By default, the circle sizes depend on the precision of each estimate (the inverse of its within-study variance), which is the weight given to each study in the fixed-effects model.

`randomsize` is for use with the `graph` option. It specifies that the size of the circles will depend on the weights in the random-effects model rather than the precision of each estimate. These random-effects weights depend on the estimate of τ^2.

The remaining options will mainly be of interest to more advanced users:

`noconstant` suppresses the constant term (intercept). This is rarely appropriate in meta-regression. *Note*: It might seem tempting to use the `noconstant` option in the cholesterol example to force the regression line through the origin, on the reasoning that an intervention that has no effect on cholesterol should have no effect on the odds of IHD. We would advise against using this option, however, both here and in most other circumstances. Using it here involves the assumption that the effect of the intervention on IHD is mediated entirely by cholesterol reduction. It also would not allow for measurement error in cholesterol reduction, which, through attenuation by errors (regression dilution bias), could lead to a nonzero intercept even when a zero intercept would be expected.

The `mm`, `reml`, and `eb` options are alternatives that specify the method of estimation of the additive (between-study) component of variance τ^2:

mm specifies the use of method of moments to estimate the additive (between-study) component of variance τ^2; this is a generalization of the DerSimonian and Laird (1986) method commonly used for random-effects meta-analysis. For speed, this is the default when the `permute()` option is specified, because it is the only noniterative method.

reml specifies the use of REML to estimate the additive (between-study) component of variance τ^2. This is the default unless the `permute()` option is specified. This revised version uses Stata's ML facilities to maximize the REML log likelihood. It will therefore not give identical results to the previous version of metareg, which used an approximate iterative method.

eb specifies the use of the "empirical Bayes" method to estimate τ^2 (Morris 1983).

knapphartung makes a modification to the variance of the estimated coefficients suggested by Knapp and Hartung (2003) and supported by Higgins and Thompson (2004), accompanied by the use of a t distribution in place of the standard normal distribution when calculating p-values and confidence intervals. This is the default unless the `permute()` option is specified.

z requests that the knapphartung modification not be applied and that the standard normal distribution be used to calculate p-values and confidence intervals. This is the default when the `permute()` option is specified with a fixed-effects model.

tau2test adds to the output two tests of $\tau^2 = 0$. The first is based on the residual heterogeneity statistic, Q_{res}. The second (not available if the mm option is also specified) is a likelihood-ratio test based on the REML log likelihood. These are two tests of the same null hypothesis (the fixed-effects model with $\tau^2 = 0$), but the alternative hypotheses are different, as are the distributions of the test statistics under the null, so close agreement of the two tests is not guaranteed. Both tests are typically of little interest because it is more helpful to quantify heterogeneity than to test for it (see section 5.3).

level(#) specifies the confidence level, as a percentage, for confidence intervals. The default is level(95) or as set by `set level`; see [U] **20.7 Specifying the width of confidence intervals**.

permute(...) calculates p-values by using a Monte Carlo permutation test. See section 6.3 below for more information about this option.

log requests the display of the iteration log during estimation of τ^2. This is ignored if the mm option is specified, because this uses a noniterative method.

maximize_options are ignored unless estimation of τ^2 is by REML. These options control the maximization process; see [R] **maximize**. They are ignored if the mm option is specified. You should never need to specify them; they are supported only in case problems in the REML estimation of τ^2 are ever reported or suspected.

6.3   Option for permutation test

The `permute()` option calculates p-values by using a Monte Carlo permutation test, as recommended by Higgins and Thompson (2004). To address multiple testing, `permute()` also calculates p-values for the most- to least-significant covariates, as the same authors also recommend.

The syntax of `permute()` is

permute(# [, univariable detail joint(*varlist1* [| *varlist2* ...])])

where # is required and specifies the number of random permutations to perform. Larger values give more precise p-values but take longer.

There are three suboptions:

univariable indicates that p-values should be calculated for a series of single covariate meta-regressions of each covariate in *varlist* separately, instead of a multiple meta-regression of all covariates in *varlist* simultaneously.

detail requests lengthier output in the format given by [R] **permute**.

joint(*varlist1* [| *varlist2* ...]) specifies that a permutation p-value should also be computed for a joint test of the variables in each varlist.

The `eform`, `level()`, and `z` options have no effect when the `permute()` option is specified.

6.4   Syntax of predict

The syntax of `predict` (see [R] **predict**) following `metareg` is

predict [*type*] *newvar* [*if*] [*in*] [, *statistic*]

statistic	description
xb	fitted values; the default
stdp	standard error of the prediction
stdf	standard error of the forecast
u	predicted random effects
ustandard	standardized predicted random effects
xbu	prediction including random effects
stdxbu	standard error of xbu
hat	leverage (diagonal elements of hat matrix)

These statistics are available both in and out of sample; type `predict ... if e(sample) ...` if wanted only for the estimation sample.

6.5 Options for predict

xb, the default, calculates the linear prediction, $\mathbf{x}_i\mathbf{b}$, that is, the fitted values *excluding* the random effects.

stdp calculates the standard error of the prediction (the standard error of the fitted values excluding the random effects).

stdf calculates the standard error of the forecast. This gives the standard deviation of the predicted distribution of the true value of *depvar* in a future study, with the covariates given by *varlist*. $\texttt{stdf}^2 = \texttt{stdp}^2 + \widehat{\tau}^2$.

u calculates the predicted random effects, u_i. These are the best linear unbiased predictions of the random effects, also known as the empirical Bayes (or posterior mean) estimates of the random effects, or as shrunken residuals.

ustandard calculates the standardized predicted random effects, i.e., the predicted random effects, u_i, divided by their (unconditional) standard errors. These may be useful for diagnostics and model checking.

xbu calculates the prediction *including* the random effects, $\mathbf{x}_i\mathbf{b} + u_i$, also known as the empirical Bayes estimates of the effects for each study.

stdxbu calculates the standard error of the prediction including random effects.

hat calculates the leverages (the diagonal elements of the projection hat matrix).

6.6 Saved results

When the permute() option is not specified, metareg saves the following in e():

Scalars

e(N)	number of observations	e(tau2)	estimate of τ^2
e(df_m)	model degrees of freedom	e(Q)	Cochran's Q
e(df_Q)	degrees of freedom for test of $Q = 0$	e(I2)	I-squared
		e(q_KH)	Knapp–Hartung variance modification factor
e(df_r)	residual degrees of freedom (if t tests used)	e(remll_c)	REML log likelihood, comparison model
e(remll)	REML log likelihood	e(tau2_0)	τ^2, constant-only model
e(chi2_c)	χ^2 for comparison test	e(chi2)	model χ^2
e(F)	model F statistic		

Macros

e(cmd)	metareg	e(depvar)	name of dependent variable
e(predict)	program used to implement predict	e(method)	REML, Method of moments, or Empirical Bayes
e(wsse)	name of wsse() variable	e(properties)	b V

Matrices

e(b)	coefficient vector	e(V)	variance–covariance matrix of estimators

Functions

e(sample)	marks estimation sample

`metareg, permute()` saves the following in `r()`:

Scalars
    `r(N)`   number of observations

Matrices
| `r(b)` | observed t statistics, T_{obs} | `r(p)` | observed proportions |
| `r(c)` | count when $\lvert T \rvert \geq \lvert T_{\mathrm{obs}} \rvert$ | `r(reps)` | number of nonmissing results |

7   Methods and formulas

The residual heterogeneity statistic, Q_{res}, is the residual weighted sum of squares from the fixed-effects model and is the same as the goodness-of-fit statistic computed by `vwls`:

$$Q_{\mathrm{res}} = \sum_i \left(\frac{y_i - \mathbf{x}_i \widehat{\boldsymbol{\beta}}}{\sigma_i} \right)^2$$

The proportion of residual variation due to heterogeneity is

$$I^2 = \max \left\{ \frac{Q_{\mathrm{res}} - (n - k)}{Q_{\mathrm{res}}}, \ 0 \right\}$$

The proportion of the between-study variance explained by the covariates (adjusted R-squared) is $R_a^2 = (\widehat{\tau}_0^2 - \widehat{\tau}^2)/\widehat{\tau}_0^2$, where $\widehat{\tau}^2$ and $\widehat{\tau}_0^2$ are the estimates of the between-study variance in models with and without the covariates, respectively.

7.1   Estimation of τ^2

Several different algorithms have been proposed for estimation of the between-study variance, τ^2, in meta-analysis (Sidik and Jonkman 2007) and meta-regression (Thompson and Sharp 1999). Three algorithms are available in this version of `metareg`. In each case, if the estimated value of τ^2 is negative, it is set to zero.

Method of moments is the only noniterative method, so it has the advantages of speed and robustness. It is the natural extension of the DerSimonian and Laird (1986) estimate commonly used in random-effects meta-analysis. The method-of-moments estimate of τ^2 is obtained by equating the observed value of Q_{res} to its expected value under the random-effects model, giving (DuMouchel and Harris 1983, eq. 3.12)

$$\widehat{\tau}_{\mathrm{MM}}^2 = \frac{Q_{\mathrm{res}} - (n + k)}{\sum_i \{ 1/\sigma_i^2 (1 - h_i) \}}$$

Here h_i is the ith diagonal element of the hat matrix $\mathbf{X}(\mathbf{X}'\mathbf{V}_0^{-1}\mathbf{X})^{-1}\mathbf{X}\mathbf{V}_0^{-1}$, where $\mathbf{V}_0 = \mathrm{diag}(\sigma_1^2, \sigma_2^2, \ldots, \sigma_n^2)$.

The iterative methods below use $\widehat{\tau}_{MM}^2$ as a starting value (this is a change from the original version of `metareg` (Sharp 1998), which used zero as a starting value).

REML estimation of τ^2 is based on maximization of the residual (or restricted) log likelihood,

$$L_R(\tau^2) = -\frac{1}{2}\sum_i\left\{\log(\sigma_i^2+\tau^2)+\frac{(y_i-\mathbf{x}_i\widehat{\boldsymbol{\beta}})^2}{\sigma_i^2+\tau^2}\right\}-\frac{1}{2}\log|\mathbf{X}'\mathbf{V}^{-1}\mathbf{X}|$$

where $\mathbf{V}=\mathrm{diag}(\sigma_1^2+\tau^2,\sigma_2^2+\tau^2,\ldots,\sigma_n^2+\tau^2)$ and $\widehat{\boldsymbol{\beta}}=(\mathbf{X}'\mathbf{V}^{-1}\mathbf{X})^{-1}\mathbf{X}'\mathbf{V}^{-1}\mathbf{y}$ (Harville 1977). This log likelihood is maximized by Stata's `ml` command, using the `d0` method, which calculates all derivatives numerically.

The "empirical Bayes" estimator of τ^2 is so named because of its introduction in an article on empirical Bayes inference by Morris (1983), although as he states, any approximately unbiased estimate of τ^2 could be used in such a setting. Thompson and Sharp (1999) found it to give substantially larger estimates of τ^2 than other methods. Others suggest it performs well in simulations based on 2×2 tables (Berkey et al. 1995; Sidik and Jonkman 2007), although this may be due to overestimation of the within-study standard errors in small studies by the conventional (Woolf) estimate rather than the properties of the empirical Bayes method itself (Sutton and Higgins 2008). It can also be considered to be a method-of-moments estimator, formed by equating the weighted sum of squares of the residuals from the random-effects model to its expected value (Knapp and Hartung 2003). It is found by iterating the following equation (Morris 1983; Berkey et al. 1995):

$$\widehat{\tau}_{\mathrm{EB}}^2=\frac{n/(n-k)\sum_i\left\{(y_i-\mathbf{x}_i\widehat{\boldsymbol{\beta}})^2/(\sigma_i^2+\widehat{\tau}_{\mathrm{EB}}^2)-\sigma_i^2\right\}}{\sum_i(\sigma_i^2+\widehat{\tau}_{\mathrm{EB}}^2)^{-1}}$$

At each iteration, $\widehat{\boldsymbol{\beta}}$ must be reestimated using a weighted least-squares regression of $\mathbf{y}$ on $\mathbf{X}$ with the weights $1/(\sigma_i^2+\widehat{\tau}_{\mathrm{EB}}^2)$.

7.2   Estimation of β

Once τ^2 has been estimated by one of the methods above, the estimated coefficients, $\widehat{\boldsymbol{\beta}}$, are obtained by a weighted least-squares regression of $\mathbf{y}$ on $\mathbf{X}$ with the weights $1/(\sigma_i^2+\widehat{\tau}^2)$. The conventional estimate of the variance–covariance matrix of the coefficients is $(\mathbf{X}'\widehat{\mathbf{V}}^{-1}\mathbf{X})^{-1}$, where $\widehat{\mathbf{V}}=\mathrm{diag}(\sigma_1^2+\widehat{\tau}^2,\sigma_2^2+\widehat{\tau}^2,\ldots,\sigma_n^2+\widehat{\tau}^2)$.

7.3   Knapp–Hartung variance modification

Knapp and Hartung (2003) proposed multiplying the conventional estimate of the variance of the coefficients given above by $\max(q,1)$, where the *Knapp–Hartung variance modification factor* is

$$q=\frac{1}{n-k}\sum_i\frac{(y_i-\mathbf{x}_i\widehat{\boldsymbol{\beta}})^2}{\sigma_i^2+\widehat{\tau}^2}$$

With the "empirical Bayes" estimator of $\widehat{\tau}^2$, $q=1$, so this modification has no effect (Knapp and Hartung 2003).

7.4   Methods and formulas for predict

The *standard error of the prediction* (stdp) is $s_{p_i} = \sqrt{\mathbf{x}_i(\mathbf{X}'\widehat{\mathbf{V}}^{-1}\mathbf{X})^{-1}\mathbf{x}_i'}$.

The *leverages*, or diagonal elements of the projection matrix (hat), are

$$h_i = s_{p_i}^2/(\sigma_i^2 + \tau^2)$$

The *standard error of the forecast* (stdf) is $s_{f_i} = \sqrt{s_{p_i}^2 + \tau^2}$.

Denote the previously estimated coefficient vector by $\mathbf{b}$, and let $\lambda_i = \hat{\tau}^2/(\sigma_i^2 + \hat{\tau}^2)$ denote the empirical Bayes shrinkage factor for the ith observation.

The *predicted random effects* (u) are $u_i = \lambda_i(y_i - \mathbf{x}_i\mathbf{b})$.

The *standardized predicted random effects* (ustandard) are

$$u_{s_j} = (y_i - \mathbf{x}_i\mathbf{b})\left/\sqrt{\sigma_i^2 + \tau^2 - s_{p_i}^2}\right.$$

The *prediction including random effects* (xbu), or empirical Bayes estimate, is

$$\mathbf{x}_i\mathbf{b} + u_i = \lambda_i y_i + (1 - \lambda_i)\mathbf{x}_i\mathbf{b}$$

The *standard error of the prediction including random effects* (stdxbu) is

$$\sqrt{\lambda_i^2(\sigma_i^2 + \tau^2) + (1 - \lambda_i^2)s_{p_i}^2}$$

8   Acknowledgments

Stephen Sharp gave permission to release this package under the same name as his original Stata package for meta-regression and to incorporate his code. Debbie Lawlor gave permission to use the example dataset on exercise for depression and provided additional unpublished data. We thank Simon Thompson for his helpful comments on the manuscript, and we thank the organizers of and participants at a meeting in Park City, Utah, in 2005 for discussions that influenced the output displayed by metareg. Finally, we wish to thank the referee for helpful comments, which led to improvements in the program and the article.

9   References

Benjamini, Y., and Y. Hochberg. 1995. Controlling the false discovery rate: A practical and powerful approach to multiple testing. *Journal of the Royal Statistical Society, Series B* 57: 289–300.

Berkey, C. S., D. C. Hoaglin, F. Mosteller, and G. A. Colditz. 1995. A random-effects regression model for meta-analysis. *Statistics in Medicine* 14: 395–411.

Davey Smith, G., M. Egger, and A. N. Phillips. 1997. Meta-analysis: Beyond the grand mean? *British Medical Journal* 315: 1610–1614.

DerSimonian, R., and N. Laird. 1986. Meta-analysis in clinical trials. *Controlled Clinical Trials* 7: 177–188.

DuMouchel, W. H., and J. E. Harris. 1983. Bayes methods for combining the results of cancer studies in humans and other species. *Journal of the American Statistical Association* 78: 293–308.

Harris, R. J., M. J. Bradburn, J. J. Deeks, R. M. Harbord, D. G. Altman, and J. A. C. Sterne. 2008. metan: Fixed- and random-effects meta-analysis. *Stata Journal* 8: 3–28. (Reprinted in this collection on pp. 29–54.)

Harville, D. A. 1977. Maximum likelihood approaches to variance component estimation and to related problems. *Journal of the American Statistical Association* 72: 320–338.

Higgins, J. P. T., and S. G. Thompson. 2004. Controlling the risk of spurious findings from meta-regression. *Statistics in Medicine* 23: 1663–1682.

Higgins, J. P. T., S. G. Thompson, J. J. Deeks, and D. G. Altman. 2002. Statistical heterogeneity in systematic reviews of clinical trials: A critical appraisal of guidelines and practice. *Journal of Health Services Research and Policy* 7: 51–61.

———. 2003. Measuring inconsistency in meta-analyses. *British Medical Journal* 327: 557–560.

Higgins, J. P. T., S. G. Thompson, and D. J. Spiegelhalter. 2009. A re-evaluation of random-effects meta-analysis. *Journal of the Royal Statistical Society, Series A* 172: 137–159.

Knapp, G., and J. Hartung. 2003. Improved tests for a random-effects meta-regression with a single covariate. *Statistics in Medicine* 22: 2693–2710.

Lawlor, D. A., and S. W. Hopker. 2001. The effectiveness of exercise as an intervention in the management of depression: Systematic review and meta-regression analysis of randomised controlled trials. *British Medical Journal* 322: 763.

Lipsey, M. W., and D. B. Wilson. 2001. *Practical Meta-Analysis*. Thousand Oaks, CA: Sage.

Manly, B. F. J. 2007. *Randomization, Bootstrap and Monte Carlo Methods in Biology*. 3rd ed. Boca Raton, FL: Chapman & Hall/CRC.

Morris, C. N. 1983. Parametric empirical Bayes inference: Theory and applications. *Journal of the American Statistical Association* 78: 47–55.

Newson, R. B., and The ALSPAC Study Team. 2003. Multiple-test procedures and smile plots. *Stata Journal* 3: 109–132.

Shaffer, J. P. 1995. Multiple hypothesis testing. *Annual Review of Psychology* 46: 561–584.

Sharp, S. 1998. sbe23: Meta-analysis regression. *Stata Technical Bulletin* 42: 16–22. Reprinted in *Stata Technical Bulletin Reprints*, vol. 7, pp. 148–155. College Station, TX: Stata Press. (Reprinted in this collection on pp. 112–120.)

Sidik, K., and J. N. Jonkman. 2007. A comparison of heterogeneity variance estimators in combining results of studies. *Statistics in Medicine* 26: 1964–1981.

Sterne, J. A. C., and G. Davey Smith. 2001. Sifting the evidence—What's wrong with significance tests? *British Medical Journal* 322: 226–231.

Storey, J. D., J. E. Taylor, and D. Siegmund. 2004. Strong control, conservative point estimation and simultaneous conservative consistency of false discovery rates: A unified approach. *Journal of the Royal Statistical Society, Series B* 66: 187–205.

Sutton, A. J., and J. P. T. Higgins. 2008. Recent developments in meta-analysis. *Statistics in Medicine* 27: 625–650.

Thompson, S. G. 1993. Controversies in meta-analysis: The case of the trials of serum cholesterol reduction. *Statistical Methods in Medical Research* 2: 173–192.

Thompson, S. G., and J. P. T. Higgins. 2002. How should meta-regression analyses be undertaken and interpreted? *Statistics in Medicine* 21: 1559–1573.

———. 2005. Can meta-analysis help target interventions at individuals most likely to benefit? *Lancet* 365: 341–346.

Thompson, S. G., and S. Sharp. 1999. Explaining heterogeneity in meta-analysis: A comparison of methods. *Statistics in Medicine* 18: 2693–2708.

Wacholder, S., S. Chanock, M. Garcia-Closas, L. El ghormli, and N. Rothman. 2004. Assessing the probability that a positive report is false: An approach for molecular epidemiology studies. *Journal of the National Cancer Institute* 96: 434–442.

Westfall, P. H., and S. S. Young. 1993. *Resampling-Based Multiple Testing: Examples and Methods for p-Value Adjustment*. New York: Wiley.

The Stata Technical Bulletin (1998)
STB-42, sbe23, pp. 16–22

Meta-analysis regression[1]

Stephen Sharp
MRC Epidemiology Unit
Institute of Metabolic Science
Cambridge, UK
stephen.sharp@mrc-epid.cam.ac.uk

The command `metareg` extends a random-effects meta-analysis to estimate the extent to which one or more covariates, with values defined for each study in the analysis, explain heterogeneity in the treatment effects. Such analysis is sometimes termed "meta-regression" (Lau, Ioannidis, and Schmid 1998). Examples of such study-level covariates might be average duration of follow-up, some measure of study quality, or, as described in this article, a measure of the geographical location of each study. `metareg` fits models with two additive components of variance, one representing the variance within units, the other the variance between units, and therefore is applicable both to the meta-analysis situation, where each unit is one study, and to other situations such as multicenter trials, where each unit is one center. Here `metareg` is explained in the meta-analysis context.

1 Background

Suppose y_i represents the treatment effect measured in study i (k independent studies, $i = 1, \ldots, k$), such as a log odds-ratio or a difference in means, v_i is the (within-study) variance of y_i, and $x_{i1}, \ldots, x_{ip}$ are measured study-level covariates. A weighted normal errors regression model is

$$Y \sim N(X\beta, V)$$

where $Y = (y_1, \ldots, y_k)^T$ is the $k \times 1$ vector of treatment effects, with ith element y_i, X is a $k \times (p + 1)$ design matrix with ith row $(1, x_{i1}, \ldots, x_{ip})$, $\beta = (\beta_0, \ldots, \beta_p)^T$ is a $(p + 1) \times 1$ vector of parameters, and V is a $k \times k$ diagonal variance matrix, with ith diagonal element v_i.

The parameters of this model can be estimated in Stata using `regress` with analytic weights $w_i = 1/v_i$. However, v_i represents the variance of the treatment effect within study i, so this model does not take into account any possible residual heterogeneity in the treatment effects *between* studies. One approach to incorporating residual heterogeneity is to include an additive between-study variance component τ^2, so the ith diagonal element of the variance matrix V becomes $v_i + \tau^2$.

The parameters of the model can then be estimated using a weighted regression with weights equal to $1/v_i + \tau^2$, but τ^2 must be explicitly estimated in order to carry out

---

1. This article uses the obsolete meta-analysis command `meta`, as well as obsolete graphics commands. The syntax described in the article still functions, although the updated syntax described in the article by Harbord and Higgins (2008) is preferred.—Ed.

the regression. `metareg` allows four alternative methods for estimation of τ^2, three of them are iterative, while one is noniterative and an extension of the moment estimator proposed for random effects meta-analysis without covariates (DerSimonian and Laird 1986).

2   Method-of-moments estimator

Maximum likelihood estimates of the β parameters are first obtained by weighted regression assuming $\widehat{\tau}^2 = 0$, and then a moment estimator of τ^2 is calculated using the residual sum of squares from the model,

$$\text{RSS} = \sum_{i=1}^{k} w_i (y_i - \widehat{y}_i)^2$$

as follows:

$$\widehat{\tau}^2_{\text{mm}} = \frac{\text{RSS} - \{k - (p+1)\}}{\sum_{i=1}^{k} w_i - \text{tr}\{V^{-1}X(X'V^{-1}X)^{-1}X'V^{-1}\}}$$

where $\widehat{\tau}^2_{\text{mm}} = 0$ if $\text{RSS} < k - (p+1)$ (DuMouchel and Harris 1983).

A weighted regression is then carried out with new weights $w_i^* = 1/\widehat{\tau}^2 + v_i$ to provide a new estimate of β. The formula for $\widehat{\tau}^2_{\text{mm}}$ in the case of no covariate reduces to the standard moment estimator (DerSimonian and Laird 1986).

3   Iterative procedures

Three other methods for estimating τ^2 have been proposed and require an iterative procedure.

Starting with $\widehat{\tau}^2 = 0$, a regression using weights $w_i^* = 1/v_i$ gives initial estimates of β. The fitted values $\widehat{y}_i$ from this model can then be used in one of three formulas for estimation of τ^2, given below:

$$\widehat{\tau}^2_{\text{ml}} = \frac{\sum_{i=1}^{k} w_i^{*2}\{(y_i - \widehat{y}_i)^2 - v_i\}}{\sum_{i=1}^{k} w_i^{*2}} \quad \text{maximum likelihood (Pocock, Cook, and Beresford}$$

1981)

$$\widehat{\tau}^2_{\text{reml}} = \frac{\sum_{i=1}^{k} w_i^{*2}\left\{\dfrac{k}{k-(p+1)}(y_i - \widehat{y}_i)^2 - v_i\right\}}{\sum_{i=1}^{k} w_i^{*2}} \quad \text{restricted maximum likelihood}$$

(Berkey et al. 1995)

$$\widehat{\tau}_{eb}^2 = \frac{\sum_{i=1}^k w_i^* \left\{ \dfrac{k}{k-(p+1)} (y_i - \widehat{y}_i)^2 - v_i \right\}}{\sum_{i=1}^k w_i^*} \quad \text{empirical Bayes (Berkey et al. 1995)}$$

In each case, if the estimated value $\widehat{\tau}^2$ is negative, it is set to zero.

Using the estimate $\widehat{\tau}^2$, new weights $w_i^* = 1/\widehat{\tau}^2 + v_i$ (or $1/v_i$ if $\widehat{\tau}$ is zero) are then calculated, and hence new estimates of β, fitted values $\widehat{y}_i$, and thence $\widehat{\tau}^2$. The procedure continues until the difference between successive estimates of τ^2 is less than a prespecified number (such as 10^{-4}). The standard errors of the final estimates of β are calculated forcing the scale parameter to be 1, since the weights are equal to the reciprocal variances.

4   Syntax

`metareg` has the usual syntax for a regression command, with the additional requirement that the user specify a variable containing either the within-study standard error or variance.

`metareg` *y varlist* $\left[\,if\,\right]$ $\left[\,in\,\right]$, $\left\{\underline{\text{wsse}}(varname) \mid \underline{\text{wsvar}}(varname) \mid \right.$

$\left. \underline{\text{wsse}}(varname)\ \underline{\text{wsvar}}(varname) \right\}$ $\left[\,\underline{\text{bse}}\text{st}(\{\text{reml} \mid \text{ml} \mid \text{eb} \mid \text{mm}\})\right.$

$\underline{\text{tole}}\text{ran}(\#)\ \underline{\text{l}}\text{evel}(\#)\ \underline{\text{noite}}\text{r}\left.\,\right]$

The command supplies estimated parameters, standard errors, Z statistics, p values and confidence intervals, in the usual regression output format. The estimated value of τ^2 is also given.

5   Options

`wsse`(*varname*) is a variable in the dataset, which contains the within-studies standard error $\sqrt{v_i}$. Either this or the `wsvar` option below (or both) must be specified.

`wsvar`(*varname*) is a variable in the dataset, which contains the within-studies variance v_i. Either this or the `wsse` option above (or both) must be specified.

Note: If both the above options are specified, the program will check that the variance is the square of the standard error for each study.

`bsest`($\{$`reml` | `ml` | `eb` | `mm`$\}$) specifies the method for estimating τ^2. The default is `reml` (restricted maximum likelihood), with the alternatives being `ml` (maximum likelihood), `eb` (empirical Bayes), and `mm` (method of moments).

`toleran`(#) specifies the difference between values of $\widehat{\tau}^2$ at successive iterations required for convergence. If # is n, the process will not converge until successive values of $\widehat{\tau}^2$ differ by less than 10^{-n}. The default is 4.

`level(#)` specifies the confidence level, as a percentage, for confidence intervals. The default is `level(95)` or as set by `set level`; see [U] **20.7 Specifying the width of confidence intervals**.

`noiter` requests that the log of the iterations in the `reml`, `ml`, or `eb` procedures be suppressed from the output.

6 Example

BCG is a vaccine widely used to give protection against tuberculosis. Colditz et al. (1994) performed a meta-analysis of all published trials, which randomized subjects to either BCG vaccine or placebo, and then had similar surveillance procedures to monitor the outcome, diagnosis of tuberculosis.

The data in `bcg.dta` are as reported by Berkey et al. (1995). Having read the file into Stata, the log odds-ratio of tuberculosis comparing BCG with placebo, and its standard error can be calculated for each study.

```
. use bcg, clear
(BCG and tuberculosis)

. describe

Contains data from bcg.dta
  obs:            13                          BCG and tuberculosis
  vars:            8
  size:          351
-----------------------------------------------------------------------------
    1. trial     str2    %9s          trial identity number
    2. lat       byte    %9.0g        absolute latitude from Equator
    3. nt        float   %9.0g        total vaccinated patients
    4. nc        float   %9.0g        total unvaccinated patients
    5. rt        int     %9.0g        tuberculosis in vaccinated
    6. rc        int     %9.0g        tuberculosis in unvaccinated
-----------------------------------------------------------------------------
Sorted by:

. list, noobs

    trial     lat       nt       nc       rt       rc
        1      44      123      139        4       11
        2      55      306      303        6       29
        3      42      231      220        3       11
        4      52    13598    12867       62      248
        5      13     5069     5808       33       47
        6      44     1541     1451      180      372
        7      19     2545      629        8       10
        8      13    88391    88391      505      499
        9      27     7499     7277       29       45
       10      42     1716     1665       17       65
       11      18    50634    27338      186      141
       12      33     2498     2341        5        3
       13      33    16913    17854       27       29

. generate logor=log((rt/(nt-rt))/(rc/(nc-rc)))
. generate selogor=sqrt((1/rc)+(1/(nc-rc))+(1/rt)+(1/(nt-rt)))
```

Note: If either `rt` or `rc` were 0, a standard approach would be to add 0.5 to each of `rt`, `rc`, `nt-rt`, and `nc-rc` for that study (Cox and Snell 1989).

A meta-analysis of the data can now be performed using the `meta` command described by Sharp and Sterne (1997, 1998).

```
. meta logor selogor, eform graph(r) id(trial) cline xlab(0.5,1,1.5) xline(1)
> boxsh(4) b2("Odds ratio - log scale")

Meta-analysis (exponential form)

         | Pooled        95% CI           Asymptotic        No. of
Method  |    Est   Lower   Upper   z_value   p_value        studies
--------+---------------------------------------------------------------
Fixed   |  0.647   0.595   0.702   -10.319    0.000           13
Random  |  0.474   0.325   0.690    -3.887    0.000

Test for heterogeneity: Q= 163.165 on 12 degrees of freedom (p= 0.000)
Moment estimate of between-studies variance =  0.366
```

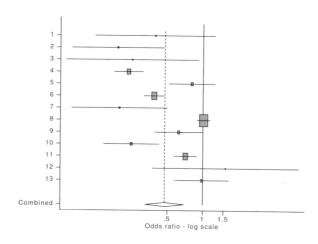

Figure 1. A meta-analysis of the BCG and Tuberculosis data

Both the graph and the statistical test indicate substantial heterogeneity between the trials, with an estimated between-studies variance of 0.366. The random effects combined estimate of 0.474, indicating a strong protective effect of BCG against tuberculosis, should not be reported without some discussion of the possible reasons for the differences between the studies (Thompson 1994).

One possible explanation for the differences in treatment effects could be that the studies were conducted at different latitudes from the equator. Berkey et al. (1995) speculated that absolute latitude, or distance of each study from the equator, may serve as a surrogate for the presence of environmental mycobacteria, which provide a certain level of natural immunity against tuberculosis. By sorting on absolute latitude, the graph obtained using `meta` shows the studies in order of increasing latitude going down the page.

```
. sort lat

. meta logor selogor, eform graph(r) id(trial) cline xlab(0.5,1,1.5) xline(1)
> boxsh(4) b2("Odds ratio - log scale")
```

(*output omitted*)

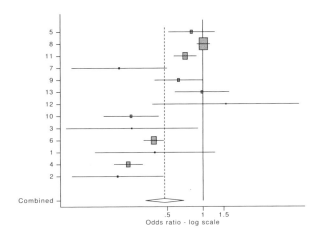

Figure 2. Same as figure 1 but sorted by latitude

The graph now suggests that BCG vaccination is more effective at higher absolute latitudes. This can be investigated further using the `metareg` command, with a REML estimate of the between-studies variance τ^2.

```
. metareg logor lat, wsse(selogor) bs(reml) noiter
Meta-analysis regression                         No of studies =    13
                                                 tau^2 method       reml
                                                 tau^2 estimate =  .0235

Successive values of tau^2 differ by less than 10^-4 - convergence achieved
------------------------------------------------------------------------------
           |       Coef.   Std. Err.       z    P>|z|     [95% Conf. Interval]
-----------+------------------------------------------------------------------
       lat |   -.0320363   .0049432    -6.481   0.000    -.0417247   -.0223479
     _cons |    .3282194   .1659807     1.977   0.048     .0029033    .6535356
------------------------------------------------------------------------------
```

This analysis shows that after allowing for additive residual heterogeneity, there is a significant negative association between the log odds-ratio and absolute latitude, i.e., the higher the absolute latitude, the lower the odds ratio, and hence the greater the benefit of BCG vaccination. The following plot of log odds-ratio against absolute latitude includes the fitted regression line from the model above. The size of the circles in the plot is inversely proportional to the variance of the log odds-ratio, so larger circles correspond to larger studies.

```
. generate invvlor=selogor^-2

. generate fit=0.328-0.032*lat

. graph logor fit lat [fw=invvlor], s(oi) c(.l)  xlab(0,10,20,30,40,50,60)
> ylab(-1.6094,-0.6931,0,0.6931) ll("Odds ratio (log scale)")
> b2("Distance from Equator (degrees of latitude)")
```

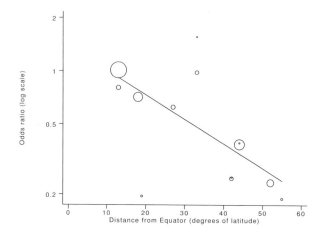

Figure 3. Plot of log odds-ratio against absolute latitude

(Note: The axes on this graph have been modified using the STAGE software)

Here a restricted maximum-likelihood method was used to estimate τ^2; the other three methods are used in turn below:

```
. metareg logor lat, wsse(selogor) bs(ml) noiter
Meta-analysis regression                            No of studies =   13
                                                    tau^2 method       ml
                                                    tau^2 estimate =  .0037
Successive values of tau^2 differ by less than 10^-4 - convergence achieved
-----------------------------------------------------------------------------
        |     Coef.    Std. Err.      z      P>|z|      [95% Conf. Interval]
--------+--------------------------------------------------------------------
    lat | -.0327447    .0033327    -9.825   0.000     -.0392767    -.0262128
  _cons |  .3725098    .1043895     3.568   0.000      .1679102     .5771093
-----------------------------------------------------------------------------
. metareg logor lat, wsse(selogor) bs(eb) noiter
Meta-analysis regression                            No of studies =   13
                                                    tau^2 method       eb
                                                    tau^2 estimate =  .1373
Successive values of tau^2 differ by less than 10^-4 - convergence achieved
-----------------------------------------------------------------------------
        |     Coef.    Std. Err.      z      P>|z|      [95% Conf. Interval]
--------+--------------------------------------------------------------------
    lat | -.0305794    .0090005    -3.398   0.001     -.0482201    -.0129388
  _cons |  .2548214    .3138935     0.812   0.417     -.3603984     .8700413
-----------------------------------------------------------------------------
```

```
. metareg logor lat, bs(mm) wsse(selogor) noiter
Warning: mm is a non-iterative method, noiter option ignored
Meta-analysis regression                        No of studies =    13
                                                tau^2 method       mm
                                                tau^2 estimate = .0480
```

```
-------------------------------------------------------------------------
          |     Coef.   Std. Err.      z    P>|z|    [95% Conf. Interval]
----------+--------------------------------------------------------------
      lat | -.0315724   .0061726   -5.115   0.000   -.0436704   -.0194744
    _cons |   .303035   .2108751    1.437   0.151   -.1102727    .7163427
-------------------------------------------------------------------------
```

The estimated value of τ^2 using a method-of-moments estimator is 0.048, compared with 0.366 before adjusting for latitude, so absolute latitude has explained almost all of the variation between the studies.

The analyses above show that the estimate of the effect of latitude is similar using all four methods. However, the estimated values of τ^2 differ considerably, with the estimate from the empirical Bayes method being largest. The restricted maximum-likelihood method corrects the bias in the maximum-likelihood estimate of τ^2. The basis for using the empirical Bayes method is less clear (Morris 1983), so this method should be used with caution. The moment-based method extends the usual random-effects meta-analysis; below `metareg` is used to fit a model with no covariate:

```
. metareg logor, bs(mm) wsse(selogor)
Meta-analysis regression                        No of studies =    13
                                                tau^2 method       mm
                                                tau^2 estimate = .3663
```

```
-------------------------------------------------------------------------
          |     Coef.   Std. Err.      z    P>|z|    [95% Conf. Interval]
----------+--------------------------------------------------------------
    _cons | -.7473923   .1922628   -3.887   0.000   -1.124221   -.3705641
-------------------------------------------------------------------------
```

Now the estimate of τ^2 is identical to that obtained earlier from `meta`, and the constant parameter is the log of the random effects pooled estimate given by `meta`.

The paper by Thompson and Sharp (1999) contains a fuller discussion both of the differences between the four methods of estimation, and other methods for explaining heterogeneity. Copies are available on request from the author.

7 Saved results

`metareg` saves the following results in the S_ macros:

S_1 k, number of studies
S_2 $\hat{\tau}^2$, estimate of between-studies variance

8   Acknowledgments

I am grateful to Simon Thompson, Ian White, and Jonathan Sterne for their helpful comments on earlier versions of this command.

9   References

Berkey, C. S., D. C. Hoaglin, F. Mosteller, and G. A. Colditz. 1995. A random-effects regression model for meta-analysis. *Statistics in Medicine* 14: 395–411.

Colditz, G. A., T. F. Brewer, C. S. Berkey, M. E. Wilson, E. Burdick, H. V. Fineberg, and F. Mosteller. 1994. Efficacy of BCG vaccine in the prevention of tuberculosis. Meta-analysis of the published literature. *Journal of the American Medical Association* 271: 698–702.

Cox, D. R., and E. J. Snell. 1989. *Analysis of Binary Data*. 2nd ed. London: Chapman & Hall.

DerSimonian, R., and N. Laird. 1986. Meta-analysis in clinical trials. *Controlled Clinical Trials* 7: 177–188.

DuMouchel, W. H., and J. E. Harris. 1983. Bayes methods for combining the results of cancer studies in humans and other species. *Journal of the American Statistical Association* 78: 293–308.

Harbord, R. M., and J. P. T. Higgins. 2008. Meta-regression in Stata. *Stata Journal* 8: 493–519. (Reprinted in this collection on pp. 85–111.)

Lau, J., J. P. A. Ioannidis, and C. H. Schmid. 1998. Summing up evidence: One answer is not always enough. *Lancet* 351: 123–127.

Morris, C. N. 1983. Parametric empirical Bayes inference: Theory and applications. *Journal of the American Statistical Association* 78: 47–55.

Pocock, S. J., D. G. Cook, and S. A. A. Beresford. 1981. Regression of area mortality rates on explanatory variables: What weighting is appropriate? *Journal of the Royal Statistical Society, Series C* 30: 286–295.

Sharp, S., and J. A. C. Sterne. 1997. sbe16: Meta-analysis. *Stata Technical Bulletin* 38: 9–14. Reprinted in *Stata Technical Bulletin Reprints*, vol. 7, pp. 100–106. College Station, TX: Stata Press.

———. 1998. sbe16.1: New syntax and output for the meta-analysis command. *Stata Technical Bulletin* 42: 6–8. Reprinted in *Stata Technical Bulletin Reprints*, vol. 7, pp. 106–108. College Station, TX: Stata Press.

Thompson, S. G. 1994. Systematic review: Why sources of heterogeneity in meta-analysis should be investigated. *British Medical Journal* 309: 1351–1355.

Thompson, S. G., and S. Sharp. 1999. Explaining heterogeneity in meta-analysis: A comparison of methods. *Statistics in Medicine* 18: 2693–2708.

Part 3

Investigating bias in meta-analysis: metafunnel, confunnel, metabias, metatrim, and extfunnel

A series of empirical studies in the medical literature have established that flaws in conducting randomized trials lead to exaggeration of intervention effect estimates and that publication and other reporting biases lead to statistically significant results being more likely to be reported. These biases can cause the results of meta-analyses to be overoptimistic. There is longstanding interest in graphical and statistical methods to detect publication bias and other biases, but the application and interpretation of the methods is controversial. Therefore, users are recommended to consult published guidance; for example, see Sterne, Egger, and Moher (2008) and Sterne et al. (2011).

Funnel plots, which can be displayed using the `metafunnel` command, have long been recommended as a graphical display that can be used to examine whether the results of a meta-analysis are affected by publication bias, because publication bias can lead to an association between study size and the size of effect and hence to an asymmetric appearance of a funnel plot. However, there are various causes of such small-study effects (Egger et al. 1997; Sterne, Gavaghan, and Egger 2000; Lau et al. 2006; Sterne et al. 2011), and these should be carefully considered when funnel plots appear asymmetric. Contour-enhanced funnel plots can be displayed using the `confunnel` command by Palmer et al. (2008). These aid interpretation by shading areas of the funnel plot corresponding to different areas of statistical significance.

Because visual assessment of funnel plots is inherently subjective, many authors have proposed statistical tests for funnel plot asymmetry. These are available in the `metabias` command. The appropriate use of such tests has been much discussed by statisticians, because the most widely used test (Egger et al. 1997) can give false-positive results in some circumstances. These issues are discussed in the article by Harbord, Harris, and Sterne (2009) describing the most recent version of `metabias`, which implements new tests for funnel plot asymmetry that aim to overcome some of the problems with the test proposed by Egger et al. (1997). The `metabias` command was originally written by Steichen (1998), and his articles describing the tests implemented in the original versions of the command are also included in this collection. Steichen also wrote the `metatrim` command, which implements the "trim and fill" approach to identifying and correcting for funnel plot asymmetry arising from publication bias (Taylor and Tweedie 1998; Duval and Tweedie 2000). This approach has also been much debated—there are suggestions that it performs poorly in the presence of substantial between-study heterogeneity (Terrin et al. 2003; Peters et al. 2007).

Augmented funnel plots may be useful both for assessing bias and to assess the impact of a hypothetical additional study on a meta-analysis (Langan et al. 2012). Crowther et al. (2012) implement these approaches in `extfunnel`. The command overlays either statistical significance contours or heterogeneity contours onto the funnel plot. The statistical significance contours define regions of the funnel plot in which a new study would have to be located to change the statistical significance of the meta-analysis. The heterogeneity contours show how a new study would affect the extent of heterogeneity in the meta-analysis. These augmented funnel plots aim to show how robust the meta-analysis is to new evidence.

1   References

Crowther, M. J., K. R. Abrams, and P. C. Lambert. 2012. Flexible parametric joint modelling of longitudinal and survival data. *Statistics in Medicine* 31: 4456–4471.

Duval, S., and R. L. Tweedie. 2000. Trim and fill: A simple funnel-plot–based method of testing and adjusting for publication bias in meta-analysis. *Biometrics* 56: 455–463.

Egger, M., G. Davey Smith, M. Schneider, and C. Minder. 1997. Bias in meta-analysis detected by a simple, graphical test. *British Medical Journal* 315: 629–634.

Harbord, R. M., R. J. Harris, and J. A. C. Sterne. 2009. sbe19_6: Updated tests for small-study effects in meta-analysis. *Stata Journal* 9: 197–210. (Reprinted in this collection on pp. 153–165.)

Langan, D., J. P. T. Higgins, W. Gregory, and A. J. Sutton. 2012. Graphical augmentations to the funnel plot assess the impact of additional evidence on a meta-analysis. *Journal of Clinical Epidemiology* 65: 511–519.

Lau, J., J. P. A. Ioannidis, N. Terrin, C. H. Schmid, and I. Olkin. 2006. The case of the misleading funnel plot. *British Medical Journal* 333: 597–600.

Palmer, T. M., J. L. Peters, A. J. Sutton, and S. G. Moreno. 2008. Contour-enhanced funnel plots for meta-analysis. *Stata Journal* 8: 242–254. (Reprinted in this collection on pp. 139–152.)

Peters, J. L., A. J. Sutton, D. R. Jones, K. R. Abrams, and L. Rushton. 2007. Performance of the trim and fill method in the presence of publication bias and between-study heterogeneity. *Statistics in Medicine* 26: 4544–4562.

Steichen, T. J. 1998. sbe19: Tests for publication bias in meta-analysis. *Stata Technical Bulletin* 41: 9–15. Reprinted in *Stata Technical Bulletin Reprints*, vol. 7, pp. 125–133. College Station, TX: Stata Press. (Reprinted in this collection on pp. 166–176.)

Sterne, J. A. C., M. Egger, and D. Moher. 2008. Addressing reporting biases. In *Cochrane Handbook for Systematic Reviews of Interventions*, ed. J. P. T. Higgins and S. Green, 297–334. Chichester, UK: Wiley.

Sterne, J. A. C., D. Gavaghan, and M. Egger. 2000. Publication and related bias in meta-analysis: Power of statistical tests and prevalence in the literature. *Journal of Clinical Epidemiology* 53: 1119–1129.

Sterne, J. A. C., A. J. Sutton, J. P. A. Ioannidis, N. Terrin, D. R. Jones, J. Lau, J. Carpenter, G. Rücker, R. M. Harbord, C. H. Schmid, J. Tetzlaff, J. J. Deeks, J. L. Peters, P. Macaskill, G. Schwarzer, S. Duval, D. G. Altman, D. Moher, and J. P. T. Higgins. 2011. Recommendations for examining and interpreting funnel plot asymmetry in meta-analyses of randomised controlled trials. *British Medical Journal* 343: d4002.

Taylor, S. J., and R. L. Tweedie. 1998. Practical estimates of the effect of publication bias in meta-analysis. *Australasian Epidemiologist* 5: 14–17.

Terrin, N., C. H. Schmid, J. Lau, and I. Olkin. 2003. Adjusting for publication bias in the presence of heterogeneity. *Statistics in Medicine* 22: 2113–2126.

The Stata Journal (2004)
4, Number 2, st0061, pp. 127–141

Funnel plots in meta-analysis

Jonathan A. C. Sterne and Roger M. Harbord
Department of Social Medicine
University of Bristol
Bristol, UK
roger.harbord@bristol.ac.uk

Abstract. Funnel plots are a visual tool for investigating publication and other bias in meta-analysis. They are simple scatterplots of the treatment effects estimated from individual studies (horizontal axis) against a measure of study size (vertical axis). The name "funnel plot" is based on the precision in the estimation of the underlying treatment effect increasing as the sample size of component studies increases. Therefore, in the absence of bias, results from small studies will scatter widely at the bottom of the graph, with the spread narrowing among larger studies. Publication bias (the association of publication probability with the statistical significance of study results) may lead to asymmetrical funnel plots. It is, however, important to realize that publication bias is only one of a number of possible causes of funnel plot asymmetry—funnel plots should be seen as a generic means of examining small study effects (the tendency for the smaller studies in a meta-analysis to show larger treatment effects) rather than a tool to diagnose specific types of bias. This article introduces the `metafunnel` command, which produces funnel plots in Stata. In accordance with published recommendations, standard error is used as the measure of study size. Treatment effects expressed as ratio measures (for example risk ratios or odds ratios) may be plotted on a log scale.

Keywords: st0061, metafunnel, funnel plots, meta-analysis, publication bias, small-study effects

1 Introduction

The substantial recent interest in meta-analysis (the statistical methods that are used to combine results from a number of different studies) is reflected in a number of user-written commands that do meta-analysis in Stata. Meta-analyses should be based on *systematic reviews* of relevant literature. A systematic review is a systematic assembly, critical appraisal, and synthesis of all relevant studies on a specific topic. The main feature that distinguishes systematic from narrative reviews is a methods section that clearly states the question to be addressed and the methods and criteria to be employed for identifying and selecting relevant studies and extracting and analyzing information (Egger, Davey Smith, and Altman 2001).

While systematic reviews and meta-analyses have the potential to produce precise estimates of treatment effects that reflect all of the relevant literature, they are not immune to bias. *Publication bias*—the association of publication probability with the statistical significance of study results—is well documented as a problem in the medical research literature (Stern and Simes 1997). Further, it has been demonstrated that randomized controlled trials for which concealment of treatment allocation is not adequate, or which are not double blind, produce estimated treatment effects that appear more beneficial (Schulz et al. 1995).

2 Funnel plots

Funnel plots are simple scatterplots of the treatment effects estimated from individual studies against a measure of study size. The name "funnel plot" is based on the precision in the estimation of the underlying treatment effect increasing as the sample size of component studies increases. Results from small studies will therefore scatter widely at the bottom of the graph, with the spread narrowing among larger studies. In the absence of bias, the plot will resemble a symmetrical, inverted funnel, as shown in the top graph of figure 1.

If there is bias, for example, because smaller studies showing no statistically significant effects (open circles in figure 1) remain unpublished, then such publication bias will lead to an asymmetrical appearance of the funnel plot with a gap in the right bottom side of the graph (middle graph of figure 1). In this situation, the combined effect from meta-analysis will overestimate the treatment's effect. The more pronounced the asymmetry, the more likely it is that the amount of bias will be substantial.

It is important to realize that publication bias is only one of a number of possible explanations for funnel plot asymmetry; these are discussed in more detail in section 2.3. For example, trials of lower quality yield exaggerated estimates of treatment effects (Schulz et al. 1995). Smaller studies are, on average, conducted and analyzed with less methodological rigor than larger studies (Egger et al. 2003), so asymmetry may also result from the overestimation of treatment effects in smaller studies of lower methodological quality (bottom graph of figure 1). Unfortunately, funnel plot asymmetry has often been equated with publication bias without consideration of its other possible explanations; for example, the help file for the `metabias` command in Stata (written in 1998) refers only to publication bias.

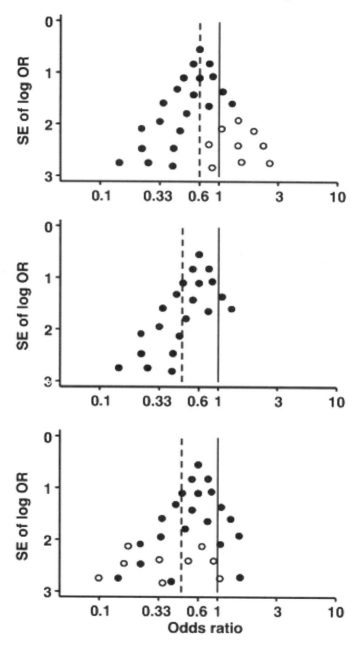

Figure 1. Hypothetical funnel plots: (top) symmetrical plot in the absence of bias (open circles indicate smaller studies showing no beneficial effects); (middle) asymmetrical plot in the presence of publication bias (smaller studies showing no beneficial effects are missing); (bottom) asymmetrical plot in the presence of bias due to low methodological quality of smaller studies (open circles indicate small studies of inadequate quality whose results are biased towards larger beneficial effects).

Although it is conventional to plot treatment effects on the horizontal axis and the measure of study size on the vertical axis, it is certainly not an error to plot the axes the other way around. Indeed, such a choice is arguably more consistent with standard statistical practice in that the variable on the vertical axis is usually hypothesized to depend on the variable on the horizontal axis. Such funnel plots can be plotted in Stata using the `metabias` command (Steichen 1998; Steichen, Egger, and Sterne 1998).

2.1 Choice of axis in funnel plots

The majority of endpoints in randomized trials of medical treatments are binary, with treatment effects most commonly expressed as ratio measures (odds ratio, risk ratio, or hazard ratio). (This may not be true of trials in other disciplines, such as psychology or social research.) The use of ratio measures is justified by empirical evidence that there is less between-trial heterogeneity in treatment effects based on ratio measures than difference measures (Deeks and Altman 2001; Engels et al. 2000). As is generally the case in meta-analysis, the *log* of the ratio measure and its standard error are used in funnel plots.

Sterne and Egger (2001) consider choice of axis in funnel plots of meta-analyses with binary outcomes. Although sample size or functions of sample size have often been used as the vertical axis, this is problematic because the precision of a treatment effect estimate is determined by both the sample size and by the number of events. Thus, studies with very different sample sizes may have the same standard error and precision and vice versa. Therefore, the shape of plots using sample size on the vertical axis is not predictable except that, in the absence of bias, it should be symmetric. After considering various possible choices of vertical axis, Sterne and Egger conclude that standard error of the treatment effect estimate is likely to be preferable in many situations. Funnel plots may also be drawn using precision (= 1/(standard error)) on the vertical axis using the `funnel2` command distributed as part of the `metaggr` package (Bradburn, Deeks, and Altman 1998). Such plots tend to emphasize differences between the largest study and the others.

2.2 Example

The trials of magnesium therapy following myocardial infarction (heart attack) are a well-known example in which the results of a meta-analysis, which appeared to provide clear evidence that magnesium therapy reduced mortality, were contradicted by subsequent larger trials that found no evidence that magnesium influenced mortality. Figure 2 is a funnel plot based on the results of 15 trials of the effect of magnesium on mortality following myocardial infarction. Because the smaller trials produced smaller odds ratios (more substantial reductions in mortality associated with magnesium therapy), the funnel plot is clearly asymmetric.

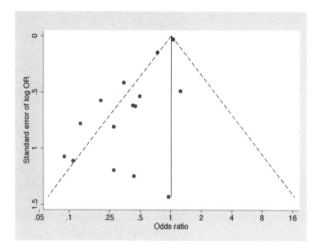

Figure 2. Funnel plot, using data from 15 trials of magnesium therapy following my-ocardial infarction.

The horizontal axis of figure 2 (treatment odds ratio) is drawn on a log scale, so that (for example) odds ratios of 2 and 0.5 are the same distance from the null value of 1 (no treatment effect). This is equivalent to plotting the log odds-ratio on the horizontal axis. The standard error of the log OR is plotted on the vertical axis. Note that the largest studies have the smallest standard errors, so to place the largest studies at the top of the graph, the vertical axis must be reversed (standard error 0 at the top).

The solid vertical line represents the summary estimate of the treatment effect, derived using fixed-effects meta-analysis. This is close to 1 because the estimated treatment odds ratios in the largest studies were close to 1. For the purposes of displaying the center of the plot in the absence of bias, calculation of the summary log odds-ratio using fixed rather than random-effects meta-analysis is preferable because the random-effects estimate gives greater relative weight to smaller studies and will, therefore, be more affected if publication bias is present (Poole and Greenland 1999).

Interpretation of funnel plots is facilitated by inclusion of diagonal lines representing the 95% confidence limits around the summary treatment effect, i.e., [summary effect estimate − (1.96 × standard error)] and [summary effect estimate + (1.96 × standard error)] for each standard error on the vertical axis. These show the expected distribution of studies in the absence of heterogeneity or of selection biases: in the absence of heterogeneity, 95% of the studies should lie within the funnel defined by these straight lines. Because these lines are not strict 95% limits, they are referred to as "pseudo 95% confidence limits".

2.3   Sources of funnel plot asymmetry

Funnel plots were first proposed as a means of detecting a specific form of bias—publication bias. However as explained earlier (see the bottom graph of figure 1), the exaggeration of treatment effects in small studies of low quality provides a plausible alternative mechanism for funnel plot asymmetry. Egger et al. (1997) list different possible reasons for funnel plot asymmetry, which are summarized in table 1.

Table 1. Potential sources of asymmetry in funnel plots

---

1. Selection biases
    Publication bias
        Location biases
            Language bias
            Citation bias
            Multiple publication bias

---

2. True heterogeneity
    Size of effect differs according to study size:
        Intensity of intervention
        Differences in underlying risk

---

3. Data irregularities
    Poor methodological design of small studies
    Inadequate analysis
    Fraud

---

4. Artifact
    Heterogeneity due to poor choice of effect measure

---

5. Chance

In addition to selective publication of studies according to their results, other possible biases affecting the selection of studies for inclusion in meta-analyses include the propensity for the results to affect the language of publication (Jüni et al. 2002); the possibility that results affect the frequency with which a study is cited and, hence, its probability of inclusion in a meta-analysis, and the multiple publication of studies with demonstrating an effect of the intervention (Tramer et al. 1997).

It is important to realize that funnel plot asymmetry need not result from bias. The studies displayed in a funnel plot may not always estimate the same underlying effect of the same intervention, and such heterogeneity in results may lead to asymmetry in funnel plots if the true treatment effect is larger in the smaller studies. For example, if a combined outcome is considered, then substantial benefit may be seen only in subjects at high risk for the component of the combined outcome which is affected by the intervention (Davey Smith and Egger 1994; Glasziou and Irwig 1995). Some interventions may have been implemented less thoroughly in larger studies, thus explaining the more positive results in smaller studies. For example, an asymmetrical funnel plot was found in a meta-analysis of trials examining the effect of inpatient comprehensive geriatric assessment programs on mortality. An experienced consultant geriatrician was more likely to be actively involved in the smaller trials and this may explain the larger treatment effects observed in these trials (Egger et al. 1997; Stuck et al. 1993).

The way in which data irregularities such as low methodological quality of smaller studies may result in funnel plot asymmetry was described earlier. Poor choice of effect measure may also result in funnel plot asymmetry; for example, it has been shown that meta-analyses in which intervention effects are measured as risk differences are more heterogeneous than those in which intervention effects are measured as risk ratios or odds ratios (Deeks and Altman 2001; Engels et al. 2000). The inappropriate use of risk differences may also result in funnel plot asymmetry—if the effect of intervention is homogeneous on the risk-ratio scale, then the risk difference will be smaller in studies that have low event rates.

2.4   Tests for funnel plot asymmetry

It is, of course, possible that an asymmetrical funnel plot arises merely by the play of chance. Statistical tests for funnel plot asymmetry have been proposed by Begg and Mazumdar (1994) and by Egger et al. (1997). These are available in the Stata command `metabias` (Steichen 1998; Steichen, Egger, and Sterne 1998). The test proposed by Egger et al. (1997) is algebraically identical to a test that there is no linear association between the treatment effect and its standard error and, hence, that there is no straight-line association in the funnel plot of treatment effect against its standard error (see Sterne, Gavaghan, and Egger [2000] for details). The corresponding fitted line may be added to the funnel plot using the `egger` option of the `metafunnel` command—see section 5 below.

2.5   Small-study effects

Funnel plot asymmetry thus raises the possibility of bias, but it is not proof of bias. It is important to note, however, that asymmetry (unless produced by chance alone) will always lead us to question the interpretation of the overall estimate of effect when studies are combined in a meta-analysis; for example, if the study size predicts the treatment effect, what treatment effect will apply if the treatment is adopted in routine practice? Sterne, Egger, and Davey Smith (2001) and Sterne, Gavaghan, and Egger (2000) have suggested that the funnel plot should be seen as a generic means of examining "small-study effects" (the tendency for the smaller studies in a meta-analysis to show larger treatment effects) rather than as a tool to diagnose specific types of bias.

When funnel plot asymmetry is found, its possible causes should be carefully considered. For example, how comprehensive was the literature search that located the trials included in the meta-analysis? Does reported trial quality differ between larger and smaller studies? Is there a plausible reason for the effect of intervention to be greater in smaller trials? It is possible that differences between smaller and larger trials are accounted for by a trial characteristic; this may be investigated using the `by()` option of the `metafunnel` command, as described in section 6 below. Explanations for heterogeneity may be investigated more formally using meta-regression (Thompson and Sharp 1999) to investigate associations between study characteristics and intervention effect estimates. For example, we might investigate evidence that studies in which reported allocation concealment is unclear or inadequate tend to result in more beneficial treatment effect estimates. Meta-regression analyses may be done using the Stata command `metareg` (Sharp 1998); however, it will not necessarily be possible to provide a definitive explanation for funnel plot asymmetry. In medical research, meta-analyses typically contain 10 or fewer trials (Sterne, Gavaghan, and Egger 2000). Power to detect associations between study characteristics and intervention effect estimates will therefore often be low, in which case it may not be possible to identify a particular study characteristic as the cause of the heterogeneity.

3 Syntax

`metafunnel` { *theta* { *se* | *var* } | *exp(theta)* { *ll ul* [*cl*] } } [*if*] [*in*]
     [, `by`(*by_var*) [`var` | `ci`] `nolines` `forcenull` `reverse` `eform` `egger`
     graph_options]

4 Description

`metafunnel` plots funnel plots. The syntax for `metafunnel` is based on the same framework as for the `meta`, `metabias`, `metacum`, and `metatrim` commands. The user provides the effect estimate as *theta* (e.g., the log odds-ratio) and a measure of theta's variability (i.e., its standard error or its variance). Alternatively, the user provides *exp(theta)* (e.g., an odds ratio), its confidence interval, and, optionally, the confidence level.

5 Options

`by`(*by_var*) displays subgroups according to the value of *by_var*. The legend displays the value labels for the levels of *by_var* if these are present; otherwise, it displays the value of each level of *by_var*.

`var` and `ci` indicate the meaning of the input variables in the same way as for the other meta-analysis commands listed above. The help file for `meta` gives a full explanation.

`nolines` specifies that pseudo 95% confidence interval lines not be included in the plot. The default is to include them.

`forcenull` forces the vertical line at the center of the funnel to be plotted at the null treatment effect of zero (1 when the treatment effect is exponentiated). The default is for the line to be plotted at the value of the fixed-effects summary estimate.

reverse inverts the funnel plot so that larger studies are displayed at the bottom of the plot with smaller studies at the top. This may also be achieved by specifying noreverse as part of the yscale(*axis_description*) graphics option.

eform exponentiates the treatment effect *theta* and displays the horizontal axis (treatment effect) on a log scale. As discussed in section 2.2, this is useful for displaying ratio measures, such as odds ratios and risk ratios.

egger adds the fitted line corresponding to the regression test for funnel plot asymmetry proposed by Egger et al. (1997) and implemented in metabias (see section 2.4). This option may not be combined with the by() option.

graph_options are any options allowed by the twoway scatter command that can be used to change the appearance of the points and add labels. If option egger is specified, the look of the fitted line can be changed using the options clstyle, clpattern, clwidth, and clcolor explained under *connect_options* in Stata's built-in help system and the graphics manual.

6   Examples

Listing the data for the 15 magnesium trials produces the following output:

```
. list trial trialnam year dead1 alive1 dead0 alive0, noobs
```

trial	trialnam	year	dead1	alive1	dead0	alive0
1	Morton	1984	1	39	2	34
2	Rasmussen	1986	9	126	23	112
3	Smith	1986	2	198	7	193
4	Abraham	1987	1	47	1	45
5	Feldstedt	1988	10	140	8	140
6	Schechter	1989	1	58	9	47
7	Ceremuzynski	1989	1	24	3	20
8	Singh	1990	6	70	11	64
9	Pereira	1990	1	26	7	20
10	Schechter 1	1991	2	87	12	68
11	Golf	1991	5	18	13	20
12	Thogersen	1991	4	126	8	114
13	LIMIT-2	1992	90	1069	118	1039
14	Schechter 2	1995	4	103	17	91
15	ISIS-4	1995	2216	26795	2103	26936

To use the metafunnel command, we first need to derive the treatment effect and its standard error for each trial. Here, we will express the treatment effects as log odds-ratios.

```
. generate or = (dead1/alive1)/(dead0/alive0)
. generate logor = log(or)
. generate selogor = sqrt((1/dead1)+(1/alive1)+(1/dead0)+(1/alive0))
```

A funnel plot can then be drawn using the following syntax, which includes the regression line corresponding to the regression test for funnel plot asymmetry proposed by Egger et al. (1997):

```
. metafunnel logor selogor, xtitle(Log odds ratio) ytitle(Standard error of log OR)
> egger
```

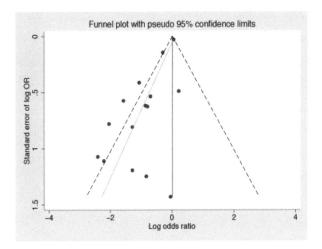

Figure 3. Funnel plot, using data from 15 trials of magnesium therapy following myocardial infarction, with log odds-ratios displayed on the horizontal axis.

By default, the subtitle "Funnel plot with pseudo 95% confidence limits" is displayed. ("Funnel plot" is displayed if the **nolines** options is specified.) This may be changed using the graphics option **subtitle**(*tinfo*).

Note that the log odds-ratio and its standard error may be derived automatically using the **metan** command. (The latest version of this command may be installed by typing **ssc install metaaggr.pkg, replace** in the Stata Command window.) Typing

```
. metan dead1 alive1 dead0 alive0, or
```

produces a meta-analysis of the effect of magnesium and creates variables _ES, containing the odds ratio in each study, and _selogES, containing the standard error of the log odds-ratio. Thus, we may derive the log odds-ratio by typing

```
. generate log_ES = log(_ES)
```

The list output below shows that variables **log_ES _selogES** are identical to variables **logor** and **selogor** derived earlier.

```
. list trial trialnam year logor selogor _ES log_ES _selogES, noobs
```

trial	trialnam year	logor	selogor	_ES	log_ES	_selogES
1	Morton 1984	-.8303483	1.247018	.4358974	-.8303483	1.247018
2	Rasmussen 1986	-1.056053	.4140706	.3478261	-1.056053	.4140706
3	Smith 1986	-1.27834	.8081392	.2784993	-1.27834	.8081392
4	Abraham 1987	-.0434851	1.42951	.9574468	-.0434851	1.42951
5	Feldstedt 1988	.2231435	.4891684	1.25	.2231435	.4891684
6	Schechter 1989	-2.40752	1.072208	.0900383	-2.40752	1.072208
7	Ceremuzynski 1989	-1.280934	1.193734	.2777778	-1.280934	1.193734
8	Singh 1990	-.695748	.5361776	.4987013	-.695748	.5361776
9	Pereira 1990	-2.208274	1.109648	.1098901	-2.208274	1.109648
10	Schechter 1 1991	-2.03816	.7807263	.1302682	-2.03816	.7807263
11	Golf 1991	-.8501509	.6184486	.4273504	-.8501509	.6184486
12	Thogersen 1991	-.7932307	.6258662	.452381	-.7932307	.6258662
13	LIMIT-2 1992	-.2993398	.1465729	.7413074	-.2993398	.1465729
14	Schechter 2 1995	-1.570789	.5740395	.2078812	-1.570789	.5740395
15	ISIS-4 1995	.0575872	.0316421	1.059278	.0575872	.0316421

The following command was used to produce figure 2 (see section 2.2), in which the horizontal axis is the treatment odds ratio, displayed on a log scale:

```
. metafunnel logor selogor, xlab(.05 .1 .25 .5 1 2 4 8 16)
> xscale(log) xtitle(Odds ratio) eform subtitle( )
> ytitle(Standard error of log OR)
```

When the `eform` option is used, the label of the horizontal axis (treatment effect, *theta*) is changed accordingly, unless there is a variable label for *theta* or the `xtitle(`*axis_title*`)` graphics option is used.

Finally, we will illustrate the use of the `by()` option by grouping the studies according to whether they were published during the 1980s or the 1990s:

```
. generate period = year
. recode period 1980/1989=1 1990/1999=2
(period: 15 changes made)
. label define periodlab 1 "1980s" 2 "1990s"
. label values period periodlab
. tab period
```

period	Freq.	Percent	Cum.
1980s	7	46.67	46.67
1990s	8	53.33	100.00
Total	15	100.00	

Using the latest version of the `metan` command (Bradburn, Deeks, and Altman 1998), we can examine the effect of magnesium separately, according to time period.

```
. metan dead1 alive1 dead0 alive0, or by(period) label(namevar=trialnam)
```

Study	OR	[95% Conf. Interval]		% Weight
1980s				
Morton	0.436	0.038	5.022	0.09
Rasmussen	0.348	0.154	0.783	0.99
Smith	0.278	0.057	1.357	0.32
Abraham	0.957	0.058	15.773	0.05
Feldstedt	1.250	0.479	3.261	0.35
Schechter	0.090	0.011	0.736	0.42
Ceremuzynski	0.278	0.027	2.883	0.14
Sub-total				
M-H pooled OR	0.437	0.267	0.714	2.36
1990s				
Singh	0.499	0.174	1.426	0.47
Pereira	0.110	0.012	0.967	0.31
Schechter 1	0.130	0.028	0.602	0.57
Golf	0.427	0.127	1.436	0.39
Thogersen	0.452	0.133	1.543	0.37
LIMIT-2	0.741	0.556	0.988	5.04
Schechter 2	0.208	0.067	0.640	0.75
ISIS-4	1.059	0.996	1.127	89.74
Sub-total				
M-H pooled OR	1.020	0.961	1.083	97.64
Overall				
M-H pooled OR	1.007	0.948	1.068	100.00

```
Test(s) of heterogeneity:
          Heterogeneity   degrees of
            statistic       freedom       P      I-squared**
1980s          7.85           6          0.250      23.5%
1990s         30.27           7          0.000      76.9%
Overall       46.61          14          0.000      70.0%
Overall Test for heterogeneity between sub-groups :
               8.50           1          0.004

** I-squared: the variation in OR attributable to heterogeneity

Significance test(s) of OR=1
1980s            z=  3.31      p = 0.001
1990s            z=  0.66      p = 0.511
Overall          z=  0.22      p = 0.829
```

The by() option of the metafunnel command is used to display separate symbols for the two time periods; the resulting funnel plot is displayed in figure 4.

```
. metafunnel logor selogor, xlab(.05 .1 .25 .5 1 2 4 8 16)
> xscale(log) xtitle(Odds ratio) eform subtitle( )
> ytitle(Standard error of log OR) by(period)
```

As demonstrated by the analysis according to time period, the larger studies were published later. Perhaps more surprisingly, the asymmetry appears to result more from the studies published during the 1990s than from those published during the 1980s.

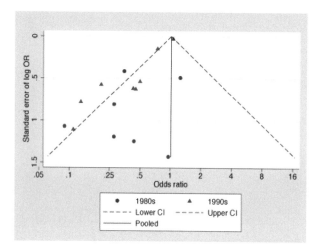

Figure 4. Funnel plot, using data from 15 trials of magnesium therapy following myocardial infarction, grouped according to date of publication.

7   Acknowledgments

Portions of the code for `metafunnel` were originally written by Thomas Steichen, who also gave helpful comments on an early version of the command. We are grateful to Nicholas J. Cox, who provided extensive programming advice.

8   References

Begg, C. B., and M. Mazumdar. 1994. Operating characteristics of a rank correlation test for publication bias. *Biometrics* 50: 1088–1101.

Bradburn, M. J., J. J. Deeks, and D. G. Altman. 1998. sbe24: metan—an alternative meta-analysis command. *Stata Technical Bulletin* 44: 4–15. Reprinted in *Stata Technical Bulletin Reprints*, vol. 8, pp. 86–100. College Station, TX: Stata Press. (Reprinted in this collection on pp. 3–28.)

Davey Smith, G., and M. Egger. 1994. Who benefits from medical interventions? Treating low risk patients can be a high risk strategy. *British Medical Journal* 308: 72–74.

Deeks, J. J., and D. G. Altman. 2001. Effect measures for meta-analysis of trials with binary outcomes. In *Systematic Reviews in Health Care: Meta-Analysis in Context*, 2nd edition, ed. M. Egger, G. Davey Smith, and D. G. Altman, 2nd ed., 313–335. London: BMJ Books.

Egger, M., G. Davey Smith, and D. G. Altman. 2001. *Systematic Reviews in Health Care: Meta-Analysis in Context*. 2nd ed. London: BMJ Books.

Egger, M., G. Davey Smith, M. Schneider, and C. Minder. 1997. Bias in meta-analysis detected by a simple, graphical test. *British Medical Journal* 315: 629–634.

Egger, M., P. Jüni, C. Bartlett, F. Holenstein, and J. A. C. Sterne. 2003. How important are comprehensive literature searches and the assessment of trial quality in systematic reviews? Empirical study. *Health Technology Assessment* 7: 1–68.

Engels, E. A., C. H. Schmid, N. Terrin, I. Olkin, and J. Lau. 2000. Heterogeneity and statistical significance in meta-analysis: An empirical study of 125 meta-analyses. *Statistics in Medicine* 19: 1707–1728.

Glasziou, P. P., and L. M. Irwig. 1995. An evidence based approach to individualizing treatment. *British Medical Journal* 311: 1356–1359.

Jüni, P., F. Holenstein, J. A. C. Sterne, C. Bartlett, and M. Egger. 2002. Direction and impact of language bias in meta-analyses of controlled trials: Empirical study. *International Journal of Epidemiology* 31: 115–123.

Poole, C., and S. Greenland. 1999. Random-effects meta-analyses are not always conservative. *American Journal of Epidemiology* 150: 469–475.

Schulz, K. F., I. Chalmers, R. J. Hayes, and D. G. Altman. 1995. Empirical evidence of bias. Dimensions of methodological quality associated with estimates of treatment effects in controlled trials. *Journal of the American Medical Association* 273: 408–412.

Sharp, S. 1998. sbe23: Meta-analysis regression. *Stata Technical Bulletin* 42: 16–22. Reprinted in *Stata Technical Bulletin Reprints*, vol. 7, pp. 148–155. College Station, TX: Stata Press. (Reprinted in this collection on pp. 112–120.)

Steichen, T. J. 1998. sbe19: Tests for publication bias in meta-analysis. *Stata Technical Bulletin* 41: 9–15. Reprinted in *Stata Technical Bulletin Reprints*, vol. 7, pp. 125–133. College Station, TX: Stata Press. (Reprinted in this collection on pp. 166–176.)

Steichen, T. J., M. Egger, and J. A. C. Sterne. 1998. sbe19.1: Tests for publication bias in meta-analysis. *Stata Technical Bulletin* 44: 3–4. Reprinted in *Stata Technical Bulletin Reprints*, vol. 8, pp. 84–85. College Station, TX: Stata Press. (Reprinted in this collection on pp. 177–179.)

Stern, J. M., and R. J. Simes. 1997. Publication bias: Evidence of delayed publication in a cohort study of clinical research projects. *British Medical Journal* 315: 640–645.

Sterne, J. A. C., and M. Egger. 2001. Funnel plots for detecting bias in meta-analysis: Guidelines on choice of axis. *Journal of Clinical Epidemiology* 54: 1046–1055.

Sterne, J. A. C., M. Egger, and G. Davey Smith. 2001. Investigating and dealing with publication and other bias. In *Systematic Reviews in Health Care: Meta-Analysis in Context*, 2nd edition, ed. M. Egger, G. Davey Smith, and D. G. Altman, 189–208. London: BMJ Books.

Sterne, J. A. C., D. Gavaghan, and M. Egger. 2000. Publication and related bias in meta-analysis: Power of statistical tests and prevalence in the literature. *Journal of Clinical Epidemiology* 53: 1119–1129.

Stuck, A. E., A. L. Siu, G. D. Wieland, J. Adams, and L. Z. Rubenstein. 1993. Comprehensive geriatric assessment: A meta-analysis of controlled trials. *Lancet* 342: 1032–1036.

Thompson, S. G., and S. Sharp. 1999. Explaining heterogeneity in meta-analysis: A comparison of methods. *Statistics in Medicine* 18: 2693–2708.

Tramer, M. R., D. J. M. Reynolds, R. A. Moore, and H. J. McQuay. 1997. Impact of covert duplicate publication on meta-analysis: A case study. *British Medical Journal* 315: 635–640.

The Stata Journal (2008)

8, Number 2, gr0033, pp. 242–254

Contour-enhanced funnel plots for meta-analysis

Tom M. Palmer

MRC Centre for Causal Analyses in Translational Epidemiology

Department of Social Medicine

University of Bristol

Bristol, UK

tom.palmer@bristol.ac.uk

Jaime L. Peters

Department of Health Sciences

University of Leicester

Leicester, UK

Alex J. Sutton

Department of Health Sciences

University of Leicester

Leicester, UK

Santiago G. Moreno

Department of Health Sciences

University of Leicester

Leicester, UK

Abstract. Funnel plots are commonly used to investigate publication and related biases in meta-analysis. Although asymmetry in the appearance of a funnel plot is often interpreted as being caused by publication bias, in reality the asymmetry could be due to other factors that cause systematic differences in the results of large and small studies, for example, confounding factors such as differential study quality. Funnel plots can be enhanced by adding contours of statistical significance to aid in interpreting the funnel plot. If studies appear to be missing in areas of low statistical significance, then it is possible that the asymmetry is due to publication bias. If studies appear to be missing in areas of high statistical significance, then publication bias is a less likely cause of the funnel asymmetry. It is proposed that this enhancement to funnel plots should be used routinely for meta-analyses where it is possible that results could be suppressed on the basis of their statistical significance.

Keywords: gr0033, confunnel, funnel plots, meta-analysis, publication bias, small-study effects

1 Introduction

Publication bias is the phenomenon where studies with uninteresting or unfavorable results are less likely to be published than those with more favorable results (Rothstein, Sutton, and Borenstein 2005). If publication bias exists, then the published literature is a biased sample of all studies, and any meta-analysis based on it will be similarly biased.

Funnel plots are commonly used to investigate publication and related biases in meta-analysis (Sterne, Becker, and Egger 2005). They consist of a simple scatterplot of each study's estimate of effect against some measure of its variability, commonly plotted on the x and y axes, respectively (although this goes against the usual convention of plotting the response variable on the y axis). In this way, the studies with the least variable effect sizes appear at the top of the funnel, and the smaller, less precise studies appear at the bottom. In the absence of publication bias, the studies will fan out in a symmetrical funnel shape around the pooled estimate, as variability due to sampling error increases down the y axis. If publication bias is present, then the funnel will appear asymmetric because of the systematic suppression of studies.

A complication in interpreting funnel plots is that funnel asymmetry could be due to factors other than publication bias, such as systematic differences in the results of large and small studies caused by confounding factors such as differential study quality; these differences are sometimes called small-study effects (Sterne and Egger 2001). The aim of the contour-enhanced funnel plot is to aid in disentangling these different causes of funnel asymmetry (Peters et al. 2008).

Funnel plots in Stata were previously described by Sterne and Harbord (2004), and there are several commands available in Stata for drawing funnel plots including `metafunnel`, `funnel` (available with `metan`), and `metabias`. These commands are described in more detail in a frequently asked question about the Stata commands available for meta-analysis; the frequently asked question can be found on Stata's web site at http://www.stata.com/support/faqs/stat/meta.html. In Stata 10, typing `help meta` displays a help file with information about the user-written commands for meta-analysis and tells which are the latest versions.

This article introduces another command for meta-analysis called `confunnel`, which produces contour-enhanced funnel plots. The concept of the contour-enhanced funnel plot is explained in the next section, followed by a description of the command syntax and options. The use of `confunnel` is demonstrated on a well-known meta-analysis example, and the use of the command is also explained in conjunction with some of the other user-written meta-analysis commands.

2   Contour-enhanced funnel plots

There is evidence that, generally, the primary driver for the suppression of studies is the level of statistical significance of study results, with studies that do not attain perceived milestones of statistical significance (i.e., $p < 0.05$ or 0.01) being less likely to be published (Easterbrook et al. 1991; Dickersin 1997; Ioannidis 1998). Despite this, no method has been previously considered to identify the areas of the funnel plot that correspond to different levels of statistical significance, to assess whether any observed asymmetry is likely caused by publication bias.

On a contour-enhanced funnel plot, contours of statistical significance are overlaid on the funnel plot (Peters et al. 2008). Adding contours of statistical significance in this way facilitates the assessment of whether the areas where studies exist are areas of statistical significance and whether the areas where studies are potentially missing correspond to areas of low statistical significance. If studies appear to be missing in areas of low statistical significance, then it is possible that the asymmetry is due to publication bias. Conversely, if the area where studies are perceived to be missing are areas of high statistical significance, then publication bias is a less likely cause of the funnel asymmetry.

There has been discussion as to which is the most informative scale for funnel plots of binary outcome meta-analyses. The consensus is that using the standard error, the variance, or their inverses is most sensible over using an alternative such as sample size (Sterne and Egger 2001; Sterne, Becker, and Egger 2005). Using the standard error on the y axis is easiest to interpret because, in this instance, the contours of statistical significance are linear, which is because they are derived from the Wald statistic for each study's effect estimate. The `confunnel` command has an option to use standard error, inverse standard error, variance, or inverse variance on the y axis.

A meta-analysis of trials investigating magnesium therapy following myocardial infarction is a well-known example in the literature where the presence of publication bias is suspected (Teo et al. 1991; ISIS-4 Collaborative Group 1995; Sterne, Bradburn, and Egger 2001). An initial meta-analysis found that magnesium therapy reduced the risk of mortality; however, a number of larger trials were subsequently published that found no evidence that magnesium therapy reduced the risk of mortality. A standard funnel plot is given for this meta-analysis in figure 1, which was generated by using the `metafunnel` command as shown in the following syntax:

```
. use magnesium
. gen logES = logor
. gen selogES = selogor
. metafunnel logES selogES
```

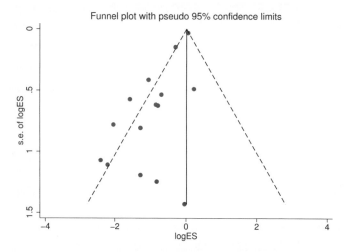

Figure 1. `metafunnel` funnel plot

When the standard error is used on the y axis of a funnel plot, it is conventional to reverse the axis so that the most precise studies are displayed at the top of the plot.

Figure 1 is compared with the equivalent funnel plot produced by `confunnel`, shown in figure 2. The addition of the contours of statistical significance makes it easier to assess the proportion of studies published in the meta-analysis at and around statistical significance. The syntax for the default `confunnel` plot, with the `sj` scheme, is

```
. confunnel logES selogES
```

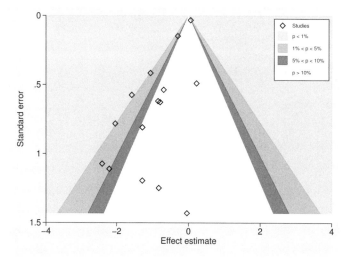

Figure 2. `confunnel` funnel plot using default options

In both figures 1 and 2, there is a strong suggestion of asymmetry in the funnel, suggesting that studies are missing on the right-hand side of the plot, but figure 2 makes it easier to assess the statistical significance of the hypothetical missing studies. The area where missing studies are perceived includes regions of both low and high statistical significance (i.e., the area crosses over the contours), suggesting studies that showed magnesium to be nonsignificantly and significantly less effective to be missing. Therefore, publication bias cannot be accepted as the only cause of funnel asymmetry if it is believed studies are being suppressed because of a mechanism based on two-sided p-values.

It is important to emphasize the differences between the pseudo 95% confidence limits produced by `metafunnel` on figure 1 and the contours of statistical significance produced by `confunnel` on figure 2 (Peters et al. 2008). The pseudo 95% confidence limits illustrate the expected 95% confidence interval about the pooled fixed-effects estimate for the meta-analysis. The pseudo-confidence limits therefore help to assess the extent of between-study heterogeneity in the meta-analysis and the asymmetry on the funnel plot. Unlike the pseudo-confidence limits, the contours of statistical significance are independent of the pooled estimate; therefore, if the pooled estimate is subject to bias, then the contours of significance will not be affected. Also, when the pooled estimate is at the null, the pseudo 95% confidence limits coincide with the two-sided 5% significance contours.

3 The confunnel command

The `confunnel` command plots contour-enhanced funnel plots for study outcome measures in a meta-analysis. Contours of statistical significance from one- or two-sided Wald tests can be plotted using shaded or dashed contour lines. Contours can be plotted along any number of chosen levels of statistical significance; by default, 1%, 5%, and 10% significance contours are plotted. As previously mentioned, `confunnel` has the choice of four y axes. The command also has been designed to be flexible, allowing the user to add extra features to the funnel plot.

3.1 Syntax

confunnel *varname1 varname2* [*if*] [*in*] [, <u>contours</u>(*numlist*)

   <u>contcolor</u>(*color*) <u>extraplot</u>(*plots*) <u>functionlowopts</u>(*options*)

   <u>functionuppopts</u>(*options*) <u>legendlabels</u>(*labels*) <u>legendopts</u>(*options*)

   <u>metric</u>(se | invse | var | invvar) <u>onesided</u>(lower | upper)

   <u>scatteropts</u>(*options*) <u>shadedcontours</u> [<u>no</u>]<u>shadedregions</u> <u>solidcontours</u>

   <u>studylab</u>(*string*) <u>twowayopts</u>(*options*) *twoway_options*]

The first variable, *varname1* is the variable corresponding to the effect estimates, often log odds-ratios, and the second variable, *varname2*, is the variable corresponding to the standard errors of the effect estimates.

3.2　Options

contours(*numlist*) specifies the significance levels of the contours to be plotted; the default is set to 1%, 5%, and 10% significance levels.

contcolor(*color*) specifies the color of the contour lines if shadedcontours is not specified.

extraplot(*plots*) specifies one or multiple additional plots to be overlaid on the funnel plot.

functionlowopts(*options*) and functionuppopts(*options*) pass options to the twoway function commands used to draw the significance contours; for example, the line widths can be changed.

legendlabels(*labels*) specifies labels to appear in the legend for extra elements added to the funnel plot.

legendopts(*options*) passes options to the plot legend.

metric(se | invse | var | invvar) specifies the metric of the y axis of the plot. se, invse, var, and invvar stand for standard error, inverse standard error, variance, and inverse variance, respectively; the default is se.

onesided(lower | upper) can be lower or upper, for lower-tailed or upper-tailed levels of statistical significance, respectively. If unspecified, two-sided significance levels are used to plot the contours.

scatteropts(*options*) specifies any of the options documented in [G] **graph twoway scatter**.

shadedcontours specifies shaded, instead of black, contour lines. Specify this option with the noshadedregions option.

[no]shadedregions suppresses or specifies shaded regions between the contours. This option provides plots that are more similar to those in the original paper by Peters et al. (2008). A plot with shadedregions is the default.

solidcontours specifies solid, instead of dashed, contour lines. Specify this option with both the shadedcontours and the noshadedregions options.

studylab(*string*) specifies the label for the scatter points in the legend. The default is "Studies".

twowayopts(*options*) specifies options passed to the twoway plotting function.

twoway_options are any of the options documented in [G] ***twoway_options***.

4　Use of confunnel

The following subsections use the meta-analysis of magnesium therapy following myocardial infarction.

4.1   Demonstration of some confunnel options

Figure 3 shows the use of the inverse standard error on the y axis; the syntax is as follows:

```
. confunnel logES selogES, metric(invse)
```

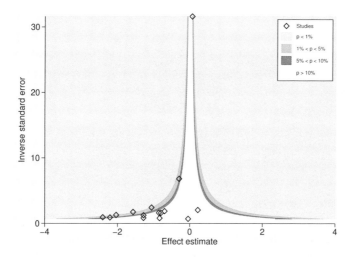

Figure 3. `confunnel` funnel plot using inverse standard error on the y axis

If there is strong evidence that studies are suppressed based on a one-sided (rather than a two-sided) significance test, this can be investigated using the `onesided()` option. This is shown in figure 4 and in the following syntax, which also depicts the contours using lines rather than shaded regions:

```
. confunnel logES selogES, onesided(lower) noshadedregions
```

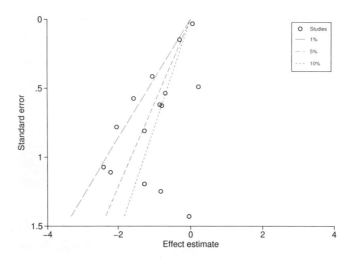

Figure 4. `confunnel` using lower tail one-sided significance regions

Unlike figure 2, in figure 4 (based on one-sided *p*-values) the area where studies are perceived missing is within the region of low statistical significance. Under this assumption, it is more reasonable to consider publication bias as the potential cause of the funnel asymmetry. In this context, the one-sided assumption implies that studies showing magnesium to be harmful are likely to be suppressed regardless of the significance of the results. Previous methods to address publication bias have made various assumptions about the sidedness of suppression; for example, the trim-and-fill method is one-sided, whereas Egger's regression test is two-sided (Duval and Tweedie 2000; Egger et al. 1997).

Figure 5 shows using variance on the *y* axis, using the shaded and solid contours options, and labeling the *x* axis with odds ratios on the funnel plot. The syntax is shown here (`confunnel` was run prior to these commands in order to see where Stata placed the tick marks on the *x* axis):

```
. local t1 = round(exp(-4)*100)/100
. local t2 = round(exp(-2)*100)/100
. local t3 = exp(0)
. local t4 = round(exp(2)*100)/100
. local t5 = round(exp(4)*100)/100
. confunnel logES selogES, metric(var) twowayopts(xtitle("Odds ratios")
> `"xlabel(-4 "`t1'" -2 "`t2'" 0 "`t3'" 2 "`t4'" 4 "`t5'")"')
```

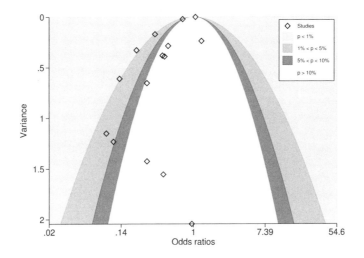

Figure 5. `confunnel` using variance on the y axis

4.2 Use of confunnel with metan, metabias, and metatrim

The `metan` command for meta-analysis (Bradburn, Deeks, and Altman 1998; Harris et al. 2008) can be used to generate the information to display the pooled fixed-effects estimate with its pseudo 95% confidence interval (or, indeed, the pooled random-effects estimate) on the `confunnel` plot; this is shown in figure 6. In this example, because the pooled log odds-ratio was very close to 0, the pseudo 95% confidence interval (for the pooled fixed-effects estimate) almost coincided with the 5% significance contours, which are symmetric about the null hypothesis. The syntax for figure 6 is as follows:

```
. capture drop logES selogES
. metan alive0 dead0 alive1 dead1, or nograph fixed
  (output omitted)
. local fixedlogES = log(r(ES))
. generate logES = log(_ES)
. rename _selogES selogES
. summarize selogES, meanonly
. local semax = r(max)
```

```
. confunnel logES selogES, extraplot(function `fixedlogES´, horizontal
> lc(black) range(0 `semax´) || function `fixedlogES´ + x*invnormal(.025),
> horizontal range(0 `semax´) lc(black) || function `fixedlogES´ +
> x*invnormal(.975), horizontal range(0 `semax´) lc(black))
> legendlabels(`"10 "F.E. & 95% C.I.""´)
```

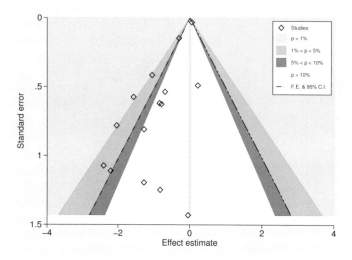

Figure 6. `confunnel` with `metafunnel` features using `metan`

Egger's test investigating possible small-study reporting bias can be represented on the funnel plot by using the information from the `metabias` command (Egger et al. 1997; Steichen 1998); this is shown in figure 7 and in the following syntax:

```
. metabias logES selogES, egger
  (output omitted)
. matrix b = e(b)
. local bias = b[1,2]
. local slope = b[1,1]
. summarize selogES, meanonly
. local semax = r(max)
. metabias alive0 dead0 alive1 dead1, harbord
  (output omitted)
. matrix c = e(b)
. local modbias = c[1,2]
. local modslope = c[1,1]
. confunnel logES selogES, contours(5 10) extraplot(function (`bias´*x + `slope´),
> horizontal range(0 `semax´) lc(black) || function (`modbias´*x + `modslope´),
> horizontal range(0 `semax´) lc(black) lp(dash))
> legendlabels(`"7 "Egger" 8 "Harbord""´)
```

Also shown on the figure is the modified Egger test using the `metabias` command (Harbord, Harris, and Sterne 2009) because Egger's test has been shown to be biased for binary outcome meta-analyses (Harbord, Egger, and Sterne 2006).

The modified Egger's test is performed on different scales from those of the axes of the funnel plot, but when all trials have a reasonable sample size with small effect estimates, it is not unreasonable to view it on a funnel plot.

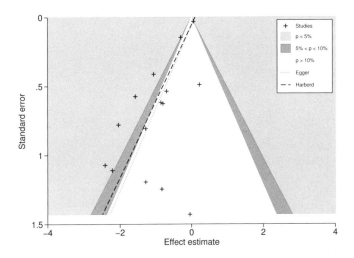

Figure 7. `confunnel` with Egger's and Harbord's regression tests using `metabias`

The trim-and-fill method (Duval and Tweedie 2000) can be applied using `metatrim` (Steichen 2000). In order to demonstrate `confunnel` displaying filled studies, a meta-analysis of the risk of lung cancer from passive smoking is used (Hackshaw, Law, and Wald 1997; Rothstein, Sutton, and Borenstein 2005). Applying the trim-and-fill method, the passive smoking meta-analysis produces seven filled studies, shown in figure 8 and described with the following syntax:

```
. use passivesmoking, clear
. local n = _N
. metan logOR selogOR, nograph
  (output omitted)
. local ES = r(ES)
. summarize selogOR, meanonly
. local semax = r(max)
. metatrim logOR selogOR, save(metatrimdata, replace)
  (output omitted)
. use metatrimdata, clear
. local nfilled = _N - `n'
. metan filled fillse, nograph
  (output omitted)
. local filledES = r(ES)
```

```
. confunnel filled fillse if _n > `nfilled´, contours(5 10)
> extraplot(scatter fillse filled if _n <= `nfilled´, m(T) mc(gs8) ||
> function `ES´, horizontal lc(black) range(0 `semax´) ||
> function `filledES´, horizontal lc(gs8) range(0 `semax´))
> legendlabels(`"7 "Filled" 8 "F.E." 9 "F.E. filled""´)
```

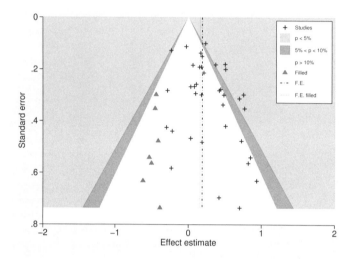

Figure 8. `confunnel` with filled studies from `metatrim`. The vertical dotted line shows the pooled log odds-ratio on the original meta-analysis, while the vertical short dash–dotted line shows the pooled estimate including the filled studies.

It is possible to consider the studies filled by trim and fill as a guide to the likely location of missing studies. With the contours added to the funnel plot containing the filled studies, it is possible to assess the projected significance of the missing studies to determine if it is reasonable to assume such studies could be suppressed by publication bias based on a p-value selection mechanism. In figure 8, trim and fill estimates that seven studies are missing, all of which indicate those exposed to passive smoking are at a reduced risk of lung cancer and all of which are in the region of $p > 0.10$. Hence, it is plausible that publication bias is the cause of the observed asymmetry in this funnel plot.

5   Discussion

The use of the contour-enhanced funnel plot, implemented with the `confunnel` command, is recommended to investigate meta-analyses where it is possible that results could be suppressed on the basis of their statistical significance. In practice, it is suspected that this could include the majority of contexts in which meta-analysis is conducted, certainly in medicine and related disciplines. Exceptions do exist, for example, where noncomparative effect sizes are combined (e.g., in a surgical case series or for incidence or prevalence data); statistical significance will often have no meaning, and in such cases the contours would not be relevant.

An issue with the interpretation of the contour-enhanced funnel plot is that the significance contours can draw the analyst into thinking that the studies should be symmetric about the null hypothesis of the Wald test, because this is the point at which the contours meet when standard error or variance is used on the y axis. But this should be avoided because the studies should form a symmetric funnel shape centered around the true underlying effect size and not the null. Because of this, it can be helpful to plot the meta-analysis pooled estimate for the data on the funnel, although the analyst should be aware that this too may be biased if publication bias is present.

In conclusion, funnel plots are a useful tool in the assessment of systematic differences between the effects in smaller and larger studies in a meta-analysis, regardless of the underlying reason for the differences. Funnel plots can be enhanced by the inclusion of contours of statistical significance, which aid in the interpretation of whether such differences in study estimates in a meta-analysis are most likely to be due to publication bias or other factors.

6   References

Bradburn, M. J., J. J. Deeks, and D. G. Altman. 1998. sbe24: metan—an alternative meta-analysis command. *Stata Technical Bulletin* 44: 4–15. Reprinted in *Stata Technical Bulletin Reprints*, vol. 8, pp. 86–100. College Station, TX: Stata Press. (Reprinted in this collection on pp. 3–28.)

Dickersin, K. 1997. How important is publication bias? A synthesis of available data. *AIDS Education and Prevention* 9: S15–S21.

Duval, S., and R. L. Tweedie. 2000. Trim and fill: A simple funnel-plot–based method of testing and adjusting for publication bias in meta-analysis. *Biometrics* 56: 455–463.

Easterbrook, P. J., J. A. Berlin, R. Gopalan, and D. R. Matthews. 1991. Publication bias in clinical research. *Lancet* 337: 867–872.

Egger, M., G. Davey Smith, M. Schneider, and C. Minder. 1997. Bias in meta-analysis detected by a simple, graphical test. *British Medical Journal* 315: 629–634.

Hackshaw, A. K., M. R. Law, and N. J. Wald. 1997. The accumulated evidence on lung cancer and environmental tobacco smoke. *British Medical Journal* 315: 980–988.

Harbord, R. M., M. Egger, and J. A. C. Sterne. 2006. A modified test for small-study effects in meta-analyses of controlled trials with binary endpoints. *Statistics in Medicine* 25: 3443–3457.

Harbord, R. M., R. J. Harris, and J. A. C. Sterne. 2009. sbe19_6: Updated tests for small-study effects in meta-analysis. *Stata Journal* 9: 197–210. (Reprinted in this collection on pp. 153–165.)

Harris, R. J., M. J. Bradburn, J. J. Deeks, R. M. Harbord, D. G. Altman, and J. A. C. Sterne. 2008. metan: Fixed- and random-effects meta-analysis. *Stata Journal* 8: 3–28. (Reprinted in this collection on pp. 29–54.)

Ioannidis, J. P. A. 1998. Effect of the statistical significance of results on the time to completion and publication of randomized efficacy trials. *Journal of the American Medical Association* 279: 281–286.

ISIS-4 Collaborative Group. 1995. ISIS-4: A randomised factorial trial assessing early oral captopril, oral mononitrate, and intravenous magnesium sulphate in 58,050 patients with suspected acute myocardial infarction. *Lancet* 345: 669–685.

Peters, J. L., A. J. Sutton, D. R. Jones, K. R. Abrams, and L. Rushton. 2008. Contour-enhanced meta-analysis funnel plots help distinguish publication bias from other causes of asymmetry. *Journal of Clinical Epidemiology* 61: 991–996.

Rothstein, H. R., A. J. Sutton, and M. Borenstein, eds. 2005. *Publication Bias in Meta-Analysis: Prevention, Assessment and Adjustments*. Chichester, UK: Wiley.

Steichen, T. J. 1998. sbe19: Tests for publication bias in meta-analysis. *Stata Technical Bulletin* 41: 9–15. Reprinted in *Stata Technical Bulletin Reprints*, vol. 7, pp. 125–133. College Station, TX: Stata Press. (Reprinted in this collection on pp. 166–176.)

———. 2000. sbe39: Nonparametric trim and fill analysis of publication bias in meta-analysis. *Stata Technical Bulletin* 57: 8–14. Reprinted in *Stata Technical Bulletin Reprints*, vol. 10, pp. 108–117. College Station, TX: Stata Press. (Reprinted in this collection on pp. 180–192.)

Sterne, J. A. C., B. J. Becker, and M. Egger. 2005. The funnel plot. In *Publication Bias in Meta-Analysis: Prevention, Assessment and Adjustments*, ed. H. R. Rothstein, A. J. Sutton, and M. Borenstein, 75–98. Chichester, UK: Wiley.

Sterne, J. A. C., M. J. Bradburn, and M. Egger. 2001. Meta-analysis in Stata. In *Systematic Reviews in Health Care: Meta-Analysis in Context*, 2nd edition, ed. M. Egger, G. Davey Smith, and D. G. Altman, 347–369. London: BMJ Books.

Sterne, J. A. C., and M. Egger. 2001. Funnel plots for detecting bias in meta-analysis: Guidelines on choice of axis. *Journal of Clinical Epidemiology* 54: 1046–1055.

Sterne, J. A. C., and R. M. Harbord. 2004. Funnel plots in meta-analysis. *Stata Journal* 4: 127–141. (Reprinted in this collection on pp. 124–138.)

Teo, K. K., S. Yusuf, R. Collins, P. H. Held, and R. Peto. 1991. Effects of intravenous magnesium in suspected acute myocardial infarction: Overview of randomised trials. *British Medical Journal* 303: 1499–1503.

The Stata Journal (2009)
9, Number 2, sbe19_6, pp. 197–210

153

Updated tests for small-study effects in meta-analyses

Roger M. Harbord
Department of Social Medicine
University of Bristol
Bristol, UK
roger.harbord@bristol.ac.uk

Ross J. Harris
Centre for Infections
Health Protection Agency
London, UK
ross.harris@hpa.org.uk

Jonathan A. C. Sterne
Department of Social Medicine
University of Bristol
Bristol, UK

Abstract. This article describes an updated version of the `metabias` command, which provides statistical tests for funnel plot asymmetry. In addition to the previously implemented tests, `metabias` implements two new tests that are recommended in the recently updated *Cochrane Handbook for Systematic Reviews of Interventions* (Higgins and Green 2008). The first new test, proposed by Harbord, Egger, and Sterne (2006, *Statistics in Medicine* 25: 3443–3457), is a modified version of the commonly used test proposed by Egger et al. (1997, *British Medical Journal* 315: 629–634). It regresses $Z/\sqrt{V}$ against $\sqrt{V}$, where Z is the efficient score and V is Fisher's information (the variance of Z under the null hypothesis). The second new test is Peters' test, which is based on a weighted linear regression of the intervention effect estimate on the reciprocal of the sample size. Both of these tests maintain better control of the false-positive rate than the test proposed by Egger at al., while retaining similar power.

Keywords: sbe19_6, metabias, meta-analysis, publication bias, small-study effects, funnel plots

1 Introduction

Publication and related biases in meta-analysis are often examined by visually checking for asymmetry in funnel plots. However, such visual interpretation is inherently subjective. Tests for funnel plot asymmetry (small-study effects [Sterne, Gavaghan, and Egger 2000]) examine whether the association between estimated intervention effects and a measure of study size (such as the standard error of the intervention effect) is greater than might be expected to occur by chance.

This update to the `metabias` command (Steichen 1998; Steichen, Egger, and Sterne 1998) implements two new tests for funnel plot asymmetry that are recommended in the chapter addressing reporting biases (Sterne, Egger, and Moher 2008) in the recent update to the *Cochrane Handbook for Systematic Reviews of Interventions* (Higgins and Green 2008). The modified version of Egger's test (Egger et al. 1997) proposed by Harbord, Egger, and Sterne (2006) still uses linear regression but is based on the efficient score and its variance, Fisher's information. The test proposed by Peters et al. (2006) is based on a weighted linear regression of the intervention effect estimate on the reciprocal of the sample size. These tests address mathematical problems that can occur with the commonly used Egger test and the rank correlation test proposed by Begg and Mazumdar (1994), which was also available in the original version of `metabias`. As with other recently updated meta-analysis commands, the syntax for `metabias` now corresponds to that for the main meta-analysis command, `metan`.

2 Syntax

`metabias` *varlist* [*if*] [*in*], egger harbord peters begg [graph nofit or rr
   level(*#*) *graph_options*]

As in the `metan` command, *varlist* corresponds to either binary data—in this order: cases and noncases for the experimental group, then cases and noncases for the control group (d_1 h_1 d_0 h_0)—or the intervention effect and its standard error (*theta se_theta*).

The Harbord and Peters tests require binary data. Although the Egger test can be used with binary data, it is recommended only for studies with continuous (numerical) outcome variables and intervention effects measured as mean differences with the format *theta se_theta*.

`by` is allowed with `metabias`; see [D] **by**.

3 Options

egger, harbord, peters, and begg specify that the original Egger test, Harbord's modified test, Peters' test, or the rank correlation test proposed by Begg and Mazumdar (1994) be reported, respectively. There is no default; one test must be chosen.

graph displays a Galbraith plot (the standard normal deviate of intervention effect estimate against its precision) for the original Egger test or a modified Galbraith plot of $Z/\sqrt{V}$ versus $\sqrt{V}$ for Harbord's modified test. There is no corresponding plot for the Peters or Begg tests.

nofit suppresses the fitted regression line and confidence interval around the intercept in the Galbraith plot.

or (the default for binary data) uses odds ratios as the effect estimate of interest.

rr specifies that risk ratios rather than odds ratios be used. This option is not available for the Peters test.

`level(#)` specifies the confidence level, as a percentage, for confidence intervals. The default is `level(95)` or as set by `set level`; see [U] **20.7 Specifying the width of confidence intervals**.

graph_options are any of the options documented in [G] **graph twoway scatter**. In particular, the options for specifying marker labels are useful.

4   Background

A funnel plot is a simple scatterplot of intervention effect estimates from individual studies against some measure of each study's size or precision (Light and Pillemer 1984; Begg and Berlin 1988; Sterne and Egger 2001). It is common to plot effect estimates on the horizontal axis and the measure of study size on the vertical axis. This is the opposite of the usual convention for twoway plots, in which the outcome (e.g., intervention effect) is plotted on the vertical axis and the covariate (e.g., study size) is plotted on the horizontal axis. The name "funnel plot" arises from the fact that precision of the estimated intervention effect increases as the size of the study increases. Effect estimates from small studies will therefore scatter more widely at the bottom of the graph, with the spread narrowing among larger studies. In the absence of bias, the plot should approximately resemble a symmetrical (inverted) funnel. The `metafunnel` command (Sterne and Harbord 2004) can be used to display funnel plots, while the `confunnel` command (Palmer et al. 2008) can be used to display "contour-enhanced" funnel plots.

Funnel plots are commonly used to assess evidence that the studies included in a meta-analysis are affected by publication bias. If smaller studies without statistically significant effects remain unpublished, this can lead to an asymmetrical appearance of the funnel plot. However, the funnel plot is better seen as a generic means of displaying *small-study effects*—a tendency for the intervention effects estimated in smaller studies to differ from those estimated in larger studies (Sterne, Gavaghan, and Egger 2000). Small-study effects may be due to reporting biases, including publication bias and selective reporting of outcomes (Chan et al. 2004), poor methodological quality leading to spuriously inflated effects in smaller studies, or true heterogeneity (when the size of the intervention effect differs according to study size) (Egger et al. 1997; Sterne, Gavaghan, and Egger 2000). Apparent small-study effects can also be artifactual, because, in some circumstances, sampling variation can lead to an association between the intervention effect and its standard error (Irwig et al. 1998). Finally, small-study effects may be due to chance; this is addressed by statistical tests for funnel plot asymmetry.

For outcomes measured on a continuous (numerical) scale, tests for funnel plot asymmetry are reasonably straightforward. Using an approach proposed by Egger et al. (1997), we can perform a linear regression of the intervention effect estimates on their standard errors, weighting by 1/(variance of the intervention effect estimate). This looks for a straight-line relationship between the intervention effect and its standard error. Under the null hypothesis of no small-study effects, such a line would be vertical on a funnel plot. The greater the association between intervention effect and standard error, the more the slope would move away from vertical. The weighting is important to ensure that the regression estimates are not dominated by the smaller studies. It is mathematically equivalent, however, to a test of zero intercept in an unweighted regression on Galbraith's radial plot (Galbraith 1988) of the standard normal deviate, defined as the effect estimate divided by its standard error, against the precision, defined as the reciprocal of the standard error; and in fact, this method is used in `metabias`. If the regression line on a Galbraith plot is constrained to pass through the origin, its slope gives the summary estimate of fixed-effects meta-analysis as suggested by Galbraith. But if the intercept is estimated, a test of the null hypothesis of zero intercept tests for no association between the effect size and its standard error.

The Egger test has been by far the most widely used and cited approach to testing for funnel plot asymmetry. Unfortunately, there are statistical problems with this approach because the standard error of the log odds-ratio is correlated with the size of the odds ratio due to sampling variability alone, even in the absence of small-study effects (Irwig et al. 1998); see Deeks, Macaskill, and Irwig (2005) for an algebraic explanation of this phenomenon. This can cause funnel plots that were plotted using log odds-ratios (or odds ratios on a log scale) to appear asymmetric and can mean that p-values from the Egger test are too small, leading to false-positive test results. These problems are especially prone to occur when the intervention has a large effect, when there is substantial between-study heterogeneity, when there are few events per study, or when all studies are of similar sizes. Therefore, a number of authors have proposed alternative tests for funnel plot asymmetry. These are reviewed in a new chapter in the recently updated *Cochrane Handbook for Systematic Reviews of Interventions* (Higgins and Green 2008), which also gives guidance on testing for funnel plot asymmetry (Sterne, Egger, and Moher 2008).

4.1 Notation

We shall be primarily concerned with meta-analysis of 2×2 tables, where each study contains an intervention group and a control group, and the outcome is binary. We shall use the notation shown in table 1 for a single 2×2 table, using the letter d to denote those who experience the event of interest and h for those who do not, with subscripts 0 and 1 to indicate the control and intervention groups, respectively. We shall concentrate on the log odds-ratio, ϕ, as the measure of intervention effect, estimated by $\phi = \log(d_1 h_0 / d_0 h_1)$. The usual estimate of the variance of the log odds-ratio is the Woolf formula (Woolf 1955), $\mathrm{Var}(\phi) = 1/d_0 + 1/h_0 + 1/d_1 + 1/h_1$, the square root of which gives the estimated standard error, $\mathrm{SE}(\phi)$.

Table 1. Notation for a single 2×2 table

	Outcome		
	Experienced event d (disease)	Did not experience event h (healthy)	Total
Group 1 (intervention)	d_1	h_1	n_1
Group 2 (control)	d_0	h_0	n_0
Total	d	h	n

The Egger test is based on a two-sided t test of the null hypothesis of zero slope in a linear regression of ϕ against $\text{SE}(\phi)$, weighted by $1/\text{Var}(\phi)$ (Sterne, Gavaghan, and Egger 2000). This is equivalent to a two-sided t test of the null hypothesis of zero intercept in an unweighted linear regression of $\phi/\text{SE}(\phi)$ against $1/\text{SE}(\phi)$, which are the axes used in the Galbraith plot.

4.2 New tests for funnel plot asymmetry

Harbord's modification to Egger's test is based on the component statistics of the score test, namely, the efficient score, Z, and the score variance (Fisher's information), V. Z is the first derivative, and V is minus the second derivative of the log likelihood with respect to ϕ evaluated at $\phi = 0$ (Whitehead and Whitehead 1991; Whitehead 1997). The intercept in a regression of $Z/\sqrt{V}$ against $\sqrt{V}$ is used as a measure of the magnitude of small-study effects, with a two-sided t test of the null hypothesis of zero intercept giving a formal test for small-study effects. This is identical to a test of nonzero slope in a regression of Z/V against $1 = \sqrt{V}$ with weights V. If all marginal totals are considered fixed, V has no sampling error and hence no correlation with Z. If, as seems more realistic, n_0 and n_1 are considered fixed but d and h are not, the correlation remains lower than that between ϕ and its variance as calculated by the Woolf formula, leading to reduced false-positive rates (Harbord, Egger, and Sterne 2006).

Using standard likelihood theory (Whitehead 1997), it can also be shown that when ϕ is small and n is large, $\phi \approx Z/V$ and $\text{Var}(\phi) \approx 1/V$. It follows that the modified test becomes equivalent to the original Egger test when all trials are large and have small effect sizes. A plot of $Z = \sqrt{V}$ against $\sqrt{V}$ is therefore similar to Galbraith's radial plot of $\phi = \text{SE}(\phi)$ against $1/\text{SE}(\phi)$, as noted by Galbraith himself (Galbraith 1988).

When the parameter of interest is the log odds-ratio, ϕ, the efficient score is

$$Z = d_1 - dn_1/n$$

and the score variance evaluated at $\phi = 0$ is

$$V = n_0 n_1 dh/n^2(n-1)$$

The formula for V given above is obtained by using conditional likelihood, conditioning on the marginal totals d and h in table 1. When the parameter of interest is the log risk-ratio, it can be shown by using standard profile likelihood arguments that $Z = (d_1 n - d n_1)/h$ and $V = n_0 n_1 d/(nh)$.

The Peters test is based on a linear regression of ϕ on $1/n$, with weights dh/n. The slope of the regression line is used as a measure of the magnitude of small-study effects, with a two-sided t test of the null hypothesis of zero slope giving a formal test for small-study effects. This is a modification of Macaskill's test (Macaskill, Walter, and Irwig 2001), with the inverse of the total sample size as the independent variable rather than total sample size. The use of the inverse of the total sample size gives more balanced type I error rates in the tail probability areas than where there is no transformation of sample size (Peters et al. 2006). For balanced trials ($n_0 = n_1$), the weights dh/n are proportional to V.

When there is little or no between-trial heterogeneity, the Harbord and Peters tests have false-positive rates close to the nominal level while maintaining similar power to the original linear regression test proposed by Egger et al. (1997) (Harbord, Egger, and Sterne 2006; Peters et al. 2006; Rücker, Schwarzer, and Carpenter 2008).

5   Example

We shall use an example taken from a systematic review of randomized trials of nicotine replacement therapies in smoking cessation (Silagy et al. 2004), restricted to the 51 trials that used chewing gum as the method of delivery.

```
. use nicotinegum
(Nicotine gum for smoking cessation)

. describe

Contains data from nicotinegum.dta
  obs:            51                          Nicotine gum for smoking cessation
  vars:            5                          8 Jan 2009 12:02
  size:          663 (99.9% of memory free)  (_dta has notes)
```

variable name	storage type	display format	value label	variable label
trialid	byte	%9.0g		
d1	int	%8.0g		Intervention successes
h1	int	%9.0g		Intervention failures
d0	int	%8.0g		Control successes
h0	int	%9.0g		Control failures

```
Sorted by:  trialid
```

A standard fixed-effects meta-analysis, with intervention effects measured as odds ratios, suggests that there was a beneficial effect of the intervention (unusually for a medical meta-analysis, the event of interest here, smoking cessation, is good news rather than bad):

```
. metan d1 h1 d0 h0, or nograph
        Study     |   OR    [95% Conf. Interval]    % Weight
------------------+-----------------------------------------------
1                 |  2.253    1.277    3.972           2.18
2                 |  1.850    0.989    3.460           1.98
3                 |  1.039    0.708    1.524           6.96
4                 |  1.416    0.599    3.350           1.21
5                 |  0.977    0.497    1.919           2.33
6                 |  4.773    1.910   11.932           0.70
7                 |  1.761    0.796    3.893           1.26
8                 |  3.159    1.138    8.768           0.69
9                 |  1.533    0.771    3.048           1.83
10                |  1.385    0.888    2.160           4.55
11                |  2.949    1.009    8.615           0.61
12                |  2.293    1.239    4.245           1.92
13                |  1.234    0.490    3.106           1.12
14                |  2.624    1.026    6.708           0.87
15                |  2.035    0.783    5.289           0.82
16                |  2.822    1.329    5.994           1.13
17                |  0.869    0.461    1.636           2.82
18                |  0.887    0.326    2.408           1.10
19                |  3.404    1.689    6.861           1.18
20                |  2.170    1.101    4.279           1.59
21                |  1.412    0.572    3.487           1.08
22                |  2.029    0.800    5.148           0.97
23                |  0.955    0.294    3.098           0.77
24                |  1.250    0.472    3.311           1.00
25                |  1.847    0.461    7.397           0.41
26                |  3.327    1.371    8.077           0.76
27                |  1.434    0.843    2.441           3.16
28                |  1.333    0.428    4.155           0.72
29                |  1.235    0.931    1.638          11.86
30                |  3.142    1.776    5.558           1.84
31                |  3.522    0.853   14.543           0.28
32                |  1.168    0.704    1.937           3.81
33                |  1.511    0.835    2.735           2.45
34                |  3.824    1.150   12.713           0.39
35                |  1.165    0.405    3.349           0.85
36                |  1.345    0.349    5.188           0.50
37                |  0.483    0.042    5.624           0.26
38                |  1.713    1.212    2.421           6.33
39                |  1.393    0.572    3.389           1.09
40                |  1.844    1.204    2.822           4.30
41                |  1.460    0.775    2.751           2.18
42                |  1.269    0.776    2.075           3.84
43                |  4.110    1.564   10.799           0.59
44                |  2.082    1.504    2.881           6.57
45                |  1.714    0.523    5.621           0.57
46                |  1.294    0.749    2.236           2.98
47                |  5.313    0.701   40.255           0.20
48                |  2.703    0.509   14.357           0.25
49                |  2.124    0.928    4.858           1.07
50                |  1.760    0.549    5.643           0.58
51                |  1.460    0.679    3.140           1.49
------------------+-----------------------------------------------
M-H pooled OR     |  1.658    1.515    1.815         100.00
------------------+-----------------------------------------------
```

```
Heterogeneity chi-squared =  62.04 (d.f. = 50) p = 0.118
I-squared (variation in OR attributable to heterogeneity) =  19.4%
Test of OR=1 : z=  10.99 p = 0.000
```

The `metan` command automatically creates the variables _ES, corresponding to the odds ratio, and _selogES, corresponding to the standard error of the log odds-ratio. We can use these to derive variables for input to the `metafunnel` command:

```
. generate logor = log(_ES)
. generate selogor = _selogES
```

We now use `metafunnel` to draw a funnel plot with the log odds-ratio, ϕ, on the horizontal axis and its standard error, SE(ϕ), on the vertical axis. The `egger` option draws a line corresponding to the weighted regression of the log odds-ratio on its standard error that is the basis of Egger's regression test; see figure 1.

```
. metafunnel logor selogor, egger
```

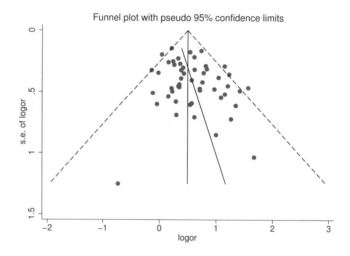

Figure 1. Funnel plot of the log odds-ratio, ϕ, against its standard error, SE(ϕ), including the fitted regression line from the standard regression (Egger) test for small-study effects

The funnel plot appears asymmetric, with smaller studies (those with larger standard errors) tending to have larger (more beneficial) odds ratios. This may suggest publication bias.

We use the `metabias` command to perform a test of small-study effects employing the commonly used Egger test.

```
. metabias d1 h1 d0 h0, egger
Note: data input format tcases tnoncases ccases cnoncases assumed.
Note: odds ratios assumed as effect estimate of interest
Note: peters or harbord tests generally recommended for binary data

Egger's test for small-study effects:
Regress standard normal deviate of intervention
effect estimate against its standard error
Number of studies =  51                          Root MSE     =   1.082
```

Std_Eff	Coef.	Std. Err.	t	P>|t|	[95% Conf. Interval]	
slope	.2832569	.1188368	2.38	0.021	.0444455	.5220683
bias	.7045941	.3566387	1.98	0.054	-.0120982	1.421286

```
Test of H0: no small-study effects            P = 0.054
```

The estimated bias coefficient is 0.705 with a standard error of 0.357, giving a p-value of 0.054. The test thus provides weak evidence for the presence of small-study effects.

The same results can be produced by using the derived variables logor and selogor:

```
. metabias logor selogor, egger
  (output omitted)
```

We now use Harbord's modified test:

```
. metabias d1 h1 d0 h0, harbord graph
Note: data input format tcases tnoncases ccases cnoncases assumed.
Note: odds ratios assumed as effect estimate of interest

Harbord's modified test for small-study effects:
Regress Z/sqrt(V) on sqrt(V) where Z is efficient score and V is score variance
Number of studies =  51                          Root MSE     =   1.092
```

Z/sqrt(V)	Coef.	Std. Err.	t	P>|t|	[95% Conf. Interval]	
sqrt(V)	.3468707	.126528	2.74	0.009	.0926032	.6011382
bias	.5273137	.3866755	1.36	0.179	-.2497398	1.304367

```
Test of H0: no small-study effects            P = 0.179
```

The estimated intercept is 0.527 with a standard error of 0.387, giving a p-value of 0.179. The modified test thus suggests little evidence for small-study effects. The modified Galbraith plot of $Z/\sqrt{V}$ versus $\sqrt{V}$ is shown in figure 2.

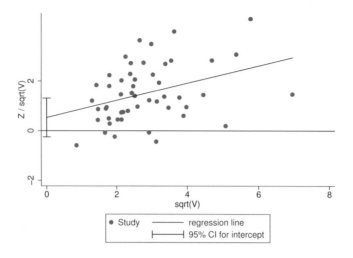

Figure 2. Modified Galbraith plot of $Z/\sqrt{V}$ versus $\sqrt{V}$

Finally, we will use Peters' test for small-study effects:

```
. metabias d1 h1 d0 h0, peters
Note: data input format tcases tnoncases ccases cnoncases assumed.
Note: odds ratios assumed as effect estimate of interest

Peter's test for small-study effects:
Regress intervention effect estimate on 1/Ntot, with weights SF/Ntot
Number of studies =  51                            Root MSE       =  .3897
```

Std_Eff	Coef.	Std. Err.	t	P>|t|	[95% Conf. Interval]	
bias	26.20225	14.58572	1.80	0.079	-3.108842	55.51334
constant	.4197904	.0776552	5.41	0.000	.2637364	.5758443

```
Test of H0: no small-study effects        P = 0.079
```

In this example, the p-value from Peters' test is closer to that from Egger's test than it is to the p-value from Harbord's test. These differing results emphasize the importance of selecting a test in advance; picking a test result from among several is strongly discouraged.

6   Saved results

For all tests, the following scalars are returned:

r(N)	number of studies
r(p_bias)	*p*-value of the bias estimate

For the regression-based tests (Harbord, Peters, and Egger), the following scalars are returned:

r(df_r)	degrees of freedom
r(bias)	estimate of bias (the constant in the regression equation for the Egger and Harbord tests, and the slope for the Peters test)
r(se_bias)	standard error of bias estimate
r(rmse)	root mean squared error of fitted regression model

For Begg's test, the following scalars are returned:

r(score)	Kendall's score $(P–Q)$
r(score_sd)	standard deviation of Kendall's score
r(p_bias_ncc)	*p*-value for Begg's test (not continuity-corrected)

7   Discussion

We have described how to use the metabias command to perform two tests for funnel plot asymmetry. These tests are among those recommended in the *Cochrane Handbook for Systematic Reviews of Interventions* (Higgins and Green 2008) because they reduce the inflation of the false-positive rate (type I error) that can occur for the Egger test, while retaining power compared with alternative tests. metabias allows only one test to be specified. Systematic reviewers should ideally specify their chosen test in advance of the analysis and should avoid choosing from among the results of several tests. Although simulation studies comparing the different tests have been reported (Harbord, Egger, and Sterne 2006; Peters et al. 2006; Rücker, Schwarzer, and Carpenter 2008), no test currently has been shown to be superior in all circumstances. A fuller discussion of these issues is available in chapter 10 (Sterne, Egger, and Moher 2008) of the *Cochrane Handbook*.

Tests for funnel plot asymmetry should not be seen as a foolproof method of detecting publication bias or other small-study effects. We recommend that tests for funnel plot asymmetry be used only when there are at least 10 studies included in the meta-analysis. Even then, power may be low. False-positive results may occur in the presence of substantial between-study heterogeneity, and no test performs well when all studies are of a similar size. Although funnel plots, and tests for funnel plot asymmetry, may alert us to a problem that needs considering when interpreting the results of a meta-analysis, they do not provide a solution to this problem.

8   Acknowledgment

We are grateful to Thomas Steichen, who wrote the original version of the `metabias` command and gave us permission to update it.

Some of the guidance in this article is based on the chapter "Addressing reporting biases" (Sterne, Egger, and Moher 2008), published in the new *Cochrane Handbook for Systematic Reviews of Interventions* (Higgins and Green 2008).

9   References

Begg, C. B., and J. A. Berlin. 1988. Publication bias: A problem in interpreting medical data. *Journal of the Royal Statistical Society, Series A* 151: 419–463.

Begg, C. B., and M. Mazumdar. 1994. Operating characteristics of a rank correlation test for publication bias. *Biometrics* 50: 1088–1101.

Chan, A.-W., A. Hróbjartsson, M. T. Haahr, P. C. Gøtzche, and D. G. Altman. 2004. Empirical evidence for selective reporting of outcomes in randomized trials: Comparison of protocols to published articles. *Journal of the American Medical Association* 291: 2457–2465.

Deeks, J. J., P. Macaskill, and L. M. Irwig. 2005. The performance of tests of publication bias and other sample size effects in systematic reviews of diagnostic test accuracy was assessed. *Journal of Clinical Epidemiology* 58: 882–893.

Egger, M., G. Davey Smith, M. Schneider, and C. Minder. 1997. Bias in meta-analysis detected by a simple, graphical test. *British Medical Journal* 315: 629–634.

Galbraith, R. F. 1988. A note on graphical presentation of estimated odds ratios from several clinical trials. *Statistics in Medicine* 7: 889–894.

Harbord, R. M., M. Egger, and J. A. C. Sterne. 2006. A modified test for small-study effects in meta-analyses of controlled trials with binary endpoints. *Statistics in Medicine* 25: 3443–3457.

Higgins, J. P. T., and S. Green, eds. 2008. *Cochrane Handbook for Systematic Reviews of Interventions*. Chichester, UK: Wiley.

Irwig, L. M., P. Macaskill, G. Berry, and P. P. Glasziou. 1998. Bias in meta-analysis detected by a simple, graphical test. Graphical test is itself biased. *British Medical Journal* 316: 470.

Light, R. J., and D. B. Pillemer. 1984. *Summing Up: The Science of Reviewing Research*. Cambridge, MA: Harvard University Press.

Macaskill, P., S. D. Walter, and L. M. Irwig. 2001. A comparison of methods to detect publication bias in meta-analysis. *Statistics in Medicine* 20: 641–654.

Palmer, T. M., J. L. Peters, A. J. Sutton, and S. G. Moreno. 2008. Contour-enhanced funnel plots for meta-analysis. *Stata Journal* 8: 242–254. (Reprinted in this collection on pp. 139–152.)

Peters, J. L., A. J. Sutton, D. R. Jones, K. R. Abrams, and L. Rushton. 2006. Comparison of two methods to detect publication bias in meta-analysis. *Journal of the American Medical Association* 295: 676–680.

Rücker, G., G. Schwarzer, and J. Carpenter. 2008. Arcsine test for publication bias in meta-analyses with binary outcomes. *Statistics in Medicine* 27: 746–763.

Silagy, C., T. Lancaster, L. Stead, D. Mant, and G. Fowler. 2004. Nicotine replacement therapy for smoking cessation. *Cochrane Database of Systematic Reviews* 3: CD000146.

Steichen, T. J. 1998. sbe19: Tests for publication bias in meta-analysis. *Stata Technical Bulletin* 41: 9–15. Reprinted in *Stata Technical Bulletin Reprints*, vol. 7, pp. 125–133. College Station, TX: Stata Press. (Reprinted in this collection on pp. 166–176.)

Steichen, T. J., M. Egger, and J. A. C. Sterne. 1998. sbe19.1: Tests for publication bias in meta-analysis. *Stata Technical Bulletin* 44: 3–4. Reprinted in *Stata Technical Bulletin Reprints*, vol. 8, pp. 84–85. College Station, TX: Stata Press. (Reprinted in this collection on pp. 177–179.)

Sterne, J. A. C., and M. Egger. 2001. Funnel plots for detecting bias in meta-analysis: Guidelines on choice of axis. *Journal of Clinical Epidemiology* 54: 1046–1055.

Sterne, J. A. C., M. Egger, and D. Moher. 2008. Addressing reporting biases. In *Cochrane Handbook for Systematic Reviews of Interventions*, ed. J. P. T. Higgins and S. Green, 297–334. Chichester, UK: Wiley.

Sterne, J. A. C., D. Gavaghan, and M. Egger. 2000. Publication and related bias in meta-analysis: Power of statistical tests and prevalence in the literature. *Journal of Clinical Epidemiology* 53: 1119–1129.

Sterne, J. A. C., and R. M. Harbord. 2004. Funnel plots in meta-analysis. *Stata Journal* 4: 127–141. (Reprinted in this collection on pp. 124–138.)

Whitehead, A., and J. Whitehead. 1991. A general parametric approach to the meta-analysis of randomized clinical trials. *Statistics in Medicine* 10: 1665–1677.

Whitehead, J. 1997. *The Design and Analysis of Sequential Clinical Trials*. Rev. 2nd ed. Chichester, UK: Wiley.

Woolf, B. 1955. On estimating the relation between blood group and disease. *Annals of Human Genetics* 19: 251–253.

The Stata Technical Bulletin (1998)
STB-41, sbe19, pp. 9–15

Tests for publication bias in meta-analysis[1]

Thomas J. Steichen
RJRT
steichen@triad.rr.com

1 Syntax

The syntax of metabias is

metabias {*theta* {*se_theta* | *var_theta*} | *exp*(*theta*) *ll ul* [*cl*]} [*if*] [*in*] [,

   by(*by_var*) graph({begg | egger}) level(*#*) {var | ci} *graph_options*]]

where the syntax construct {*a* | *b* | ...} means choose one and only one of {*a, b, ...*}.

2 Description

metabias performs the Begg and Mazumdar (1994) adjusted rank correlation test for publication bias and performs the Egger et al. (1997) regression asymmetry test for publication bias. As options, it provides a funnel graph of the data or the regression asymmetry plot.

The Begg adjusted rank correlation test is a direct statistical analogue of the visual funnel graph. Note that both the test and the funnel graph have low power for detecting publication bias. The Begg and Mazumdar procedure tests for publication bias by determining if there is a significant correlation between the effect estimates and their variances. metabias carries out this test by, first, standardizing the effect estimates to stabilize the variances and, second, performing an adjusted rank correlation test based on Kendall's τ.

The Egger et al. regression asymmetry test and the regression asymmetry plot tend to suggest the presence of publication bias more frequently than the Begg approach. The Egger test detects funnel plot asymmetry by determining whether the intercept deviates significantly from zero in a regression of standardized effect estimates against their precision.

The user provides the effect estimate, *theta*, to metabias as a log risk-ratio, log odds-ratio, or other direct measure of effect. Along with *theta*, the user supplies a measure of *theta*'s variability (i.e., its standard error, *se_theta*, or its variance, *var_theta*). Alternatively, the user may provide the exponentiated form, *exp*(*theta*), (i.e., a risk ratio or odds ratio) and its confidence interval, (*ll, ul*).

---

1. This article describes the original version of the metabias command, which is now obsolete. Syntax for the current version of the command is described in the article by Harbord, Harris, and Sterne (2009).—Ed.

The funnel graph plots *theta* versus *se_theta*. Guide lines to assist in visualizing the funnel are plotted at the variance-weighted (fixed effects) meta-analytic effect estimate and at pseudo confidence interval limits about that effect estimate (i.e., at *theta* $\pm$ $z \times se\_theta$, where z is the standard normal variate for the confidence level specified by option `level()`). Asymmetry on the right of the graph (where studies with high standard error are plotted) may give evidence of publication bias.

The regression asymmetry graph plots the standardized effect estimates, *theta/ se_theta*, versus precision, $1/se\_theta$, along with the variance-weighted regression line and the confidence interval about the intercept. Failure of this confidence interval to include zero indicates asymmetry in the funnel plot and may give evidence of publication bias. Guide lines at $x = 0$ and $y = 0$ are plotted to assist in visually determining if zero is in the confidence interval.

`metabias` will perform stratified versions of both the Begg and Mazumdar test and the Egger regression asymmetry test when option `by(`*by_var*`)` is specified. Variable *by_var* indicates the categorical variable that defines the strata. The procedure reports results for each strata and for the stratified tests. The graphs, if selected, plot only the combined unstratified data.

3   Options

`by(`*by_var*`)` requests that the stratified tests be carried out with strata defined by *by_var*.

`graph(begg)` requests the Begg funnel graph showing the data, the fixed-effects (variance-weighted) meta-analytic effect, and the pseudo confidence interval limits about the meta-analytic effect.

`graph(egger)` requests the Egger regression asymmetry plot showing the standardized effect estimates versus precision, the variance-weighted regression line, and the confidence interval about the intercept.

`level(`#`)` specifies the confidence level, as a percentage, for confidence intervals. The default is `level(95)` or as set by `set level`; see [U] **20.7 Specifying the width of confidence intervals**.

`var` indicates that *var_theta* was supplied on the command line instead of *se_theta*. Option `ci` should not be specified when option `var` is specified.

`ci` indicates that *exp(theta)* and its confidence interval, (*ll*, *ul*), were supplied on the command line instead of *theta* and *se_theta*. Option `var` should not be specified when option `ci` is specified.

graph_options are those allowed with `graph, twoway`. For `graph(begg)`, the default *graph_options* include `connect(lll.)`, `symbol(iiio)`, and `pen(3552)` for displaying the meta-analytic effect, the pseudo confidence interval limits (two lines), and the data points, respectively. For `graph(egger)`, the default *graph_options* include `connect(.ll)`, `symbol(oid)`, and `pen(233)` for displaying the data points,

regression line, and the confidence interval about the intercept, respectively. Setting
`t2title(.)` blanks out the default `t2title`.

4   Input variables

The effect estimates (and a measure of their variability) can be provided to `metabias`
in any of three ways:

1. The effect estimate and its corresponding standard error (the default method):

   . `metabias` *theta se_theta* ...

2. the effect estimate and its corresponding variance (note that option `var` must be
   specified):

   . `metabias` *theta var_theta* , `var` ...

3. the risk (or odds) ratio and its confidence interval (note that option `ci` must be
   specified):

   . `metabias` *exp(theta) ll ul* , `ci` ...

   where *exp(theta)* is the risk (or odds) ratio, *ll* is the lower limit and *ul* is the upper
   limit of the risk ratio's confidence interval.

When input method 3) is used, *cl* is an optional input variable that contains the
confidence level of the confidence interval defined by *ll* and *ul*:

   . `metabias` *exp(theta) ll ul cl* , `ci` ...

If *cl* is not provided, `metabias` assumes that each confidence interval is at the 95%
confidence level. *cl* allows the user to provide the confidence level, by study, when the
confidence intervals are not at the default level or are not all at the same level. Values
of *cl* can be provided with or without a decimal point. For example, 90 and 0.90 are
equivalent and may be mixed (e.g., 90, 0.95, 80, 0.90).

5   Explanation

Meta-analysis has become a popular technique for numerically synthesizing information
from published studies. One of the many concerns that must be addressed when performing a meta-analysis is whether selective publication of studies could lead to bias in
the meta-analytic conclusions. In particular, if the probability of publication depends
on the results of the study—for example, if reporting large or statistically significant
findings increase the chance of publication—then the possibility of bias exists.

An initial approach used to assess the likelihood of publication bias was the funnel
graph (Light and Pillemer 1984). The funnel graph plotted the outcome measure (effect size) of the component studies against the sample size (a measure of variability).

The approach assumed that all studies in the analysis were estimating the same effect, therefore the estimated effects should be distributed about the unknown true effect level and their spread should be proportional to their variances. This suggested that, when plotted, small studies should be widely spread about the average effect and the spread should narrow as sample sizes increase. If the graph suggested a lack of symmetry about the average effect, especially if small, negative studies were absent, then publication bias was assumed to exist.

Evaluation of a funnel graph was a very subjective process, with bias—or lack of bias—being in the eye of the beholder. Begg and Mazumdar (1994) noted this and observed that the presence of publication bias induced a skewness in the plot and a correlation between the effect sizes and their variances. They proposed that a *formal test for publication bias*, which is implemented in this insert, could be constructed by examining this correlation. The proposed test evaluates the significance of the Kendall's rank correlation between the standardized effect sizes and their variances.

Recently, Egger et al. (1997) proposed an alternative, regression-based test for detecting skewness in the funnel plot and, by extension, for detecting publication bias in the data. This numerical measure of funnel plot asymmetry also constitutes a *formal test for publication bias* and is implemented in this insert. The proposed test evaluates whether the intercept deviates significantly from zero in a regression of standardized effect estimates against their precision. The test is motivated by the observation that, under assumptions of a nonzero underlying effect and a lack of publication bias, 1) small studies would have both a near-zero precision (since precision is predominantly a function of sample size) and a near-zero standardized effect (because of division by a correspondingly large standard error), while 2) large studies would have both a large precision and a large standardized effect (because of division by a small standard error). Therefore the standardized effects would scatter about a regression line (approximately) through the origin that has a slope which estimates both the size and direction of the underlying effect. Under conditions of publication bias and asymmetry in the funnel plot, the subsample of small studies will differ systematically from the subsample of larger studies and the regression line will fail to go through the origin. The size of the intercept provides a measure of asymmetry—the larger the deviation from zero the greater the asymmetry. The direction of the intercept provides information on the form of the bias—a positive intercept indicates that the effect estimated from the smaller studies is greater than the effect estimated from the larger studies. Conversely, a negative intercept indicates that the effect estimated from the smaller studies is less than the effect estimated from the larger studies.

6   Begg's test

This section paraphrases the mathematical development and discussion in the Begg and Mazumdar paper (the paper also includes a detailed examination of the operating characteristics of this test and examples based on medical data).

Let (t_i, v_i), $i = 1, \ldots, k$, be the estimated effect sizes and sample variances from k studies in a meta-analysis. To construct the adjusted rank correlation test, calculate the standardized effect sizes

$$t_i^* = \frac{(t_i - \bar{t})}{(v_i^*)^{1/2}}$$

where

$$\bar{t} = \frac{\sum_{j=1}^{k} t_j v_j^{-1}}{\sum_{j=1}^{k} v_j^{-1}}$$

is the variance-weighted average effect size, and

$$v_i^* = v_i - \left(\sum_{j=1}^{k} v_j^{-1} \right)^{-1}$$

is the variance of $t_i - \bar{t}$.

Correlate the standardized effect sizes, t_i^*, with the sample variances, v_i, using Kendall's rank correlation procedure and examine the p value. A significant correlation is interpreted as providing strong evidence of publication bias.

In their examples, Begg and Mazumdar use the normalized Kendall rank correlation test statistic for data that have no ties, $z = (P - Q)/\{k(k-1)(2k+5)/18\}^{1/2}$, where P is the number of pairs of studies ranked in the same order with respect to t^* and v and Q is the number of pairs ranked in the opposite order. This statistic does not apply a continuity correction. The authors remark that the denominator should be modified if there are tied observations in either t_i^* or v_i but, instead, apparently break ties in their sample data by adding a small constant. The `metabias` procedure implemented in this insert invokes a modification of Stata's `ktau` procedure to calculate the correct statistic, whether ties exist or not, and presents the z and p values with and without the continuity correction.

Begg and Mazumdar (1994) report that the principal determinant of the power of this test is the number of component studies in the meta-analysis (as opposed to the sample sizes of the individual studies). Additionally, the power will increase with a wider range in variance (sample size) and with a smaller underlying effect size. The authors state that the test is fairly powerful for a meta-analysis of 75 component studies, only moderately powerful for one of 25 component studies, and weak when there are few component studies. They advise that "the test must be interpreted with caution in small meta-analyses. In particular, [publication] bias cannot be ruled out if the test is not significant."

A *stratified test* can also be constructed. Let $P_l - Q_l$ be the numerator of the unstratified test statistic for the lth subgroup and d_l be the square of the corresponding denominator (i.e., the variance of $P_l - Q_l$). The stratified test statistic, without continuity correction, is defined as

$$z_s = \frac{\sum_l (P_l - Q_l)}{\left(\sum_l d_l \right)^{1/2}}$$

The `metabias` procedure implemented in this insert calculates the correct stratified statistic, whether ties exist or not, and presents the z_s and p_s values with and without the continuity correction.

Begg and Mazumdar assume that the sampling distribution of t is normal, i.e., $t \sim N(\delta, v_i)$, where δ is the common effect size to be estimated and the v_i are the variances, which depend on the sample sizes of the individual component studies. They argue that the normality assumption is reasonable because t is "invariably a summary estimate of some parameter, and as such will possess an asymptotic normal distribution in most circumstances." The subsequent asymptotic-normality assumption for z_s inherently follows from this argument.

7 Egger's test

This section paraphrases the method development and discussion in the Egger et al. paper. (The paper also provides an empirical evaluation, based on only eight examples from the medical literature, of the ability of the regression asymmetry test to correctly predict whether a meta-analysis of smaller studies will be concordant with the results of a subsequent large trial.)

Let (t_i, v_i), $i = 1, \ldots, k$, be the estimated effect sizes and sample variances from k studies in a meta-analysis. Define the standardized effect size as $t_i^* = t_i / v_i^{1/2}$, the precision as $s^{-1} = 1/v_i^{1/2}$, and the weight as $w_i = 1/v_i$. (In this form of standardization, t^* is a *standard normal deviate* and is designated as such in the Egger paper.) Fit t^* to s^{-1} using standard weighted linear regression with weights w and linear equation: $t^* = \alpha + \beta s^{-1}$. A significant deviation from zero of the estimated intercept, $\hat{\alpha}$, is interpreted as providing evidence of asymmetry in the funnel plot and of publication bias in the sampled data.

Egger et al. (1997) fit both weighted and unweighted regression lines and select the results of the analysis yielding the intercept with the larger deviation from zero. This insert implements only the weighted analysis.

Egger et al. (1997) do not provide a formal analysis of coverage (i.e., nominal significance level) or power for this test, though they do provide a number of assertions about power. First, they state that "[i]n contrast to the overall test of heterogeneity, the test for funnel plot asymmetry assesses a specific type of heterogeneity and provides a more powerful test in this situation." Second, they state that "[i]n some situations... power is gained by weighting the analysis." Lastly, in a comparison to the Begg and Mazumdar test, they state that "the linear regression approach may be more powerful than the rank correlation test." Egger et al. note, though, that "any analysis of heterogeneity depends on the number of trials included in a meta-analysis, which is generally small, and this limits the statistical power of the test."

Although the paper provides no formal analysis in support of these assertions, an empirical evaluation based on eight examples from the medical literature is reported. This evaluation assessed the ability of the regression asymmetry test to correctly predict whether a meta-analysis of smaller studies will be concordant with the results of a subsequent large trial. For these eight examples, the test detected bias in 3 of 4 cases where a meta-analysis disagreed with a subsequent large trial and indicated no bias in all 4 cases where the meta-analysis agreed with the subsequent large trial. In contrast, the Begg and Mazumdar test was significant for only 1 of the 4 discordant cases (but like Egger's test, for none of the concordant cases). Nonetheless, eight example cases are too few to be statistically convincing and the test remains unvalidated. Further, the lack of coverage analysis leaves open the question of false-positive claims of asymmetry and publication bias. Interestingly, if the Egger's publication bias test is too liberal (a concern that the author of this insert holds) that translates into conservativeness at the meta-analysis level since the bias test will suggest too frequently that caution is needed in interpreting the results of the meta-analysis.

An approximate *stratified test* can be constructed using logic similar to that of Begg and Mazumdar (although Egger et al. did not do so). Let a_l be the intercept from the regression equation for the lth subgroup and v_l^a be the variance of a_l. The stratified test statistic is defined as

$$z_s = \frac{\sum_l a_l / v_l^a}{\left(\sum_l 1/v_l^a \right)^{1/2}}$$

and is assumed to be distributed asymptotically normal. In this form, the stratified estimate is simply the variance-weighted, fixed effect meta-analysis of the intercepts. This stratified test is implemented in this insert.

8   Examples

Begg and Mazumdar illustrated their method with examples from the literature. The first example examined the association between Chlamydia trachomatis and oral contraceptive use derived from 29 case-control studies (Cottingham and Hunter 1992). `metabias` is invoked as follows:

```
. metabias logor varlogor, var graph(egger)
```

Option `var` is used because the data were provided as log odds-ratios and variances and this avoids the, admittedly, small step of generating the standard errors. The optional Egger graph is also requested. `metabias` provides the following analysis:

```
Tests for Publication Bias
Begg´s Test
  adj. Kendall´s Score (P-Q) =      85
           Std. Dev. of Score = 53.30 (corrected for ties)
            Number of Studies =     29
                           z =   1.59
                    Pr > |z| =  0.111
                           z =   1.58 (continuity corrected)
                    Pr > |z| =  0.115 (continuity corrected)

Egger´s Test

--------------------------------------------------------------------------
Std_Eff |      Coef.   Std. Err.        t    P>|t|     [95% Conf. Interval]
--------+-----------------------------------------------------------------
  slope |   .5107122   .0266415    19.170    0.000     .4560484     .565376
   bias |   .8016095   .2961195     2.707    0.012     .1940226    1.409196
--------------------------------------------------------------------------
```

The noncontinuity-corrected test statistic, $z = 1.59$ ($p = 0.111$), differs substantially from that reported by Begg and Mazumdar, $z = 1.76$ ($p = 0.08$). It differs for two reasons: first, the `metabias` procedure corrected the standard deviation of Kendall's score for ties; and second, Begg and Mazumdar apparently carried out their calculation on data that differs slightly from the data they report in their appendix.

The difference in data is apparent when comparing the funnel graph in the published paper to that generated by `metabias`. The published graph suggests that the observation at (*logor*, *varlogor*) = (0.41, 0.162) incorrectly overlays observation (0.41, 0.083); that it, it was incorrectly entered as (0.41, 0.083). Recalculation of the test statistic with ties broken, and with the data modified to match the published graph, yields the published results.

Begg and Mazumdar report that their p of 0.08 is "strongly suggestive of publication bias." Correction of the data and calculation of the test statistic to account for the ties, as shown above, weakens this conclusion. Application of the continuity correction further weakens the conclusion. Nonetheless, with only 29 component studies, the test is expected to have only moderate power at best, and the existence of publication bias cannot be ruled out.

In contrast, the Egger's bias coefficient, $bias = 0.802$ ($P > |t| = 0.012$), strongly indicates the presence of asymmetry and publication bias. Further, the sign of the coefficient (positive) suggests that small studies overestimate the effect (or, alternatively, that negative and/or nonsignificant small studies are not included in the Cottingham and Hunter dataset). The slope coefficient, 0.511, which is an estimate of *theta* (that in a weak sense might be considered to be adjusted for the effects of publication bias), is smaller than the effects estimated from meta-analysis of these data using either fixed-effects (*theta* = 0.655) or random-effects (*theta* = 0.716). These differences in effect estimates are consistent with those expected when small, negative studies are excluded.

The Egger plot (figure 1), requested via the `graph(egger)` option, graphically shows this test and points out that the analysis is dominated by one large, very precise study. The plot also shows that the data near the origin are systematically elevated.

The Begg funnel graph of the data (figure 2), which could have been selected with the `graph(begg)` option, provides additional support for this interpretation.

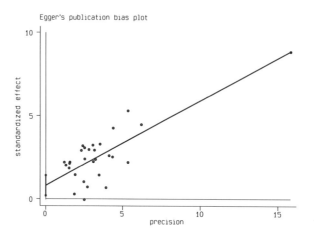

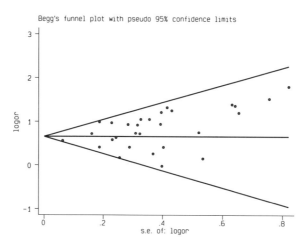

Figure 1. Figure 1 and figure 2

Most of the data points in the Begg plot fall above the meta-analytic effect estimate and there is a visible void in the lower right of the funnel, that is, in the region of low effect and high variance. This is the region where studies most likely to be subject to publication bias would appear. It is notable, though, that since the meta-analytic effect estimate and most of the individual component effect estimates are substantially above zero, the effect of publication bias, if any, would be to inflate the estimate rather than to lead to an incorrect conclusion about the existence of an effect.

Begg and Mazumdar's third example called for the use of the stratified test. These data examined the association between chlorination by-products in drinking water and cancer occurrence, with studies stratified by the site of the cancer (Morris et al. 1992). `metabias` is invoked as follows:

```
. metabias effect variance, var by(site)
```

Use of option `by(site)` informs `metabias` that the stratified tests are to be carried out and that variable `site` is to be used to define the strata. Results are provided in table format, presenting the statistics for each strata and then for the overall stratified tests:

```
Tests for Publication Bias
-------------------------------------------------------------------------------
         |    |    Begg´s          Begg´s        cont. corr. |    Egger´s
  site   | n  | score   s.d.      z      p         z      p  |  bias      p
---------+----+----------------------------------------------+---------------
 Bladder | 7  |   7     6.658    1.05   0.293     0.90   0.368 |  0.07    0.928
   Brain | 2  |   1     1.000    1.00   0.317     0.00   1.000 |  4.71      .
  Breast | 4  |   2     2.944    0.68   0.497     0.34   0.734 |  4.13    0.002
   Colon | 7  |  -1     6.658   -0.15   0.881     0.00   1.000 |  4.36    0.003
 ColoRect| 8  |   0     8.083    0.00   1.000    -0.12   1.000 |  5.33    0.273
 Esophagu| 5  |   4     4.082    0.98   0.327     0.73   0.462 |  1.85    0.456
  Kidney | 4  |   2     2.944    0.68   0.497     0.34   0.734 |  2.31    0.426
   Liver | 4  |   2     2.944    0.68   0.497     0.34   0.734 | -0.78    0.727
    Lung | 5  |   6     4.082    1.47   0.142     1.22   0.221 |  1.06    0.324
Pancreas | 6  |   5     5.323    0.94   0.348     0.75   0.452 |  1.55    0.001
  Rectum | 6  |   1     5.323    0.19   0.851     0.00   1.000 |  4.39    0.103
 Stomach | 6  |   5     5.323    0.94   0.348     0.75   0.452 |  2.02    0.042
---------+----+----------------------------------------------+---------------
 overall | 64 |  34    17.301    1.97   0.049     1.91   0.056 |  2.51    0.000
-------------------------------------------------------------------------------
```

The Begg and Mazumdar results provide no evidence of publication bias in any of the small site-specific strata, yet the stratified test statistic, $z_s = 1.97$ ($p = 0.049$) provides strong evidence that publication bias exists in the chlorinated drinking water and cancer literature. (These results also differ slightly from those published by Begg and Mazumdar in that the published score for the Pancreas strata is 6, leading to an overall score of 35 and slightly different test statistics for this strata and the overall statistic. Results for all other strata agree.) Again, the Egger test provides a stronger indication of the possible presence of publication bias in this literature. Four site-specific strata (Breast, Colon, Pancreas, and Stomach) reach statistical significance and the p value for the overall test is more significant than that of Begg's test, 0.000 versus 0.049. All but one of the individual bias coefficients are positive, as is the overall bias coefficient, suggesting that the small studies in this Morris et al. dataset are overestimating the effect (or that the negative and/or nonsignificant small studies are not included).

9 Saved results

`metabias` saves the following results:

S_1	number of studies	S_5	Begg's p value, continuity corrected
S_2	Begg's score	S_6	Egger's bias coefficient
S_3	s.d. of Begg's score	S_7	Egger's p value
S_4	Begg's p value	S_8	overall effect (log scale)

10 References

Begg, C. B., and M. Mazumdar. 1994. Operating characteristics of a rank correlation test for publication bias. *Biometrics* 50: 1088–1101.

Cottingham, J., and D. Hunter. 1992. Chlamydia trachomatis and oral contraceptive use: A quantitative review. *Genitourinary Medicine* 68: 209–216.

Egger, M., G. Davey Smith, M. Schneider, and C. Minder. 1997. Bias in meta-analysis detected by a simple, graphical test. *British Medical Journal* 315: 629–634.

Harbord, R. M., R. J. Harris, and J. A. C. Sterne. 2009. sbe19_6: Updated tests for small-study effects in meta-analysis. *Stata Journal* 9: 197–210. (Reprinted in this collection on pp. 153–165.)

Light, R. J., and D. B. Pillemer. 1984. *Summing Up: The Science of Reviewing Research.* Cambridge, MA: Harvard University Press.

Morris, R. D., A. M. Audet, I. F. Angelillo, T. C. Chalmers, and F. Mosteller. 1992. Chlorination, chlorination by-products, and cancer: A meta-analysis. *American Journal of Public Health* 82: 955–963.

The Stata Technical Bulletin (1998)
STB-44, sbe19_1, pp. 3–4

Tests for publication bias in meta-analysis[1]

Thomas J. Steichen
RJRT
steichen@triad.rr.com

Matthias Egger
Institute of Social and Preventive Medicine
University of Bern, Switzerland
egger@ispm.unibe.ch

Jonathan A. C. Sterne
Department of Social Medicine
University of Bristol Bristol, UK

1 Modification of the metabias program

This insert documents four changes to the `metabias` program (Steichen 1998). First, the weighted form of the Egger et al. (1997) regression asymmetry test for publication bias has been replaced by the unweighted form. Second, an error has been corrected in the calculation of the asymmetry test p values for individual strata in a stratified analysis. Third, error trapping has been modified to capture or report problem situations more completely and accurately. Fourth, the labeling of the Begg funnel graph has been changed to properly title the axes when the `ci` option is specified. None of these changes affects the program syntax or operation.

The first change was made because, while there is little theoretical justification for the weighted analysis, justification for the unweighted analysis is straightforward. As before, let (t_i, v_i), $i = 1, \ldots, k$, be the estimated effect sizes and sample variances from k studies in a meta-analysis. Egger et al. defined the standardized effect size as $t_i^* = t_i/v_i^{1/2}$, and the precision as $s^{-1} = 1/v_i^{1/2}$. For the unweighted form of the asymmetry test, they fit t^* to s^{-1} using standard linear regression and the equation $t^* = \alpha + \beta s^{-1}$. A significant deviation from zero of the estimated intercept, $\widehat{\alpha}$, is then interpreted as providing evidence of asymmetry in the funnel plot and of publication bias in the sampled data.

Jonathan Sterne (private communication to Matthias Egger) noted that this "unweighted" asymmetry test is merely a reformulation of a standard weighted regression of the original effect sizes, t_i, against their standard errors, $v_i^{1/2}$, where the weights are the usual $1/v_i$. It follows then that the "weighted" asymmetry test is merely a weighted regression of the original effect sizes against their standard errors, but with weights $1/v_i^2$. This form has no obvious theoretical justification.

We note further that the "unweighted" asymmetry test weights the data in a manner consistent with the weighting of the effect sizes in a typical meta-analysis (i.e., both use

---

1. This article describes the original version of the `metabias` command, which is now obsolete. Syntax for the current version of the command is described in the article by Harbord, Harris, and Sterne (2009).—Ed.

the inverse variances). Thus, bias is detected using the same weighting metric as in the meta-analysis.

For these reasons, this insert restricts `metabias` to the unweighted form of the Egger et al. regression asymmetry test for publication bias.

The second change to `metabias` is straightforward. A square root was inadvertently left out of the formula for the p value of the asymmetry test that is calculated for an individual stratum when option `by()` is specified. This formula has been corrected. Users of this program should repeat any stratified analyses they performed with the original program. Please note that unstratified analyses were not affected by this error.

The third change to `metabias` extends the error-trapping capability and reports previously trapped errors more accurately and completely. A noteworthy aspect of this change is the addition of an error trap for the `ci` option. This trap addresses the situation where epidemiological effect estimates and associated error measures are provided to `metabias` as risk (or odds) ratios and corresponding confidence intervals. Unfortunately, if the user failed to specify option `ci` in the previous release, `metabias` assumed that the input was in the default (*theta*, *se_theta*) format and calculated incorrect results. The current release checks for this situation by counting the number of variables on the command line. If more than two variables are specified, `metabias` checks for the presence of option `ci`. If `ci` is not present, `metabias` assumes it was accidentally omitted, displays an appropriate warning message, and proceeds to carry out the analysis as if `ci` had been specified.

Warning: The user should be aware that it remains possible to provide *theta* and its variance, *var_theta*, on the command line without specifying option `var`. This error, unfortunately, cannot be trapped and will result in an incorrect analysis. Though only a limited safeguard, the program now explicitly indicates the data input option specified by the user, or alternatively, warns that the default data input form was assumed.

The fourth change to `metabias` has effect only when options `graph(begg)` and `ci` are specified together. `graph(begg)` requests a funnel graph. Option `ci` indicates that the user provided the effect estimates in their exponentiated form, $exp(theta)$—usually a risk or odds ratio, and provided the variability measures as confidence intervals, (*ll*, *ul*). Since the funnel graph always plots *theta* against its standard error, `metabias` correctly generated *theta* by taking the log of the effect estimate and correctly calculated *se_theta* from the confidence interval. The error was that the axes of the graph were titled using the variable name (or variable label, if available) and did not acknowledge the log transform. This was both confusing and wrong and is corrected in this release. Now when both `graph(begg)` and `ci` are specified, if the variable name for the effect estimate is `RR`, the y axis is titled "log[RR]" and the x axis is titled "s.e. of: log[RR]". If a variable label is provided, it replaces the variable name in these axis titles.

2 References

Egger, M., G. Davey Smith, M. Schneider, and C. Minder. 1997. Bias in meta-analysis detected by a simple, graphical test. *British Medical Journal* 315: 629–634.

Harbord, R. M., R. J. Harris, and J. A. C. Sterne. 2009. sbe19_6: Updated tests for small-study effects in meta-analysis. *Stata Journal* 9: 197–210. (Reprinted in this collection on pp. 153–165.)

Steichen, T. J. 1998. sbe19: Tests for publication bias in meta-analysis. *Stata Technical Bulletin* 41: 9–15. Reprinted in *Stata Technical Bulletin Reprints*, vol. 7, pp. 125–133. College Station, TX: Stata Press. (Reprinted in this collection on pp. 166–176.)

The Stata Technical Bulletin (2000)
STB-57, sbe39, pp. 8–14

Nonparametric trim and fill analysis of publication bias in meta-analysis

Thomas J. Steichen

RJRT

steichen@triad.rr.com

Abstract. This insert describes `metatrim`, a command implementing the Duval and Tweedie nonparametric "trim and fill" method of accounting for publication bias in meta-analysis. Selective publication of studies, which may lead to bias in estimating the overall meta-analytic effect and in the inferences derived, is of concern when performing a meta-analysis. If publication bias appears to exist, then it is desirable to consider what the unbiased dataset might look like and then to reestimate the overall meta-analytic effect after any apparently "missing" studies are included. Duval and Tweedie's "nonparametric 'trim and fill' method" is an approach designed to meet these objectives.

Keywords: meta-analysis, publication bias, nonparametric, data augmentation

1 Syntax

`metatrim` {*theta* { *se_theta* | *var_theta* } | *exp(theta) ll ul* [*cl*]} [*if*] [*in*]
  [, {<u>var</u>|ci} <u>r</u>effect <u>p</u>rint <u>estimat</u>({<u>run</u> | <u>linear</u> | <u>quadratic</u>}) <u>ef</u>orm
  <u>graph</u> <u>f</u>unnel <u>l</u>evel(*#*) <u>id</u>var(*varname*) <u>s</u>ave(*filename* [, replace])
  graph_options]

where {*a* | *b* | ...} means choose one and only one of {*a, b, . . .*}.

2 Description

`metatrim` performs the Duval and Tweedie (2000) nonparametric "trim and fill" method of accounting for publication bias in meta-analysis. The method, a rank-based data-imputation technique, formalizes the use of funnel plots, estimates the number and outcomes of missing studies, and adjusts the meta-analysis to incorporate the imputed missing data. The authors claim that the method is effective and consistent with other adjustment techniques. As an option, `metatrim` provides a funnel plot of the filled data.

The user provides the effect estimate, *theta*, to `metatrim` as a log risk-ratio, log odds-ratio, or other direct measure of effect. Along with *theta*, the user supplies a measure of *theta*'s variability (that is, its standard error, *se_theta*, or its variance, *var_theta*). Alternatively, the user may provide the exponentiated form, *exp(theta)*, (that is, a risk ratio or odds ratio) and its confidence interval, (*ll, ul*).

The funnel plot graphs *theta* versus *se_theta* for the filled data. Imputed observations are indicated by a square around the data symbol. Guide lines to assist in visualizing the center and width of the funnel are plotted at the meta-analytic effect estimate and at pseudo-confidence-interval limits about that effect estimate (that is, at *theta* $\pm$ z $\times$ *se_theta*, where z is the standard normal variate for the confidence level specified by option `level()`).

3 Options

`var` indicates that *var_theta* was supplied on the command line instead of *se_theta*. Option `ci` should not be specified when option `var` is specified.

`ci` indicates that *exp(theta)* and its confidence interval, (*ll*, *ul*), were supplied on the command line instead of *theta* and *se_theta*. Option `var` should not be specified when option `ci` is specified.

`reffect` specifies an analysis based on random-effects meta-analytic estimates. The default is to base calculations on fixed-effects meta-analytic estimates.

`print` requests that the weights used in the filled meta-analysis be listed for each study, together with the individual study estimates and confidence intervals. The studies are labeled by name if the `idvar()` option is specified, or by number otherwise.

`estimat({run | linear | quadratic})` specifies the estimator used to determine the number of points to be trimmed in each iteration. The user is cautioned that the `run` estimator, R_0, is nonrobust to an isolated negative point, and that the `quadratic` estimator, Q_0, may not be defined when the number of points in the dataset is small. The `linear` estimator, L_0, is stable in most situations and is the default.

`eform` requests that the results in the final meta-analysis, and in the `print` option, be reported in exponentiated form. This is useful when the data represent odds ratios or relative risks.

`graph` requests that point estimates and confidence intervals be plotted. The estimate and confidence interval in the graph are derived using fixed- or random-effects meta-analysis, as specified by option `reffect`.

`funnel` requests a filled funnel graph be displayed showing the data, the meta-analytic estimate, and pseudo confidence-interval limits about the meta-analytic estimate. The estimate and confidence interval in the graph are derived using fixed or random-effects meta-analysis, as specified by option `reffect`.

`level(#)` specifies the confidence level, as a percentage, for confidence intervals. The default is `level(95)` or as set by `set level`; see [U] **20.7 Specifying the width of confidence intervals**.

`idvar(`*varname*`)` indicates the character variable used to label the studies.

save(*filename*[, replace]) saves the filled data in a separate Stata data file. The
   filename is assumed to have extension .dta (an extension should not be provided
   by the user). If *filename* does not exist, it is created. If *filename* exists, an error will
   occur unless replace is also specified. Only three variables are saved: a study id
   variable and two variables containing the filled *theta* and *se_theta* values. The study
   id variable, named id in the saved file, is created by metatrim; but when option
   idvar() is specified, it is based on that id variable. The filled *theta* and *se_theta*
   variables are named filled and sefill in the saved file.

graph_options are those allowed with graph, twoway, except ylabel(), symbol(),
   xlog, ytick and gap are not recognized by graph. For funnel, the default
   graph_options include connect(lll..), symbol(iiioS), and pen(35522) for displaying the meta-analytic effect, the pseudo confidence interval limits (two lines),
   and the data points, respectively.

4 Specifying input variables

The individual effect estimates (and a measure of their variability) can be provided to
metatrim in any of three ways:

1. The effect estimate and its corresponding standard error (the default method):

   . metatrim *theta se_theta* . . .

2. The effect estimate and its corresponding variance (note that option var must be
   specified):

   . metatrim *theta var_theta*, var . . .

3. The risk (or odds) ratio and its confidence interval (note that option ci must be
   specified):

   . metatrim *exp(theta) ll ul*, ci . . .

   where *exp(theta)* is the risk (or odds) ratio, *ll* is the lower limit and *ul* is the upper
   limit of the risk ratio's confidence interval.

   When input method 3 is used, *cl* is an optional input variable that contains the
   confidence level of the confidence interval defined by *ll* and *ul*:

   . metatrim *exp(theta) ll ul cl*, ci . . .

   If *cl* is not provided, metatrim assumes that a 95% confidence level was reported for
each study. *cl* allows the user to combine studies with diverse or non-95% confidence
levels by specifying the confidence level for each study not reported at the 95% level.
Note that option level() does not affect the default confidence level assumed for the
individual studies. Values of *cl* can be provided with or without a decimal point. For
example, 90 and .90 are equivalent and may be mixed (i.e., 90, .95, 80, .90, etc.). Missing
values within *cl* are assumed to indicate a 95% confidence level.

Note that data in binary count format can be converted to the effect format used in `metatrim` by use of program `metan` (Bradburn, Deeks, and Altman 1998). `metan` automatically creates and adds variables for *theta* and *se_theta* to the raw dataset, naming them `_ES` and `_seES`. These variables can be provided to `metatrim` using the default input method.

5 Explanation

Meta-analysis is a popular technique for numerically synthesizing information from published studies. One of the many concerns that must be addressed when performing a meta-analysis is whether selective publication of studies could lead to bias in estimating the overall meta-analytic effect and in the inferences derived from the analysis. If publication bias appears to exist, then it is desirable to consider what the unbiased dataset might look like and then to reestimate the overall meta-analytic effect after any apparently "missing" studies are included. Duval and Tweedie's "nonparametric 'trim and fill' method" is designed to meet these objectives and is implemented in this insert.

An early, visual approach used to assess the likelihood of publication bias and to provide a hint of what the unbiased data might look like was the funnel graph (Light and Pillemer 1984). The funnel graph plotted the outcome measure (effect size) of the component studies against the sample size (a measure of variability). The approach assumed that all studies in the analysis were estimating the same effect. Therefore, the effect estimates should be distributed about the unknown true effect level and their spread should be proportional to their variances. This suggested that, when plotted, small studies should be widely spread about the average effect, and the spread should narrow as sample sizes increase, resulting in a symmetric, funnel-shaped graph. If the graph revealed a lack of symmetry about the average effect (especially if small, negative studies appeared to be absent) then publication bias was assumed to exist.

Evaluation of a funnel graph was a very subjective process, with bias—or lack of bias—residing in the eye of the beholder. Begg and Mazumdar (1994) noted this and observed that the presence of publication bias induced skewness in the plot and a correlation between the effect sizes and their variances. They proposed that a formal test of publication bias could be constructed by examining this correlation. More recently, Egger et al. (1997) proposed an alternative, regression-based test for detecting skewness in the funnel plot and, by extension, for detecting publication bias in the data. Their numerical measure of funnel plot asymmetry also constitutes a formal test of publication bias. Stata implementations of both the Begg and Mazumdar procedure and the Egger et al. procedure were provided in `metabias` (Steichen 1998; Steichen, Egger, and Sterne 1998).

However, neither of these procedures provided estimates of the number or characteristics of the missing studies, and neither provided an estimate of the underlying (unbiased) effect. There exist a number of methods to estimate the number of missing studies, model the probability of publication, and provide an estimate of the underlying effect size. Duval and Tweedie list some of these and note that all "are complex and highly computer-intensive to run" and, for these reasons, have failed to find acceptance among meta-analysts. They offer their new method as "a simple technique that seems to meet many of the objections to other methods."

The following sections paraphrase some of the mathematical development and discussion in the Duval and Tweedie paper.

6   Estimators of the number of suppressed studies

Let (Y_j, v_j^2), $j = 1, \ldots, n$, be the estimated effect sizes and within-study variances from n observed studies in a meta-analysis, where all such studies attempt to estimate a common global "effect size" Δ. Define the random-effects model used to combine the Y_j as

$$Y_j = \Delta + \beta_j + \varepsilon_j$$

where $\beta_j \sim N(0, \tau^2)$ accounts for heterogeneity between studies, and $\varepsilon_j \sim N(0, \sigma_j^2)$ is the within-study variability of study j. For a fixed-effects model, assume $\tau^2 = 0$.

Further, in addition to n observed studies, assume that there are k_0 relevant studies that are not observed due to publication bias. Both the value of k_0, that is, the number of unobserved studies, and the effect sizes of these unobserved studies are unknown and must be estimated.

Now, for any collection X_i, $i = 1, \ldots, N$ of random variables, each with a median of zero and sign generated according to an independent set of Bernoulli variables taking values -1 and 1, let r_i denote the rank of $|X_i|$ and

$$W_N^+ = \sum_{X_i > 0} r_i$$

be the sum of the ranks associated with positive X_i. Then W_N^+ has a Wilcoxon distribution.

Assume that among these N random variables, k_0 were suppressed, leaving n observed values. Furthermore, assume that the *suppression has taken place in such a way that the k_0 values of the X_i with the most extreme negative ranks have been suppressed.* (Note: Duval and Tweedie call this their key assumption and present it italicized, as done here, for emphasis. Further, they label the model for an overall set of studies defined in this way as a *suppressed Bernoulli model* and state that it might be expected to lead to a truncated funnel plot.)

Rank again the n observed $|X_i|$ as r_i^* running from 1 to n. Let $\gamma^* \geq 0$ denote the length of the rightmost run of ranks associated with positive values of the observed X_i; that is, if h is the index of the most negative of the X_i and r_h^* is its absolute rank, then $\gamma^* = n - r_h^*$. Define the "trimmed" rank test statistic for the observed n values as

$$T_n = \sum_{X_i > 0} r_i^*$$

Note that though the distributions of γ^* and T_n depend on k_0, the dependence is omitted in this notation. Based on these quantities, define three estimators of k_0, the number of suppressed studies:

$$R_0 = \gamma^* - 1,$$

$$L_0 = \frac{4T_n - n(n+1)}{2n - 1}$$

and

$$Q_0 = n - 1/2 - \sqrt{2n^2 - 4T_n + 1/4}$$

Duval and Tweedie provide the mean and variance of each estimator as follows (the reader should refer to the original paper for the derivation):

$$E(R_0) = k_0, \qquad \text{Var}(R_0) = 2k_0 + 2$$

$$E(L_0) = k_0 - k_0^2/(2n - 1), \qquad \text{Var}(L_0) = 16\,\text{Var}(T_n)/(2n - 1)^2$$

where

$$\text{Var}(T_n) = \{n(n+1)(2n+1) + 10k_0^3 + 27k_0^2 + 17k_0 - 18nk_0^2 - 18nk_0 + 6n^2k_0\}/24$$

and

$$E(Q_0) \approx k_0 + \frac{2\,\text{Var}(T_n)}{\{(n - 1/2)^2 - k_0(2n - k_0 - 1)\}^{3/2}},$$

$$\text{Var}(Q_0) \approx \frac{4\,\text{Var}(T_n)}{(n - 1/2)^2 - k_0(2n - k_0 - 1)}$$

The authors also report that for n large and k_0 of a smaller order than n, then asymptotically:

$$E(R_0) = k_0$$
$$E(L_0) \sim k_0$$
$$E(Q_0) \sim k_0 + 1/6$$
$$\text{Var}(R_0) = o(n)$$
$$\text{Var}(L_0) \sim n/3$$
$$\text{Var}(Q_0) \sim n/3$$

These results suggest that L_0 and Q_0 should have similar behavior, but the authors report that in practice Q_0 is often larger, sometimes excessively so. They also note that L_0 generally has smaller mean square error than Q_0 when $k_0 \geq n/4 - 2$.

Duval and Tweedie remark that the R_0 *run* estimator is rather conservative and nonrobust to the presence of a relatively isolated negative term at the end of the sequence of ranks. They suggest that the estimators based on T_n seem more robust to such a departure from the suppressed Bernoulli hypothesis. They also note that the Q_0 *quadratic* estimator is defined only when $T_n < n^2/2 + 1/16$, and that simulations show this to be violated quite frequently when the number of studies, n, is small and when the number of suppressed studies, k_0, is large relative to n. These concerns leave the L_0 *linear* estimator as the best all around choice.

Because only whole studies can be trimmed, the estimators are rounded in practice to the nearest nonnegative integer, as follows:

$$R_0^+ = \max(0, R_0)$$

$$L_0^+ = \left\{ \max\left(0, L_0 + \frac{1}{2}\right) \right\}$$

$$Q_0^+ = \left\{ \max\left(0, Q_0 + \frac{1}{2}\right) \right\}$$

where (x) is the integer part of x.

7 The iterative trim and fill algorithm

Because the global "effect size" Δ is unknown, the number and position of any missing studies is correlated with the true value of Δ. Therefore, Duval and Tweedie developed an iterative algorithm to estimate these values simultaneously. The algorithm can be used with any of the three estimators of k_0 defined in the previous section (the `metatrim` program allows the user to specify which one is to be used through the `estimat()` option). Likewise, either a fixed-effects or random-effects meta-analysis model can be used to estimate $\widehat{\Delta}^{(l)}$ within each iteration (l) of the algorithm (the default model in `metatrim` is fixed effects, but random effects is used when option `reffect` is specified). Note that the `meta` program of Sharp and Sterne (1997, 1998) is called by `metatrim` to carry out the meta-analysis calculations.

The algorithm proceeds as follows:

1. Starting with values Y_i, estimate $\widehat{\Delta}^{(1)}$ using the chosen meta-analysis model. Construct an initial set of centered values

$$Y_i^{(1)} = Y_i - \widehat{\Delta}^{(1)}, \quad i = 1, \ldots, n$$

and estimate $\widehat{k}_0^{(1)}$ using the chosen estimator for k_0 applied to the set of values $Y_i^{(1)}$.

2. Let l be the current step number. Remove $\widehat{k}_0^{(l-1)}$ values from the right end of the original Y_i and estimate $\widehat{\Delta}^{(l)}$ based on this trimmed set of $n - \widehat{k}_0^{(l-1)}$ values: $\{Y_1, \ldots, Y_{n-\widehat{k}_0^{(l-1)}}\}$. Construct the next set of centered values

$$Y_i^{(l)} = Y_i - \widehat{\Delta}^{(l)}, \quad i = 1, \ldots, n$$

and estimate $\widehat{k}_0^{(l)}$ using the chosen estimator for k_0 applied to the set of values $Y_i^{(l)}$.

3. Increment l and repeat step 2 until an iteration L where $\widehat{k}_0^{(L)} = \widehat{k}_0^{(L-1)}$. Assign this common value to be the estimated value $\widehat{k}_0$. Note that in this iteration it will also be true that $\widehat{\Delta}^{(L)} = \widehat{\Delta}^{(L-1)}$.

4. Augment (that is, "fill") the dataset Y with the $\widehat{k}_0$ imputed symmetric values

$$Y_j^* = 2\widehat{\Delta}^{(L)} - Y_{n-j+1}, \quad j = 1, \ldots, \widehat{k}_0$$

and imputed standard errors

$$\sigma_j^* = \sigma_{n-j+1}, \quad j = 1, \ldots, \widehat{k}_0$$

Estimate the "trimmed and filled" value of Δ using the chosen meta-analysis method applied to the full augmented dataset $\{Y_1, \ldots, Y_n, Y_1^*, \ldots, Y_{\widehat{k}_0}^*\}$.

Conceptually, this algorithm starts with the observed data, iteratively trims (that is, removes) extreme positive studies from the dataset until the remaining studies do not show detectable deviation from symmetry, fills (that is, imputes into the original dataset) studies that are left-side mirrored reflections (about the center of the trimmed data) of the trimmed studies and, finally, repeats the meta-analysis on the filled dataset to get "trimmed and filled" estimates. Each filled study is assigned the same standard error as the trimmed study it reflects in order to maintain symmetry within the filled dataset.

8 Example

The method is illustrated with an example from the literature that examines the association between Chlamydia trachomatis and oral contraceptive use derived from 29 case–control studies (Cottingham and Hunter 1992). Analysis of these data with the publication bias tests of Begg and Mazumdar ($p = 0.115$) and Egger et al. ($p = 0.016$), as provided in `metabias`, suggests that publication bias may affect the data. To examine the potential impact of publication bias on the interpretation of the data, `metatrim` is invoked as follows:

```
. metatrim logor varlogor, reffect funnel var
```

The random-effects model and display of the optional funnel graph are requested via options `reffect` and `funnel`. Option `var` is required because the data were provided as log odds-ratios and variances. By default, the linear estimator, L_0, is used to estimate k_0, as no other estimator was requested. `metatrim` provides the following output:

```
Note: option "var" specified.
Meta-analysis
         | Pooled      95% CI        Asymptotic       No. of
Method | Est    Lower   Upper  z_value  p_value   studies
-------+--------------------------------------------------
Fixed  |  0.655   0.571   0.738   15.359    0.000     29
Random |  0.716   0.595   0.837   11.594    0.000
Test for heterogeneity: Q= 37.034 on 28 degrees of freedom (p= 0.118)
Moment-based estimate of between studies variance =  0.021

Trimming estimator: Linear
Meta-analysis type: Random-effects model
iteration | estimate    Tn    # to trim      diff
----------+------------------------------------------
        1 |   0.716     285        5          435
        2 |   0.673     305        6           40
        3 |   0.660     313        7           16
        4 |   0.646     320        7           14
        5 |   0.646     320        7            0
Filled
Meta-analysis
         | Pooled      95% CI        Asymptotic       No. of
Method | Est    Lower   Upper  z_value  p_value   studies
-------+--------------------------------------------------
Fixed  |  0.624   0.542   0.705   14.969    0.000     36
Random |  0.655   0.531   0.779   10.374    0.000
Test for heterogeneity: Q= 49.412 on 35 degrees of freedom (p= 0.054)
Moment-based estimate of between studies variance =  0.031
```

metatrim first calls program meta to perform and report a standard meta-analysis of the original data, showing both the fixed- and random-effects results. These initial results are always reported as *theta* estimates, regardless of whether the data were provided in exponentiated form.

metatrim next reports the trimming estimator and type of meta-analysis model to be used in the iterative process, then displays results at each iteration. The estimate column shows the value of $\widehat{\Delta}^{(l)}$ at each iteration. As expected, its value at iteration 1 is the same as shown for the random-effects method in the meta-analysis panel, and then decreases in successive iterations as values are trimmed from the data. Column Tn reports the T_n statistic, column # to trim reports the successive estimates $\widehat{k}_0^{(l)}$ and column diff reports the sum of the absolute differences in signed ranks between successive iterations. The algorithm stops when diff is zero.

metatrim finishes with a call to program meta to report an analysis of the trimmed and filled data. Observe that there are now 36 studies, composed of the $n = 29$ observed studies plus the additional $\widehat{k}_0 = 7$ imputed studies. Also note that the estimate of $\widehat{\Delta}$ reported as the random effects pooled estimate for the 36 studies is not the same as the value $\widehat{\Delta}^{(5)}$ shown in the fifth (and final) line of the iteration panel. These values usually differ when the random-effects model is used (because the addition of imputed values change the estimate of τ^2) but are identical always when the fixed-effects model is used.

In summary, `metatrim` adds 7 "missing" studies to the dataset, moving the random-effects summary estimate from $\widehat{\Delta} = 0.716, 95\%$ CI: $(0.595, 0.837)$ to $\widehat{\Delta} = 0.655, 95\%$ CI: $(0.531, 0.779)$. The new estimate, though slightly lower, remains statistically significant; correction for publication bias does not change the overall interpretation of the dataset. Addition of "missing" studies results in an increased variance between studies, the estimate rising from 0.021 to 0.031, and increased evidence of heterogeneity in the dataset, $p = 0.118$ in the observed data versus $p = 0.054$ in the filled data. As expected, when the trimmed and filled dataset is analyzed with the publication bias tests of Begg and Mazumdar (1994) and Egger et al. (1997) (not shown), evidence of publication bias is no longer observed ($p = 0.753$ and $p = 0.690$, respectively).

The funnel plot (figure 1), requested via the `funnel` option, graphically shows the final filled estimate of Δ (as the horizontal line) and the augmented data (as the points), along with pseudo confidence-interval limits intended to assist in visualizing the funnel. The plot indicates the imputed data by a square around the data symbol. The filled dataset is much more symmetric than the original data and the plot shows no evidence of publication bias.

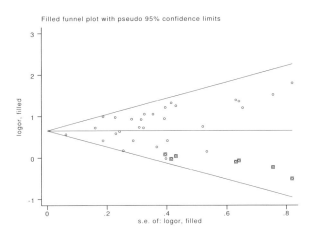

Figure 1. Funnel plot for analysis of Cottingham and Hunter data

Additional options that can be specified include `print` to show the weights, study estimates and confidence intervals for the filled dataset, `eform` to request that the results be reported in exponentiated form in the final meta-analysis and in the `print` option be reported in exponentiated form (this is useful when the data represent odds ratios or relative risks), `graph` to graphically display the study estimates and confidence intervals for the filled dataset, and `save(`*filename*`)` to save the filled data in a separate Stata dataset.

9   Remarks

The Duval and Tweedie method is based on the observation that an unbiased selection of studies that estimate the same thing should be symmetric about the underlying common effect (at least within sampling error). This implies an expectation that the number of studies, and the magnitudes of those studies, should also be roughly equivalent both above and below the common effect value. It is, therefore, reasonable to apply a nonparametric approach to test these assumptions and to adjust the data until the assumptions are met. The price of the nonparametric approach is, of course, lower power (and a concomitant expectation that one may under-adjust the data).

Duval and Tweedie use the symmetry argument in a somewhat roundabout way, choosing to first trim extreme positive studies until the remaining studies meet symmetry requirements. This makes sense when the studies are subject only to publication bias, since trimming should preferably toss out the low-weight, but extreme studies. Nonetheless, if other biases affect the data, in particular if there is a study that is high-weight and extremely positive relative to the remainder of the studies, then the method could fail to function properly. The user must remain alert to such possibilities.

Duval and Tweedie's final step—giving imputed reflections of the trimmed studies—has no effect on the final trimmed point estimate in a fixed effects analysis but does cause the confidence interval of the estimate to be smaller than that from the trimmed or original data. One could question whether this "increased" confidence is warranted.

The random-effects situation is more complex, as both the trimmed point estimate and confidence interval width are affected by filling, with a tendency for the filled data to yield a point estimate between the values from the original and trimmed data. When the random-effects model is used, the confidence interval of the filled data is typically smaller than that of either the trimmed or original data.

Experimentation suggests that the Duval and Tweedie method trims more studies than may be expected; but because of the increase in precision induced by the imputation of studies during filling, changes in the "significance" of the results occur less often than expected. Thus the two operations (trimming, which reduces the point estimate, and filling, which increases the precision) seem to counter each other.

Another phenomenon noted is a tendency for the heterogeneity of the filled data to be greater than that of the original data. This suggests that the most likely studies to be trimmed and filled are those that are most responsible for heterogeneity. The generality of this phenomenon and its impact on the analysis have not been investigated.

Duval and Tweedie provide a reasonable development based on accepted statistics; nonetheless, the number and the magnitude of the assumptions required by the method are substantial. If the underlying assumptions hold in a given dataset, then, as with many methods, it will tend to under- rather than overcorrect. This is an acceptable situation in my view (whereas "overcorrection" of publication bias would be a critical flaw).

This author presents the program as an *experimental* tool only. Users must assess for themselves both the amount of correction provided and the reasonableness of that correction. Other tools to assess publication bias issues should be used in tandem. `metatrim` should be treated as merely one of an arsenal of methods needed to fully assess a meta-analysis.

10   Saved results

`metatrim` does not save values in the system `S_#` macros, nor does it return results in `r()`.

11   Note

The command `meta` (Sharp and Sterne 1997, 1998) should be installed before running `metatrim`.

12   References

Begg, C. B., and M. Mazumdar. 1994. Operating characteristics of a rank correlation test for publication bias. *Biometrics* 50: 1088–1101.

Bradburn, M. J., J. J. Deeks, and D. G. Altman. 1998. sbe24: metan—an alternative meta-analysis command. *Stata Technical Bulletin* 44: 4–15. Reprinted in *Stata Technical Bulletin Reprints*, vol. 8, pp. 86–100. College Station, TX: Stata Press. (Reprinted in this collection on pp. 3–28.)

Cottingham, J., and D. Hunter. 1992. Chlamydia trachomatis and oral contraceptive use: A quantitative review. *Genitourinary Medicine* 68: 209–216.

Duval, S., and R. L. Tweedie. 2000. A nonparametric "trim and fill" method of accounting for publication bias in meta-analysis. *Journal of the American Statistical Association* 95: 89–98.

Egger, M., G. Davey Smith, M. Schneider, and C. Minder. 1997. Bias in meta-analysis detected by a simple, graphical test. *British Medical Journal* 315: 629–634.

Light, R. J., and D. B. Pillemer. 1984. *Summing Up: The Science of Reviewing Research.* Cambridge, MA: Harvard University Press.

Sharp, S., and J. A. C. Sterne. 1997. sbe16: Meta-analysis. *Stata Technical Bulletin* 38: 9–14. Reprinted in *Stata Technical Bulletin Reprints*, vol. 7, pp. 100–106. College Station, TX: Stata Press.

———. 1998. sbe16.1: New syntax and output for the meta-analysis command. *Stata Technical Bulletin* 42: 6–8. Reprinted in *Stata Technical Bulletin Reprints*, vol. 7, pp. 106–108. College Station, TX: Stata Press.

Steichen, T. J. 1998. sbe19: Tests for publication bias in meta-analysis. *Stata Technical Bulletin* 41: 9–15. Reprinted in *Stata Technical Bulletin Reprints*, vol. 7, pp. 125–133. College Station, TX: Stata Press. (Reprinted in this collection on pp. 166–176.)

Steichen, T. J., M. Egger, and J. A. C. Sterne. 1998. sbe19.1: Tests for publication bias in meta-analysis. *Stata Technical Bulletin* 44: 3–4. Reprinted in *Stata Technical Bulletin Reprints*, vol. 8, pp. 84–85. College Station, TX: Stata Press. (Reprinted in this collection on pp. 177–179.)

The Stata Journal (2012)
12, Number 4, gr0054, pp. 605–622

Graphical augmentations to the funnel plot to assess the impact of a new study on an existing meta-analysis

Michael J. Crowther
Department of Health Sciences
University of Leicester
Leicester, UK
michael.crowther@le.ac.uk

Dean Langan
Clinical Trials Research Unit (CTRU)
University of Leeds
Leeds, UK
d.p.langan@leeds.ac.uk

Alex J. Sutton
Department of Health Sciences
University of Leicester
Leicester, UK
ajs22@le.ac.uk

Abstract. Funnel plots are currently advocated to investigate the presence of publication bias (and other possible sources of bias) in meta-analysis. A previously described augmentation to the funnel plot—to aid its interpretation in assessing publication biases—is the addition of statistical contours indicating regions where studies would have to be for a given level of significance, as implemented in the Stata package `confunnel` by Palmer et al. (2008, *Stata Journal* 8: 242–254).

In this article, we describe the implementation of a new range of overlay augmentations to the funnel plot, many described in detail recently by Langan et al. (2012, *Journal of Clinical Epidemiology* 65: 511–519). The purpose of these overlays is to display the potential impact a new study would have on an existing meta-analysis, providing an indication of the robustness of the meta-analysis to the addition of new evidence. Thus these overlays extend the use of the funnel plot beyond assessments of publication biases. Two main graphical displays are described: 1) statistical significance contours, which define regions of the funnel plot where a new study would have to be located to change the statistical significance of the meta-analysis; and 2) heterogeneity contours, which show how a new study would affect the extent of heterogeneity in a given meta-analysis.

We present the `extfunnel` command, which implements the methods of Langan et al. (2012, *Journal of Clinical Epidemiology* 65: 511–519), and, furthermore, we extend the graphical displays to illustrate the impact a new study has on lower and upper confidence interval values and the confidence interval width of the pooled meta-analytic result. We also describe overlays for the impact of a future study on user-defined limits of clinical equivalence. We implement inverse-variance weighted methods by using both explicit formulas for contour lines and a simulation approach optimized in Mata.

Keywords: gr0054, extfunnel, funnel plots, meta-analysis, graphs

1   Introduction

The funnel plot is now a standard graphical tool for the investigation of publication biases and the extent of heterogeneity in meta-analyses. In its simplest form, a funnel plot is simply an x–y scatterplot of the individual study estimates versus some measure of estimate precision and study sample size. Asymmetry in such a plot can be an indication that publication bias is present.

An extensive set of Stata tools has been developed to facilitate the generation of funnel plots; for example, `metafunnel` and `metabias` produce funnel plot displays with various augmentations, such as a line for the pooled effect size. For more details about these commands, see Sterne and Harbord (2004); for further general guidance on the use of graphical tools in meta-analyses, see Anzures-Cabrera and Higgins (2010). An example of a funnel plot of trials of treatment for antidepressant versus placebo is presented in figure 1 (see Moreno et al. [2009] for further details). Here the plot is highly asymmetric, which is indicative of possible publication bias.

```
. use antid
. metafunnel ES seES, noline xtitle("Standardized Effect Size")
> ytitle("Standard Error")
note: default data input format (theta, se_theta) assumed
```

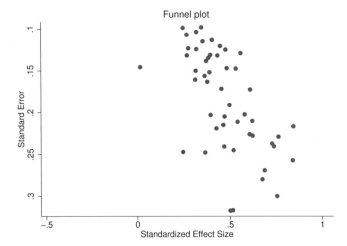

Figure 1. Funnel plot of new-generation antidepressant versus placebo for depression

A recent augmentation to the funnel plot by Peters et al. (2008) is the inclusion of statistical contours indicating regions in which studies would have to be for a given level of statistical significance. This feature is intended to aid the assessment of whether funnel plot asymmetry is likely due to publication biases or other causes. This has also been implemented in Stata as the `confunnel` command by Palmer et al. (2008).

In this article, we present a range of further graphical overlays for the funnel plot, illustrating the potential impact a new study may have when added to an existing meta-analysis. These overlays are similar to the contours for aiding the assessment of publication bias (the previous contours focus on the significance of individual studies, whereas the contours presented here focus on inferences relating to meta-analysis). However, they have a very different purpose that expands the uses of the funnel plot and is broadly applicable across meta-analyses of intervention trials, studies on the accuracy of diagnostic tests, etiological observational studies, etc. The overlays include statistical significance contours, which highlight regions where a new study would have to lie to change the statistical significance of the summary estimate of the present meta-analysis, and heterogeneity contours, which show how a new study would affect the heterogeneity of the meta-analysis. A full description of these overlays, their algebraic derivations (where possible), and applications are available in Langan et al. (2012).

We then extend the displays of Langan et al. (2012) to illustrate how a new study would affect the lower or upper confidence interval and the confidence interval width of the pooled meta-analytic estimate. We also show how these may be of particular interest with regard to diagnostic tests. Furthermore, we adapt the approach to develop a graphical display to assess the impact of a future study on user-defined limits of clinical equivalence described, for example, in Sutton et al. (2007).

The methodology is summarized in section 2, followed by a description of the command syntax in section 3. The use of `extfunnel` is shown in examples in section 4, with some additional features described in section 5. We conclude with a discussion in section 6.

2   Methodology

2.1   Statistical significance contours

When a study is added to an existing meta-analysis, it is interesting to investigate how this new study affects the statistical significance and direction of the pooled estimate and serves as an indication of the robustness of the original meta-analysis. By choosing plausible ranges for the new study's effect size and associated standard error, we can add each combination to the meta-analysis, which is subsequently re-meta-analyzed, and finally record the statistical significance of the new pooled estimate. By plotting the new study's effect size and standard error, color coded by the now updated meta-analysis' statistical significance, we obtain contour regions illustrating the robustness of the meta-analysis. We describe and illustrate the approach under fixed- and random-effects models using the inverse-variance weighted method.

Fixed effect

Under a fixed-effects model, explicit formulas for the contours can be derived based on the inverse-variance method. For further details, see Langan et al. (2012).

Random effect

Under a random-effects model, each new effect estimate and standard error spanning the entire range of the funnel plot must be combined with the original meta-analysis and analyzed separately to calculate the statistical significance level and direction of the updated pooled estimate. This is because explicit contour lines cannot be derived under a random-effects model. The computational issues of this are discussed in section 3.3.

2.2 Heterogeneity contours

A further addition to the standard funnel plot is the overlay of heterogeneity contours. These contours serve to illustrate how the between-study heterogeneity would be affected by the addition of a new study. This can be illustrated using either the between-study variance τ^2 or the I^2 statistic.

2.3 Alternative targets of inference

Because of the well-documented limitations of focusing exclusively on p-values, one can take an approach to inference that focuses more on effect-size estimation. For example, the impact of a new study on the lower or upper limits of the confidence interval of the pooled meta-analytic effect size may be of interest. For instance, in a diagnostic study where accuracy estimation rather than differences between tests is of primary concern, we may be interested in the effect a new study would have on the lower bound of the meta-analytic 95% confidence interval for sensitivity. Similarly, the confidence interval width may be of interest, and how a new study would affect the precision of the pooled estimate can be considered.

2.4 Limits of clinical equivalence

An approach to inference that aims to combine clinical information with statistical information is the use of user-specified limits of clinical equivalence. Within these limits, the two interventions are considered equivalent. Although defining the limits is of course subjective, external information can be obtained to inform the defined limits. Eight distinct scenarios are described in figure 2, detailing the characteristics of the confidence interval in relation to the limits of clinical equivalence and providing the subsequent interpretation. Further discussion on the limits of clinical equivalence can be found in Parmar, Ungerleider, and Simon (1996) and Sutton et al. (2007).

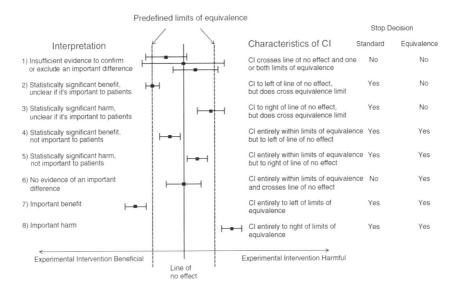

Figure 2. Limits of clinical equivalence (image adapted from Sutton et al. [2007])

3   The extfunnel command

3.1   Syntax

extfunnel *varname1* *varname2* [*if*] [*in*] [, fixedi randomi cpoints(*#*)
    null(*#*) isquared(*numlist*) tausquared(*numlist*) measure(lci|uci|ciwidth)
    loe(*numlist*) loeline newstudycontrol(*#*) newstudytreatment(*#*) or rr
    xrange(*numlist*) yrange(*numlist*) sumd sumdposition(*#*) prediction
    nonullline nopooledline noshading noscatter nometan
    label([namevar=*namevar*], [yearvar=*yearvar*]) eform
    scheme(grayscale|color) addplot(*string*) level(*#*) *twoway_options*]

The first variable, *varname1*, must correspond to the effect estimates assumed to be normally distributed (for example, log odds-ratios) with *varname2*, the associated standard errors of the effect estimates. extfunnel requires metan to be installed.

3.2   Options

fixedi specifies a fixed-effects model by using the inverse-variance method. This is the default.

randomi specifies a random-effects model by using the method of DerSimonian and Laird, with the estimate of heterogeneity being taken from the inverse-variance fixed-effects model.

cpoints($\#$) specifies the number of points to evaluate either the shaded statistical significance contours or the heterogeneity contours. The default numbers for fixed- and random-effects meta-analyses are 3500 and 100, respectively. When a random meta-analysis is invoked, the maximum number of contour points is 500. A larger number of cpoints() results in a smoother graph but takes longer to compute (see section 3.3 for more details).

null($\#$) is the value of the null hypothesis for the effect estimate. The default is null(0). This is the value that lci, uci, or ciwidth is compared with when measure() is specified. If lci or uci is specified, the value of null() is compared with the lower or upper confidence interval value, respectively, of the updated meta-analyses and is color coded depending on whether the updated estimate is less than or greater than the null(). If ciwidth is specified, then the width confidence interval of the updated meta-analyses is compared with the value defined by null().

isquared(*numlist*) specifies the values that define the I^2 contours. *numlist* must be of maximum length 5 and should have elements in the range 0–100.

tausquared(*numlist*) specifies the values that define the between-study variance (τ^2) contours. *numlist* must be a vector of maximum length 5 and should have elements in the range 0–infinity.

measure(lci|uci|ciwidth) defines the target of inference, which can be one of lci, uci, or ciwidth.

loe(*numlist*) defines the limits of clinical equivalence. The default legend assumes a beneficial and detrimental effect in specific directions. The legend can be relabeled using legend(order(1 "text1" 2 "text2" ...)). For further details, see Sutton et al. (2007).

loeline displays the limits of clinical equivalence.

newstudycontrol($\#$) defines the number of patients in the control arm of a new trial. newstudycontrol() and newstudytreatment() (explained next) defined together produce a statistical significance contour graph, whereby each possible permutation of results is calculated and analyzed within the appropriate meta-analysis model. Odds ratios and risk ratios are supported.

newstudytreatment($\#$) defines the number of patients in the treatment arm of a new trial.

or specifies to use log odds-ratios. This is the default; alternatively, rr can be specified for risk ratios. or is valid only when newstudycontrol() and newstudytreatment() are specified.

rr specifies to use log risk-ratios. rr is valid only when newstudycontrol() and newstudytreatment() are specified.

xrange(*numlist*) defines the range of effect estimates used to create the shaded contours.

yrange(*numlist*) defines the range of standard errors used to create the shaded contours.

sumd displays the summary diamond.

sumdposition(#) defines the vertical coordinate where the summary diamond should be placed. The default is sumdposition(0).

prediction displays a prediction interval for a new trial based on the current meta-analysis. The y-axis position is defined by sumdposition().

nonullline suppresses the display of the vertical line of no effect.

nopooledline suppresses the display of the vertical line at the pooled effect estimate.

noshading suppresses the display of shaded regions.

noscatter suppresses the display of the scatter of original study effects.

nometan suppresses the display of original meta-analysis results using metan.

label([namevar=*namevar*], [yearvar=*yearvar*]) labels the variable by its name, year, or both. This is a metan option. Either option or both options may be left blank. For the table display, the overall length of the label is restricted to 20 characters.

eform exponentiates the x-axis labels (valid only when the input variables are log transformed, for example, log odds-ratios or log risk-ratios).

scheme(grayscale|color) specifies the color scheme of the graph. The default is scheme(grayscale). scheme(color) can be useful to distinguish areas when loe() is specified.

addplot(*string*) allows additional twoway plots to be overlayed on the extfunnel plot.

level(#) specifies the confidence level, as a percentage, for confidence intervals. The default is level(95).

twoway_options; see [G-3] ***twoway_options***.

3.3 Computational details

As discussed in section 2.1, the inverse-variance weighted method, when applied in a random-effects setting, results in no closed-form expression to calculate the boundary contours between regions of statistical significance for the updated meta-analysis. Using either the default range of the x and y axes or the ranges entered in `xrange()` and `yrange()`, each x and y range is split at n points, where n is defined by `cpoints()`. So for example, `cpoints(500)` would result in $500 \times 500 = 250{,}000$ individual meta-analyses being conducted, with each having its statistical significance calculated. Sets of four adjacent points are then analyzed for the same statistical significance and are color coded into three categories (`Sig. effect < NULL`, `Nonsig. effect`, `Sig. effect > NULL`).

Conducting 250,000 meta-analyses will be computationally intensive. For example, if `metan` was used and specified method `randomi`, with the `nograph` and `notable` options, it would take approximately 24 hours on an Intel Core 2 Duo 3.0 GHz desktop computer to execute the `metan` command and analyze the pooled estimate and confidence interval. For this reason, the inverse-variance weighted method has been implemented in Mata, whereby all $n \times n$ random-effects meta-analyses are conducted simultaneously. When $n = 500$, this reduces computation time to approximately 90 seconds (60 seconds of which involves building the `twoway` graph). Furthermore, attempting to plot thousands of individual `twoway area` graphs on the same graph would cause an overflow of Stata's string limits; therefore the data are prepared row by row into areas of the same statistical significance.

This implementation is also used to create the fixed and random contour graphs when the target of inference is `lci`, `uci`, `ciwidth`, or the limits of clinical equivalence (although closed-form expressions may be obtainable in some contexts). Note that `extfunnel` can be quite memory intensive.

4 Example uses of extfunnel

In this section, we detail some features of `extfunnel`.

4.1 Statistical significance contours

Fixed effect

Figure 3 shows a forest plot from a fixed-effects meta-analysis of four trials investigating the change in Epworth score for an oral appliance versus continuous positive airways pressure for treating obstructive sleep apnea. For further details, see Lim et al. (2006).

```
. use epworth
. metan ES seES, fixedi notable texts(220) xlabel(-6, -4, -2, 2, 0, 2, 4, 6)
> xtitle("Effect Size")
```

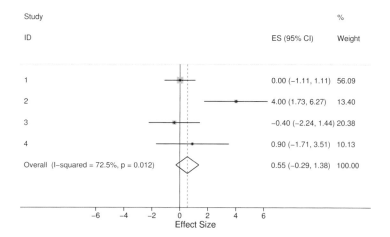

Figure 3. Forest plot of an oral appliance versus continuous positive airways pressure with outcome in the Epworth score

A nonstatistically significant (at the 5% level) summary effect estimate of 0.55 (95% CI; $[-0.29, 1.38]$) was found. By default, extfunnel displays the results from a metan call on the original meta-analysis. This can be suppressed by specifying the nometan option.

```
. use epworth

. extfunnel ES seES
Original meta-analysis results:
            Study    |    ES    [95% Conf. Interval]   % Weight
   ------------------+------------------------------------------------
   1                 |   0.000    -1.109    1.109        56.09
   2                 |   4.000     1.730    6.270        13.40
   3                 |  -0.400    -2.240    1.440        20.38
   4                 |   0.900    -1.711    3.511        10.13
   ------------------+------------------------------------------------
   I-V pooled ES     |   0.546    -0.285    1.376       100.00
   ------------------+------------------------------------------------

     Heterogeneity chi-squared =   10.91 (d.f. = 3) p = 0.012
     I-squared (variation in ES attributable to heterogeneity) =   72.5%

     Test of ES=0 : z=   1.29 p = 0.198
Building graph:
```

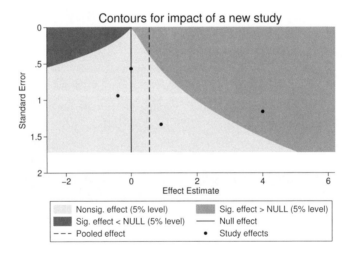

Figure 4. `extfunnel` plot of an oral appliance versus continuous positive airways pressure with outcome in the Epworth score

Using all the default options in this way produces figure 4, which presents contours for areas in which a new study would have to lie for the pooled result to be significant at the 5% level. Regions for significant effects in both directions are present on this plot. It is clear that the existing meta-analysis may not be robust to the impact of a new study. Given that one of the four studies lies in the region where a new study would have to be to change the summary estimate in favor of the continuous positive airways pressure, it is certainly plausible that a new study could alter the conclusion of the existing meta-analysis.

Random effect

Figure 5 illustrates a meta-analysis of Sanchi versus control for the treatment of ischemic stroke. For further details, see Chen et al. (2008). The log risk-ratios and associated standard errors are combined with a random-effects meta-analysis model. To illustrate the pixel-by-pixel approach, we compare `extfunnel` calls with the default `cpoints(100)` and the maximum `cpoints(500)`. Figure 5 illustrates the improved smoothness of the contours when the maximum `cpoints()` is used. The user-written `grc1leg` is used to combine the two `extfunnel` graphs.

```
. use sanchi
. quietly extfunnel ES seES, randomi xlabel(0.1 0.25 0.5 1 2)
> name(graph1, replace) eform title("100 intervals")
. quietly extfunnel ES seES, randomi cpoints(500) xlabel(0.1 0.25 0.5 1 2)
> eform name(graph2, replace) title("500 intervals")
. grc1leg graph1 graph2
```

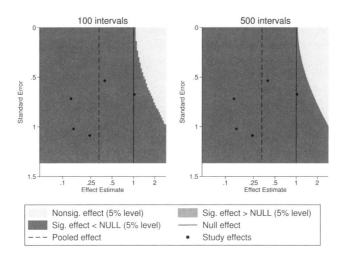

Figure 5. Sanchi versus control for acute ischemic stroke with outcome as proportion of patients with no neurological improvement

Under a random-effects model, the pooled estimate obtained from the original meta-analysis was 0.327 (95% CI; [0.153, 0.699]), with a between-study variance estimate of heterogeneity, $\tau^2 = 0.187$. The original meta-analysis shows a statistically significant reduction in the proportion of patients in the treatment group who had no improvement. From figure 5, it is clear all studies lie in the region of statistical significance with a beneficial treatment effect, which itself dominates the graph area; therefore, the meta-analysis could be considered relatively robust to the addition of a single new trial.

4.2   Heterogeneity contours

We detail example plots where we investigate the effect a new study has on estimates
of heterogeneity. Heterogeneity statistics include the between-study variance, τ^2, and
the I^2 statistic. We again use the dataset from Chen et al. (2008) investigating Sanchi
versus control for acute ischemic stroke.

τ^2 contours

Use of the `tausquared()` option can create a contour plot illustrating the effect a new
study would have on τ^2. In figure 6, the current estimate of $\tau^2 = 0.187$ is shown,
illustrating the current level of heterogeneity in the original meta-analysis. The other
contour lines represent combinations of effect estimates and standard errors that would
be required in the new study to alter the estimate of τ^2 to each particular value. Up
to five distinct contour values of τ^2 can be plotted. In figure 6, we display the summary diamond and a 95% prediction interval (the interval that the underlying effect
is estimated to lie within 95% of the time based on the existing meta-analysis) using
the `sumd` and `prediction` options, but we suppress the display of the shaded statistical
significance contours by invoking the `noshading` option.

```
. use sanchi
. program drop _all
. extfunnel ES seES, tausquared(0.04 0.1 0.187 0.4 0.5) yrange(0 1.5) noshading
> nometan prediction sumd sumdposition(0.2)
Building graph:
```

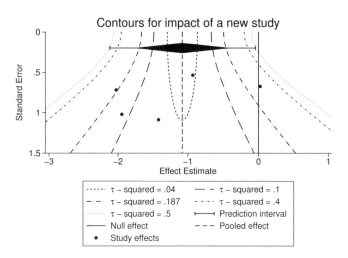

Figure 6. τ^2 heterogeneity contours; Sanchi versus control for acute ischemic stroke

If a new study lies within the region defined by the contours at $\tau^2 = 0.187$, then the estimate of τ^2 would be reduced. Similarly, if a new study lies outside the region, then τ^2 would be increased. If the new study lies on a contour defined by $\tau^2 = 0.5$, then τ^2 would be increased to 0.5 on inclusion of the new study. Mental interpolation between contours can provide a guide to the effect a new study would have on the estimate of τ^2. The prediction interval indicates the likely range of effect sizes (but ignores sampling error) for a new study and thus indicates which region of the plot is most relevant.

Similarly, the effect on the estimates of the I^2 statistic can be investigated using `isquared()`. A comparison of the τ^2 and I^2 contours, including additional examples, can be found in Langan et al. (2012).

4.3 Confidence intervals

Let us now investigate the robustness of the sensitivity of a diagnostic test. The data used in this example come from Geersing et al. (2009), who conducted a meta-analysis examining point-of-care D-dimer tests for detecting venous thromboembolism—more specifically, seven studies evaluating the Clearview Simplify D-dimer test. Analyzing sensitivity (an assessment of specificity could also be conducted using exactly the same approach) under a fixed-effects model produced a pooled sensitivity of 0.853 (95% CI; [0.817, 0.883]). We can investigate the effect of a new study on the lower confidence interval by specifying `measure(lci)` and `null(1.495)`, where invlogit(1.495) = 0.817. Analyses are conducted and plotted on the logit scale.

```
. use simplify

. extfunnel logitsens se_logitsens, fixedi measure(lci) null(1.495) nometan
> xlabel(1 2 3) yrange(0 1) cpoints(500) sumd sumdposition(0.9)
> xtitle("logit(Sensitivity)")

Building graph:
```

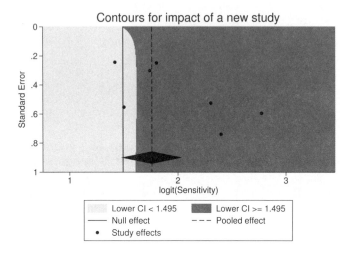

Figure 7. Target of inference—lower confidence interval

From figure 7, we can see where a new study's effect size and standard error would have to lie either to increase or to decrease the lower confidence interval. Similar plots can be produced when the upper confidence interval is of interest or when the confidence interval width is investigated.

4.4   Limits of clinical equivalence

We now illustrate how, through the use of the limits of clinical equivalence, `extfunnel` can graphically display a range of scenarios (described in figure 2) for the updated meta-analysis. We illustrate the method using the Epworth score dataset once more, defining the limits of clinical equivalence to be $(-0.25, 0.25)$.

```
. use epworth

. extfunnel ES seES, fixedi loe(-0.25 0.25) nometan sumd sumdposition(1.6)
> yrange(0 2) nopooledline loeline cpoints(400)
Building graph:
```

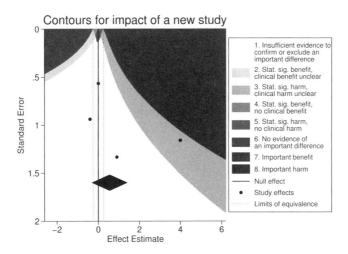

Figure 8. Limits of equivalence

In figure 8, the funnel plot is divided into the eight distinct regions defined in the legend (following the descriptions in figure 2). Here all existing studies lie in either scenario 1 (CI crosses the line of no effect and both limits of equivalence) or scenario 3 (CI lies to the right of the line of no effect but does cross the equivalence line). Given the large proportion of the plot around the pooled estimate that represents scenario 1, changing inferences with one further new study is unlikely unless it is very large (precise) or has an extreme effect size. Note that figure 4 can be considered a special case of figure 8 because figure 8 subdivides further the regions defined in figure 4. We recover figure 4 by grouping areas of the graph where the effect is greater than the null and is statistically significant (areas 3, 5, and 8); by grouping areas where the effect is nonstatistically significant (areas 1 and 6); and by grouping areas where the effect is less than the null and is statistically significant (areas 2, 4, and 7).

5   Additional feature

If the measure of interest is an odds ratio or rate ratio, then the effect of a new study with an explicit sample size can be investigated directly through invoking the `newstudytreatment()` and `newstudycontrol()` options. For example, if 100 patients are in both treatment and control arms, we can directly calculate all possible combinations of 2×2 cell counts, combining each unique new study estimate with the original meta-analysis to produce a graph such as figure 9. This process can be computationally intensive, because it has not been optimized in Mata but relies on `metan`. We use a meta-analysis from a Cochrane review investigating the use of antibiotics versus control to treat the common cold, where the outcome is the alleviation of symptoms within seven days. More details can be found in Arroll and Kenealy (2002). A fixed-effects meta-analysis of the existing studies had a nonstatistically significant pooled odds ratio of 0.796 (95% CI; [0.587, 1.080]).

```
. use colddata

. extfunnel logor selogor, fixedi newstudycontrol(100) newstudytreatment(100)
> nometan eform xlabel(0.01 0.2 1 5 100)

Building graph:
```

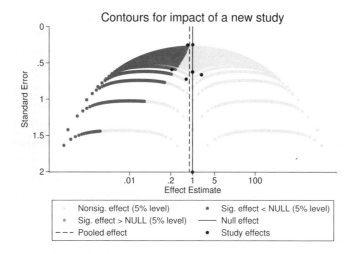

Figure 9. All possible results from the addition of a new study with 100 patients in each arm

Figure 9 shows the possible results, under a fixed-effects model, of combining a trial of size 200 with the original meta-analysis. Given that two of the original studies lie in a region of statistical significance showing a beneficial treatment effect, the original meta-analysis could be considered to be lacking robustness to the influence of a new trial.

6 Discussion

We have presented and described the `extfunnel` command, which provides a variety of graphical means of establishing the robustness of a meta-analysis to the inclusion of a new study. The simulation approach implemented in Mata has optimized the re-meta-analyses, providing an efficient and powerful tool encompassing a variety of scenarios.

We envisage that meta-analysts, trialists, and editors of portfolios of systematic reviews will find this display useful when reporting their meta-analyses, designing new studies, and prioritizing updates of existing meta-analyses, respectively (Langan et al. 2012).

The limitations of the approach include no accounting for change in the statistical model. We assume that the original meta-analysis is analyzed with the same fixed or random framework as the updated meta-analysis. Furthermore, only inverse-variance weighted methods are available because, for example, the Mantel–Haenszel approach is prohibitive: it takes into account all cell frequencies in a study's 2×2 contingency table, which cannot be displayed on a two-dimensional graph. Finally, the plots only consider the inclusion of one further study; an approach that could display the impact of multiple studies would be welcomed.

7 Acknowledgments

We thank Julian Higgins for helpful feedback on the program and the article, and Rob Herbert and Manuela Ferreira for invaluable suggestions for improvement and testing of the command. Michael Crowther is funded by a National Institute for Health Research methodology fellowship (RP-PG-0407-10314).

8 References

Anzures-Cabrera, J., and J. P. T. Higgins. 2010. Graphical displays for meta-analysis: An overview with suggestions for practice. *Research Synthesis Methods* 1: 66–80.

Arroll, B., and T. Kenealy. 2002. Antibiotics for the common cold. *Cochrane Database of Systematic Reviews* 3: CD000247.

Chen, X., M. Zhou, Q. Li, J. Yang, Y. Zhang, D. Zhang, S. Kong, D. Zhou, and L. He. 2008. Sanchi for acute ischaemic stroke. *Cochrane Database of Systematic Reviews* 4: CD006305.

Geersing, G. J., K. J. M. Janssen, R. Oudega, L. Bax, A. W. Hoes, J. B. Reitsma, and K. G. M. Moons. 2009. Excluding venous thromboembolism using point of care D-dimer tests in outpatients: A diagnostic meta-analysis. *British Medical Journal* 339: b2990.

Langan, D., J. P. T. Higgins, W. Gregory, and A. J. Sutton. 2012. Graphical augmentations to the funnel plot assess the impact of additional evidence on a meta-analysis. *Journal of Clinical Epidemiology* 65: 511–519.

Lim, J., T. J. Lasserson, J. Fleetham, and J. Wright. 2006. Oral appliances for obstructive sleep apnoea. *Cochrane Database of Systematic Reviews* 1: CD004435.

Moreno, S. G., A. J. Sutton, E. H. Turner, K. R. Abrams, N. J. Cooper, T. M. Palmer, and A. E. Ades. 2009. Novel methods to deal with publication biases: Secondary analysis of antidepressant trials in the FDA trial registry database and related journal publications. *British Medical Journal* 339: b2981.

Palmer, T. M., J. L. Peters, A. J. Sutton, and S. G. Moreno. 2008. Contour-enhanced funnel plots for meta-analysis. *Stata Journal* 8: 242–254. (Reprinted in this collection on pp. 139–152.)

Parmar, M. K. B., R. S. Ungerleider, and R. Simon. 1996. Assessing whether to perform a confirmatory randomized clinical trial. *Journal of the National Cancer Institute* 88: 1645–1651.

Peters, J. L., A. J. Sutton, D. R. Jones, K. R. Abrams, and L. Rushton. 2008. Contour-enhanced meta-analysis funnel plots help distinguish publication bias from other causes of asymmetry. *Journal of Clinical Epidemiology* 61: 991–996.

Sterne, J. A. C., and R. M. Harbord. 2004. Funnel plots in meta-analysis. *Stata Journal* 4: 127–141. (Reprinted in this collection on pp. 124–138.)

Sutton, A. J., N. J. Cooper, D. R. Jones, P. C. Lambert, J. R. Thompson, and K. R. Abrams. 2007. Evidence-based sample size calculations based upon updated meta-analysis. *Statistics in Medicine* 26: 2479–2500.

Part 4

Multivariate meta-analysis: metandi, mvmeta

There is great interest in meta-analyses of studies that estimate the accuracy of diagnostic tests. Such meta-analyses are inherently bivariate because of the trade-off between sensitivity and specificity. Harbord et al. (2007) noted that two different proposed methods for meta-analyses of test-accuracy studies (Rutter and Gatsonis 2001; Reitsma et al. 2005) were identical in many circumstances. These methods are implemented in the `metandi` command.

Multivariate random-effects meta-analysis—implemented in the `mvmeta` command (White 2009)—can be used in many circumstances, including modeling each outcome separately in a clinical trial, exploring treatment effects on both disease outcome and costs, and examining the shape of an association of a quantitative exposure with a disease outcome.

White (2011) presented updates to `mvmeta`. The main addition is the ability to fit multivariate meta-regression models. `mvmeta` also allows more covariance structures, including Riley's overall correlation model (Riley, Thompson, and Abrams 2008). Command speed was increased through use of Mata, and additional postestimation features are available, including I^2 statistics, and confidence intervals for between-study standard deviations and correlations. `mvmeta` can also report 'the best' intervention. In part 6, White (Forthcoming) describes how to use `mvmeta` to fit network meta-analysis models through his `network` suite of commands.

1   References

Harbord, R. M., J. J. Deeks, M. Egger, P. Whiting, and J. A. C. Sterne. 2007. A unification of models for meta-analysis of diagnostic accuracy studies. *Biostatistics* 8: 239–251.

Reitsma, J. B., A. S. Glas, A. W. S. Rutjes, R. J. P. M. Scholten, P. M. Bossuyt, and A. H. Zwinderman. 2005. Bivariate analysis of sensitivity and specificity produces informative summary measures in diagnostic reviews. *Journal of Clinical Epidemiology* 58: 982–990.

Riley, R. D., J. R. Thompson, and K. R. Abrams. 2008. An alternative model for bivariate random-effects meta-analysis when the within-study correlations are unknown. *Biostatistics* 9: 172–186.

Rutter, C. M., and C. A. Gatsonis. 2001. A hierarchical regression approach to meta-analysis of diagnostic test accuracy evaluations. *Statistics in Medicine* 20: 2865–2884.

White, I. R. 2009. Multivariate random-effects meta-analysis. *Stata Journal* 9: 40–56. (Reprinted in this collection on pp. 232–248.)

———. 2011. Multivariate random-effects meta-regression: Updates to mvmeta. *Stata Journal* 11: 255–270. (Reprinted in this collection on pp. 249–264.)

———. Forthcoming. Network meta-analysis. *Stata Journal* (Reprinted in this collection on pp. 321–354.)

The Stata Journal (2009)
9, Number 2, st0163, pp. 211–229

metandi: Meta-analysis of diagnostic accuracy using hierarchical logistic regression

Roger M. Harbord
Department of Social Medicine
University of Bristol
Bristol, UK
roger.harbord@bristol.ac.uk

Penny Whiting
Department of Social Medicine
University of Bristol
Bristol, UK

Abstract. Meta-analysis of diagnostic test accuracy presents many challenges. Even in the simplest case, when the data are summarized by a 2×2 table from each study, a statistically rigorous analysis requires hierarchical (multilevel) models that respect the binomial data structure, such as hierarchical logistic regression. We present a Stata package, metandi, to facilitate the fitting of such models in Stata. The commands display the results in two alternative parameterizations and produce a customizable plot. metandi requires either Stata 10 or above (which has the new command xtmelogit), or Stata 8.2 or above with gllamm installed.

Keywords: st0163, metandi, metandiplot, diagnosis, meta-analysis, sensitivity and specificity, hierarchical models, generalized mixed models, gllamm, xtmelogit, receiver operating characteristic (ROC), summary ROC, hierarchical summary ROC

1 Introduction

There are several existing user-written commands in Stata that are intended primarily for meta-analysis (see Sterne et al. [2007] for an overview). There is increasing interest in systematic reviews and meta-analyses of data from diagnostic accuracy studies (Deeks 2001b; Devillé et al. 2002; Tatsioni et al. 2005; Gluud and Gluud 2005; Mallett et al. 2006; Gatsonis and Paliwal 2006), which presents many additional challenges compared to more traditional meta-analysis applications, such as controlled trials. In particular, diagnostic accuracy cannot be adequately summarized by one measure; two measures are typically used, most often sensitivity and specificity or, alternatively, positive and negative likelihood ratios, and the two are correlated (Deeks 2001a). Meta-analysis of diagnostic accuracy therefore requires different and more complex methods than traditional meta-analysis applications, even in the simplest situation where the data from each primary study are summarized as a 2×2 table of test results against true disease status, both of which have been dichotomized. In addition, substantial between-study heterogeneity is commonplace, and the models must account for this (Lijmer, Bossuyt, and Heisterkamp 2002).

Several methods of meta-analyzing diagnostic accuracy data have been proposed, of which two are statistically rigorous: the hierarchical summary receiver operating characteristic (HSROC) model (Rutter and Gatsonis 2001) and the bivariate model (Reitsma et al. 2005). In the absence of covariates, these turn out to be different parameterizations of the same model (Harbord et al. 2007; Arends et al. 2008).

The bivariate model can be fit in Stata by using the user-written `gllamm` command, as pointed out by Coveney (2004). In Stata 10, the same model can be fit considerably faster by using the new `xtmelogit` command. In either case, however, some data preparation is required, the syntax is complex (particularly for `gllamm`), and the output is not easy to interpret.

In this article, we present a new Stata command, `metandi`, to facilitate the fitting of these hierarchical logistic regression models for meta-analysis of diagnostic test accuracy. The `metandi` command fits the model and displays the estimates in both the HSROC and bivariate parameterizations. `metandi` also displays some familiar summary measures (sensitivity and specificity, positive and negative likelihood ratios, and the diagnostic odds ratio). However, these simple summary measures fail to describe the expected trade-off between sensitivity and specificity, which is best illustrated graphically. We have therefore included a command, `metandiplot`, to simplify the plotting of graphical summaries of the fitted model, namely, the summary receiver operating characteristic (SROC) curve and the prediction region, and also to plot the summary point and its confidence region.

The name `metandi` was chosen to indicate that, like `metan` (Bradburn, Deeks, and Altman 1998), `metandi` takes the cell counts of 2×2 tables as input but is designed for meta-analysis of diagnostic accuracy.

`metandi` is not intended to provide a comprehensive package for diagnostic meta-analysis by itself; other plots are also useful, such as forest plots showing within-study estimates and confidence intervals for sensitivity and specificity separately (Deeks 2001b).

Section 4 of this article introduces an example dataset, which we will use to illustrate the commands. Section 3 then gives some background on methods and models that have been proposed for meta-analysis of diagnostic accuracy. Sections 4 and 5 illustrate the output of `metandi` and `metandiplot` on the example dataset. Section 6, which assumes somewhat greater knowledge of both statistics and Stata, gives examples of the use of `predict` after `metandi` for model checking and identification of influential studies. Finally, sections 7 and 8, which are intended mainly as reference material, detail the formal syntax of the commands, and the methods and formulas used.

2   Example: Lymphangiography for diagnosis of lymph node metastasis

We shall illustrate the use of the `metandi` package on data from 17 studies of lymphangiography for the diagnosis of lymph node metastasis in women with cervical cancer. Lymphangiography is one of three imaging techniques in the meta-analysis of Scheidler et al. (1997), and these data have been frequently used as an example for methodological papers on meta-analysis of diagnostic accuracy (Rutter and Gatsonis 2001; Macaskill 2004; Reitsma et al. 2005; Harbord et al. 2007). These data are provided in the auxiliary file `scheidler_LAG.dta`. The total number of patients in each study ranges from 21 to 300. There is one observation in the dataset for each study.

The data needed for meta-analysis consist of the number of true positives (tp), false positives (fp), false negatives (fn), and true negatives (tn).

Figure 1 shows a SROC plot of these data, generated by the official Stata commands given below. An SROC plot is similar to a conventional ROC plot (see, e.g., [R] **roc**) in that it plots sensitivity (true-positive rate) against specificity (true-negative rate), but here each symbol represents a different study rather than a different threshold within the same study. It therefore makes no sense to connect the points with a line, but it can be useful to indicate the size of each study by the symbol size. (It might be preferable to use an ellipse or rectangle to separately indicate the number of people with [tp + fn] and without [tn + fp] the disease of interest, but this is hard to achieve within the current Stata graphics system.) By convention, the specificity is plotted on a reversed scale (or equivalently, the false-positive rate is plotted on a conventional scale).

```
. use scheidler_LAG
(Lymphangiography for diagnosing lymph node metastases)
. generate sens = tp/(tp+fn)
. generate spec = tn/(tn+fp)
. label variable sens "Sensitivity"
. label variable spec "Specificity"
. local opts "xscale(reverse) xla(0(.2)1) yla(0(.2)1, nogrid) aspect(1) nodraw"
. scatter sens spec [fw=tp+fp+fn+tn], m(Oh) `opts´ name(sroccirc)
. scatter sens spec, mlabel(studyid) m(i) mlabpos(0) `opts´ name(sroclab)
. graph combine sroccirc sroclab, xsize(4.5) scale(*1.5)
```

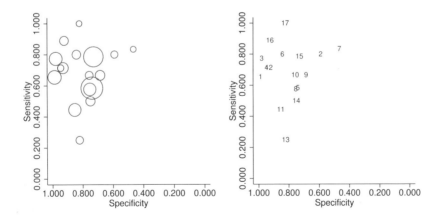

Figure 1. SROC plot of the lymphangiography data. Left panel: Studies indicated by circles sized according to the total number of individuals in each study. Right panel: Studies indicated by study ID numbers.

3   Models for meta-analysis of diagnostic accuracy

Several statistical methods for meta-analysis of data from diagnostic test accuracy studies have been proposed that account for the correlation between sensitivity and specificity (Moses, Shapiro, and Littenberg 1993; Rutter and Gatsonis 2001; Reitsma et al. 2005).

Moses, Shapiro, and Littenberg (1993) proposed a method of generating an SROC curve by using simple linear regression. This method has frequently been used, but the assumptions of simple linear regression are not met, and the method is therefore approximate. There is also uncertainty as to the most appropriate weighting of the regression (Walter 2002; Rutter and Gatsonis 2001).

Two more-complex but statistically rigorous approaches have been proposed that overcome the limitations of the linear regression method: the HSROC model (Rutter and Gatsonia 2001) and the bivariate model (Reitsma et al. 2005). Both approaches are based on hierarchical models, i.e., both approaches involve statistical distributions at two levels. At the lower level, they model the cell counts in the 2×2 tables by using binomial distributions and logistic (log-odds) transformations of proportions. Although their motivation is distinct and they allow covariates to be added to the models in different ways, it has been shown that the two models are equivalent when no covariates are fit, as well as in certain models including covariates (Harbord et al. 2007; Arends et al. 2008).

3.1   HSROC model

The HSROC model (Rutter and Gatsonis 2001) assumes that there is an underlying ROC curve in each study with parameters α and β that characterize the accuracy and asymmetry of the curve. The 2×2 table for each study then arises from dichotomizing at a positivity threshold, θ. The parameters α and θ are assumed to vary between studies; both are assumed to have normal distributions as in conventional random-effects meta-analysis. The accuracy parameter has a mean of Λ (capital lambda) and a variance of σ_α^2, while the positivity parameter θ has a mean of Θ (capital theta) and a variance of σ_θ^2. Because estimation of the shape parameter, β, requires information from more than one study, it is assumed constant across studies. When no covariates are included in an HSROC model, there are therefore five parameters: Λ, Θ, β, σ_α^2, and σ_θ^2.

3.2   Bivariate model

The bivariate model (Reitsma et al. 2005) models the sensitivity and specificity more directly. It assumes that their logit (log-odds) transforms have a bivariate normal distribution between studies. The logit-transformed sensitivities are assumed to have a mean of μ_A and a variance of σ_A^2, while the logit-transformed specificities have a mean of μ_B and a variance of σ_B^2. The trade-off between sensitivity and specificity is allowed for by also including a correlation, ρ_{AB}, that is expected to be negative. The bivariate model, like the HSROC model, therefore has five parameters when no covariates are included: μ_A, μ_B, σ_A^2, σ_B^2, and ρ_{AB}.

4   metandi output

The output from running `metandi` on the lymphangiography data is shown below (the `nolog` option suppresses the iteration log and is used here merely to save space):

```
. use scheidler_LAG, clear
(Lymphangiography for diagnosing lymph node metastases)

. metandi tp fp fn tn, nolog

True  positives: tp                      False positives: fp
False negatives: fn                      True  negatives: tn

Meta-analysis of diagnostic accuracy

Log likelihood    = -91.391372                  Number of studies =        17
```

	Coef.	Std. Err.	z	P>|z|	[95% Conf. Interval]	
Bivariate						
E(logitSe)	.7266321	.1544626			.4238909	1.029373
E(logitSp)	1.638955	.2505372			1.147911	2.129999
Var(logitSe)	.1249622	.1306738			.0160943	.9702552
Var(logitSp)	.8232703	.4055446			.3135009	2.161952
Corr(logits)	.2387873	.4557706			-.6067877	.8308258
HSROC						
Lambda	2.187142	.3086554			1.582189	2.792096
Theta	.0705698	.3271092			-.5705525	.7116921
beta	.9426366	.5764601	1.64	0.102	-.1872044	2.072478
s2alpha	.7946708	.5114529			.2250873	2.805586
s2theta	.1220778	.1082908			.0214569	.6945553
Summary pt.						
Se	.6740658	.0339356			.6044139	.7367944
Sp	.8373927	.0341147			.7591292	.8937849
DOR	10.65029	3.296352			5.806411	19.53509
LR+	4.145361	.9181013			2.685598	6.398582
LR-	.389225	.0452324			.3099427	.4887875
1/LR-	2.569208	.2985712			2.045879	3.226402

```
Covariance between estimates of E(logitSe) & E(logitSp)    .0045838
```

The bivariate and HSROC parameter estimates are displayed along with their standard errors and approximate 95% confidence intervals in the standard Stata format. The bivariate location parameters, μ_A and μ_B, are denoted by E(logitSe) and E(logitSp); the variance parameters, σ_A^2 and σ_B^2, are shown as Var(logitSe) and Var(logitSp); and the correlation, σ_{AB}, is shown as Corr(logits). The HSROC parameters are denoted by using the notation of Rutter and Gatsonis (2001) given in section 3.1, spelling out Greek letters with capital initials for the capital Greek letters Λ and Θ, and showing σ_α^2 and σ_θ^2 as s2alpha and s2theta.

z statistics and p-values are not given for most of the parameters because parameter values of zero do not correspond to null hypotheses of interest. The exception is the HSROC shape (asymmetry) parameter, β (beta), where $\beta = 0$ corresponds to a symmetric ROC curve in which the diagnostic odds ratio does not vary along the curve.

The output also gives summary values and confidence intervals for the sensitivity (Se) and specificity (Sp) (back-transformed from E(logitSe) and E(logitSp)), as well as values for the diagnostic odds ratio (DOR) and the positive and negative likelihood ratios (LR+ and LR−) at the summary point. The summary likelihood ratios will not, in general, be the same as would be obtained by first calculating the likelihood ratios for each study and meta-analyzing these. Such an approach has been deprecated in favor of the approach implemented here (Zwinderman and Bossuyt 2008). A summary value for the inverse of the negative likelihood ratio (1/LR−) is also given, because larger values of the inverse of the negative likelihood ratio indicate a more accurate test, and comparing this with the positive likelihood ratio can indicate whether a positive or negative test result has greater impact on the odds of disease.

Finally, the output shows the covariance between $\widehat{\mu}_A$ and $\widehat{\mu}_B$. This is needed to draw confidence and prediction regions, and is included to make it easier to do so in external software, such as the Cochrane Collaboration's Review Manager 5 (Nordic Cochrane Centre 2007).

❏ **Technical note**

On rare occasions, during model fitting, gllamm may report an error, such as "convergence not achieved: try with more quadrature points" or (less transparently) "log likelihood cannot be computed". Increasing the number of integration points beyond metandi's default of 5 by using the nip() option (e.g., nip(7)) may resolve this.

❏

5   metandiplot

The metandiplot command produces a graph of the model fit by metandi, which must be the last estimation-class command executed. For convenience, the metandi command has a plot option, which produces the same graph. If metandiplot is not followed by a varlist, then the study-specific estimates (shown by the circles in figure 2) are not included in the graph. The metandiplot command has options to alter the default appearance of the graph or to turn off any of the plot elements. These options are not available when using the plot option to metandi. metandiplot can be run many times with different options without refitting the model with metandi.

```
. metandiplot tp fp fn tn
```

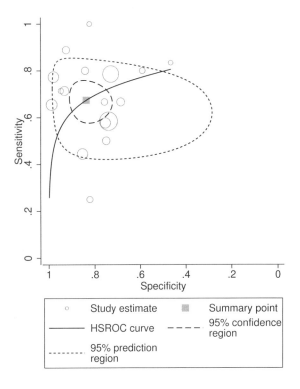

Figure 2. Plot of fitted model from `metandiplot`

The resulting graph (figure 2) shows the following summaries, together with circles showing the individual study estimates:

- A summary curve from the HSROC model

- A summary operating point, i.e., summary values for sensitivity and specificity

- A 95% confidence region for the summary operating point

- A 95% prediction region (confidence region for a forecast of the true sensitivity and specificity in a future study)

The default is to include all the summaries listed above, which can result in a rather cluttered graph, so options are included to remove any of the elements; for example, `predopts(off)` turns off the prediction region. See section 7.2 for more information about `metandiplot` options.

By default, the summary HSROC curve is displayed only for sensitivities and specificities at least as large as the smallest study-specific estimates if a varlist is included.

The shape of the prediction region is dependent on the assumption of a bivariate normal distribution for the random effects and should therefore not be overinterpreted; it is intended to give a visual representation of the extent of between-study heterogeneity, which is often considerable.

6   predict after metandi

Many of Stata's standard postestimation tools will not work after `metandi` or will not work as expected, because `metandi` temporarily reshapes the data before fitting the model.

The notable exception is `predict`, which can be used to obtain posterior predictions (empirical Bayes estimates) of the sensitivity and specificity in each study (`mu`), as well as various statistics that can be useful for detecting outliers (e.g., `ustd`) and influential observations (`cooksd`).

The help file provides basic commands for examining diagnostics. We take the opportunity here to provide slightly more customized displays.

Empirical Bayes estimates give the best estimate of the true sensitivity and specificity in each study, and these estimates will be "shrunk" toward the summary point compared with the study-specific estimates shown in figure 1.

```
. predict eb
(option mu assumed; posterior predicted Se & Sp)

. metandiplot, addplot(scatter eb1 eb0, msymbol(o))
> legend(label(5 "Empirical Bayes"))
```

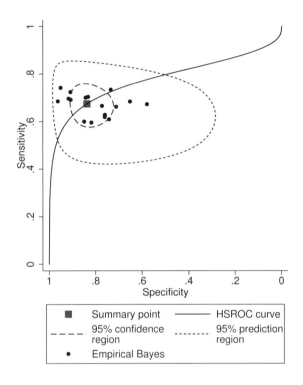

Figure 3. Empirical Bayes estimates

Comparing figure 3 with figure 2 shows that the shrinkage is generally greater for sensitivity than for specificity in this example, reflecting both the smaller variance of sensitivity (on the logit scale) and the fact that most studies have fewer participants with disease than without disease, leading to more precise estimates of specificity than of sensitivity.

Cook's distance is a measure of the influence of a study on the model parameters and can be used to check for particularly influential studies. Cook's distance is calculated using `gllapred` and so is available in Stata 10 only if the `gllamm` option was used with `metandi`. `gllapred` calculates Cook's distance to measure influence on all model parameters including the variance parameters (Skrondal and Rabe-Hesketh 2004, sec. 8.6.6). To check for outliers, standardized predicted random effects can be interpreted as standardized study-level residuals.

```
. metandi tp fp fn tn, gllamm nolog
  (output omitted)
. predict cooksd, cooksd
(Cook´s distance may take a few seconds...)
. predict ustd_Se ustd_Sp, ustd
. local opts "mlabel(studyid) mlabpos(0) m(i) nodraw"
. scatter cooksd studyid,`opts´ name(cooksd)
. scatter ustd_Se ustd_Sp, xscale(rev) xla(, grid) xline(0) yline(0) `opts´
> name(ustd)
. graph combine cooksd ustd, xsize(5) scale(*1.5)
```

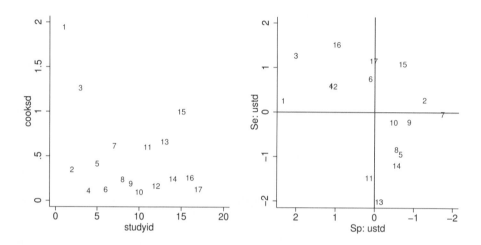

Figure 4. Left panel: Cook's distance. Right panel: Standardized residuals (standardized predicted random effects).

Figure 4 shows both Cook's distance and the standardized residuals. (The residual corresponding to specificity has been plotted on a reversed axis to correspond with the convention for ROC plots used in figure 1.) These two graphs are best read in combination. Cook's distance shows which studies are influential, while the standardized residuals give some insight into why. According to Skrondal and Rabe-Hesketh (2004), a typical cutpoint for declaring a value of Cook's D to be "large" is four times the number of parameters divided by the number of clusters (here studies). (Definitions of Cook's D differ, hence so does the cutpoint—the definition used by Stata in [R] **regress postestimation** divides by the number of parameters.) Because there are five parameters in this model, this suggests a cutpoint of 20 divided by the number of studies for interpreting Cook's D after `metandi`, giving $20/17 \approx 1.2$ for the lymphangiography meta-analysis. Here, study 1 is particularly influential, followed by study 3. Studies 1 and 3 have high standardized residuals for specificity, leading to influence on both the mean and variance of logit-transformed specificity. Study 13 has a large (negative) standardized residual for sensitivity but does not appear to be so influential as judged by its Cook's distance. Further investigation of the effect of individual studies on the model could be undertaken by refitting the model and leaving out each study in turn.

7 Syntax and options for commands

7.1 The metandi command

Syntax

metandi *tp fp fn tn* [*if*] [*in*] [, <u>p</u>lot gllamm force ip(g|m) <u>nip</u>(#)
<u>nobivariate</u> <u>noh</u>sroc <u>nos</u>ummarypt <u>detail</u> <u>level</u>(#) <u>trace</u> <u>nolog</u>]

   by is allowed with metandi; see [D] **by**.

Options

plot requests a plot of the results on an SROC plot. This is a convenience option equivalent to executing the metandiplot command after metandi with the same list of variables, *tp*, *fp*, *fn*, and *tn* (and the same *if* and *in* qualifiers, if specified). Greater control of the plot is available through the options of the metandiplot command when issued as a separate command after metandi.

gllamm specifies that the model be fit using gllamm. This is the default in Stata 8 and 9, so the option is of use only in Stata 10, in which the model is fit using xtmelogit by default.

force forces metandi to attempt to fit data where one or more studies have $tp + fn = 0$ (or $tn + fp = 0$), i.e., where there are no individuals that are positive (negative) for the reference standard. Without this option, metandi exits with an error when such data exist. Problems may be encountered in fitting such data, particularly when the model is fit using xtmelogit. Sensitivity (specificity) cannot be estimated within such studies, so they are not included in the plot produced by metandiplot.

ip(g|m) specifies the quadrature (numerical integration) method used to integrate out the random effects: ip(g), the default, gives Cartesian product quadrature, while ip(m) gives spherical quadrature, which is available in gllamm but not in xtmelogit. Spherical quadrature can be more efficient, though its properties are less well known and it can sometimes cause the adaptive quadrature step to take longer to converge. See Rabe-Hesketh, Skrondal, and Pickles (2005).

nip(#) specifies the number of integration points used for quadrature. Higher values should result in greater accuracy but typically at the expense of longer execution times. Specifying too small a value can lead to convergence problems or even failure of adaptive quadrature; if you receive the error "log likelihood cannot be computed", try increasing nip(). For Cartesian product quadrature, nip() specifies the number of points for each of the two random effects; the default is nip(5). For spherical quadrature, `nip()` specifies the degree, d, of the approximation; the default is `nip(9)`, and the only values currently supported by `gllamm` are 5, 7, 9, 11, and 15. These defaults give approximately the same accuracy, because degree d for spherical quadrature approximately corresponds in accuracy to $(d+1)/2$ points per random effect for Cartesian product quadrature (Rabe-Hesketh, Skrondal, and Pickles 2005, app. B).

`nobivariate`, `nohsroc`, and `nosummarypt` suppress reporting of the bivariate parameter estimates, the HSROC parameter estimates, or the summary point estimates, respectively.

`detail` displays the output of all `gllamm` or `xtmelogit` commands issued.

`level(#)` specifies the confidence level, as a percentage, for confidence intervals. The default is `level(95)` or as set by `set level`; see [U] **20.7 Specifying the width of confidence intervals**.

`trace` adds a display of the current parameter vector to the iteration log.

`nolog` suppresses display of the iteration log.

7.2   The metandiplot command

Syntax

`metandiplot` [*tp fp fn tn*] [*if*] [*in*] [*weight*] [, notruncate level(#)
    predlevel(*numlist*) npoints(#) *subplot_options* addplot(*plot*)
    twoway_options]

Options

`notruncate` specifies that the HSROC curve will not be truncated outside the region of the data. By default, the HSROC curve is not shown when the sensitivity or specificity is less than its smallest study estimate.

`level(#)` specifies the confidence level, as a percentage, for the confidence contour. The default is `level(95)` or as set by `set level`; see [U] **20.7 Specifying the width of confidence intervals**.

`predlevel(`*numlist*`)` specifies the levels, as a percentage, for the prediction contour(s). The default is one contour at the same probability level as the confidence region. Up to five prediction contours are allowed.

`npoints(#)` specifies the number of points to use in drawing the outlines of the confidence and prediction regions. The default is `npoints(500)`.

subplot_options, which are `summopts()`, `confopts()`, `predopts()`, `curveopts()`, and `studyopts()`, control the display of the summary point, confidence contour, prediction contour(s), HSROC curve, and study symbols, respectively. The options within each set of parentheses are simply passed through to the appropriate `twoway` plot. Any of the plots can be turned off by specifying, for example, `summopts(off)`.

`addplot`(*plot*) allows adding additional `graph twoway` plots to the graph; see [G] ***addplot_option***. For example, empirical Bayes predictions could be generated by using `predict` after `metandi` and then added to the graph. See section 6.

twoway_options are most of the options documented in [G] ***twoway_options***, including options for titles, axes, labels, schemes, and saving the graph to disk. However, the `by()` option is not allowed.

7.3 The predict command after metandi

Syntax

`predict` [*type*] *newvarlist* [*if*] [*in*] [, *statistic*]

statistic	description
`mu`	posterior predicted (empirical Bayes) sensitivity and specificity; the default
`u`	posterior means (empirical Bayes predictions, BLUPs) of random effects
`sdu`	posterior standard deviations of random effects
`ustd`	standardized posterior means of random effects
`linpred`	linear predictor with empirical Bayes predictions plugged in: linpred = xb + u
`cooksd`	Cook's distance for each study; available only when model was fit using `gllamm`

Most of the above statistics require *newvarlist* to consist of two new variables to store them: one for the statistic associated with sensitivity and one for the statistic associated with specificity. If *newvarlist* contains only one *newvar*, the statistics associated with sensitivity and specificity will be stored in *newvar*1 and *newvar*0, respectively. `cooksd`, however, is computed once for each study and therefore requires only one *newvar*. See section 6 for examples.

7.4   Saved results

`metandi` saves the following results in `e()`:

Scalars
  `e(N)`   number of studies   `e(ll)`   log likelihood

Macros
  `e(cmd)`   `metandi`   `e(predict)`   program used to implement
  `e(tpfpfntn)`   names of *tp fp fn tn*   predict
                      variables   `e(properties)` b V
  `e(cmd)`   `metareg`

Matrices
  `e(b)`   bivariate coefficient vector   `e(V)`   variance–covariance matrix of
  `e(b_hsroc)`   HSROC coefficient vector   the bivariate estimators
                                                   `e(V_hsroc)`   variance–covariance matrix of
                                                                       the HSROC estimators

Functions
  `e(sample)`   marks estimation sample

8   Methods and formulas

It is possible to use routines for linear mixed models to fit an approximate version
of the bivariate model obtained by using empirical logit transforms of the estimated
sensitivity and specificity in each study together with their estimated standard errors
(Reitsma et al. 2005). However, the small cell counts common in diagnostic accuracy
studies can lead to poor performance of such approximations. Generalized mixed models, in particular, hierarchical (mixed-effects) logistic regression, can handle the binomial
nature of the data directly and are therefore preferable (Chu and Cole 2006; Riley et al.
2007).

Such models are complex to fit, however, because they require numerical integration
(quadrature) to integrate out the random effects. `metandi` uses `gllamm` or `xtmelogit`
to fit the bivariate model by using adaptive quadrature, then transforms the parameter
estimates to those of the HSROC model by using the delta method (Cox 1998).

Because the bivariate model can sometimes prove difficult to fit, some care has been
taken to provide good starting values. First, two separate univariate models are fit
to sensitivity and specificity. These provide excellent starting values for the two mean
and two variance parameters of the bivariate model. A reasonable starting value for
the correlation parameter is obtained from the correlation between the posterior means
(empirical Bayes predictions) of the two univariate models.

We now give the mathematical forms of the bivariate and HSROC models in the
absence of covariates. See Rutter and Gatsonis (2001); Reitsma et al. (2005); and Harbord et al. (2007) for information on the models with covariates, which are not currently
supported by `metandi`.

8.1 The bivariate model

Following Reitsma et al. (2005), we denote the sensitivity in the ith study by p_{Ai} and the specificity by p_{Bi}, and base analysis on their logit transforms:

$$\mu_{Ai} = \mathrm{logit}(p_{Ai})$$

$$\mu_{Bi} = \mathrm{logit}(p_{Bi})$$

(We use the letter μ where Reitsma et al. (2005) used θ to avoid a clash of notation with the HSROC model defined in the next section.)

The bivariate model is a random-effects model in which the logit transforms of the true sensitivity and true specificity in each study have a bivariate normal distribution across studies, thereby allowing for the possibility of correlation between them (Reitsma et al. 2005):

$$\begin{pmatrix} \mu_{Ai} \\ \mu_{Bi} \end{pmatrix} \sim N \left\{ \begin{pmatrix} \mu_A \\ \mu_B \end{pmatrix}, \Sigma_{AB} \right\} \quad \text{with} \quad \Sigma_{AB} = \begin{pmatrix} \sigma_A^2 & \sigma_{AB} \\ \sigma_{AB} & \sigma_B^2 \end{pmatrix}$$

8.2 The HSROC model

The HSROC model (Rutter and Gatsonis 2001) was originally formulated in terms of the probability, π_{ij}, that a patient in study i with disease status j has a positive test result, where $j = 0$ for a patient without the disease and $j = 1$ for a patient with the disease. Therefore, sensitivity $p_{Ai} = \pi_{i1}$ and specificity $p_{Bi} = 1 - \pi_{i0}$.

The HSROC model for study i takes the form

$$\mathrm{logit}(\pi_{ij}) = (\theta_i + \alpha_i X_{ij}) \exp(-\beta X_{ij}) \tag{1}$$

where $X_{ij} = -1/2$ for those without disease ($j = 0$) and $+1/2$ for those with disease ($j = 1$). Both θ_i and α_i are allowed to vary between studies. In the model without covariates fit by `metandi`, they are assumed to have independent normal distributions with $\theta_i \sim N(\Theta, \sigma_\theta^2)$ and $\alpha_i \sim N(\Lambda, \sigma_\alpha^2)$. The model is nonlinear in the parameter β and therefore cannot be fit in `gllamm` directly.

We can rewrite (1) as two separate equations for the logit transforms of sensitivity p_{Ai} and specificity p_{Bi}, thus connecting to the parameters μ_{Ai} and μ_{Bi} of the bivariate model above:

$$\mu_{Ai} = \mathrm{logit}(p_{Ai}) = b^{-1}(\theta_i + \frac{1}{2}\alpha_i)$$

$$\mu_{Bi} = \mathrm{logit}(p_{Bi}) = -b(\theta_i - \frac{1}{2}\alpha_i)$$

This tells us that μ_{Ai} and μ_{Bi} are linear combinations of two random variables, θ_i and α_i, with independent normal distributions, and that they therefore must have a bivariate normal distribution. Some straightforward further algebra gives the explicit relationship between the parameters of the two models (Harbord et al. 2007; Arends et al.

2008), enabling HSROC parameter estimates to be obtained by transforming the bivariate parameter estimates. Standard errors for the transformed parameter estimates are obtained by the delta method, which gives the same standard errors that would be obtained from standard maximum-likelihood methods if the HSROC model were fit directly (Cox 1998).

8.3   Methods and formulas for metandiplot

HSROC curve

The HSROC model gives rise to an SROC curve by allowing the threshold parameter, θ_i, to vary while holding the accuracy parameter, α_i, fixed at its mean, Λ. The expected sensitivity for a given specificity is then given by (Rutter and Gatsonis 2001; Macaskill 2004)

$$\text{logit(sensitivity)} = \Lambda e^{-\beta/2} - e^{-\beta}\text{logit(specificity)}$$

Bivariate confidence and prediction regions

Confidence and prediction regions in SROC space can be constructed by using the estimates from the bivariate model (Reitsma et al. 2005; Harbord et al. 2007). An elliptical joint confidence region for μ_A and μ_B is most easily specified by using a parametric representation (Douglas 1993)

$$\mu_A = \widehat{\mu}_A + s_A\, c\, \cos t \tag{2}$$

$$\mu_B = \widehat{\mu}_B + s_B\, c\, \cos(t + \arccos r) \tag{3}$$

where s_A and s_B are the estimated standard errors of $\widehat{\mu}_A$ and $\widehat{\mu}_B$, r is the estimate of their correlation, and varying t from 0 to 2π generates the boundary of the ellipse. The constant c has been called the boundary constant of the ellipse (Alexandersson 2004); $c = \sqrt{2f_{2,n-2;\alpha}}$, where n is the number of studies and $f_{2,n-2;\alpha}$ is the upper $100\alpha\%$ point of the F distribution with degrees of freedom 2 and $n-2$ (Douglas 1993; Chew 1966). This ellipse is then back-transformed to conventional ROC space to give a confidence region for the summary operating point.

A prediction region giving the region that has a given probability (e.g., 95%) of including the *true* sensitivity and specificity of a future study is generated similarly. The covariance matrix for the true logit sensitivity and logit specificity in a future study is

$$\Sigma_{AB} + \text{Var}\begin{pmatrix}\widehat{\mu}_A \\ \widehat{\mu}_B\end{pmatrix}$$

In practice, both terms are estimated by fitting the model to the data. The parameters s_A, s_B, and r in (2) and (3) can then be replaced by the corresponding quantities derived from this covariance matrix to give the prediction ellipse in logit ROC space, which is then back-transformed to a prediction region for the true sensitivity and specificity of a future study in conventional ROC space.

8.4 Methods and formulas for predict

If `metandi` fit the model by using `gllamm`, then `predict` after `metandi` uses `gllapred`; see Rabe-Hesketh, Skrondal, and Pickles (2004). If `metandi` fit the model by using `xtmelogit`, `predict` after `metandi` uses the prediction facilities of `xtmelogit`; see [XT] **xtmelogit postestimation**.

9 Acknowledgments

Joseph Coveney first worked out how to fit the bivariate model by using `gllamm` and posted the syntax on Statalist in response to a query from Ben Dwamena; our thanks to Joe for generous email correspondence. We thank the authors of `gllamm` for all their work, and Sophia Rabe-Hesketh in particular for helpful email correspondence. Our thanks also to Susan Mallett and Jon Deeks for useful feedback on earlier versions of `metandi`.

10 References

Alexandersson, A. 2004. Graphing confidence ellipses: An update of ellip for Stata 8. *Stata Journal* 4: 242–256.

Arends, L. R., T. H. Hamza, J. C. van Houwelingen, M. H. Heijenbrok-Kal, M. G. M. Hunink, and T. Stijnen. 2008. Bivariate random effects meta-analysis of ROC curves. *Medical Decision Making* 28: 621–638.

Bradburn, M. J., J. J. Deeks, and D. G. Altman. 1998. sbe24: metan—an alternative meta-analysis command. *Stata Technical Bulletin* 44: 4–15. Reprinted in *Stata Technical Bulletin Reprints*, vol. 8, pp. 86–100. College Station, TX: Stata Press. (Reprinted in this collection on pp. 3–28.)

Chew, V. 1966. Confidence, prediction, and tolerance regions for the multivariate normal distribution. *Journal of the American Statistical Association* 61: 605–617.

Chu, H., and S. R. Cole. 2006. Bivariate meta-analysis of sensitivity and specificity with sparse data: A generalized linear mixed model approach. *Journal of Clinical Epidemiology* 59: 1331–1332.

Coveney, J. 2004. Re: st: bivariate random effects meta-analysis of diagnostic test. Statalist archive. Available at http://www.stata.com/statalist/archive/2004-04/msg00820.html.

Cox, C. 1998. Delta method. In *Encyclopedia of Biostatistics*, ed. P. Armitage and T. Colton, 1125–1127. New York: Wiley.

Deeks, J. J. 2001a. Systematic reviews of evaluations of diagnostic and screening tests. In *Systematic Reviews in Health Care: Meta-Analysis in Context*, 2nd edition, ed. M. Egger, G. Davey Smith, and D. G. Altman, 248–282. London: BMJ Books.

————. 2001b. Systematic reviews in health care: Systematic reviews of evaluations of diagnostic and screening tests. *British Medical Journal* 323: 157–162.

Devillé, W. L., F. Buntinx, L. M. Bouter, V. M. Montori, H. C. de Vet, D. A. W. M. van der Windt, and P. D. Bezemer. 2002. Conducting systematic reviews of diagnostic studies: Didactic guidelines. *BMC Medical Research Methodology* 2: 9.

Douglas, J. B. 1993. Confidence regions for parameter pairs. *American Statistician* 47: 43–45.

Gatsonis, C. A., and P. Paliwal. 2006. Meta-analysis of diagnostic and screening test accuracy evaluations: Methodologic primer. *American Journal of Roentgenology* 187: 271–281.

Gluud, C., and L. L. Gluud. 2005. Evidence based diagnostics. *British Medical Journal* 330: 724–726.

Harbord, R. M., J. J. Deeks, M. Egger, P. Whiting, and J. A. C. Sterne. 2007. A unification of models for meta-analysis of diagnostic accuracy studies. *Biostatistics* 8: 239–251.

Lijmer, J. G., P. M. Bossuyt, and S. H. Heisterkamp. 2002. Exploring sources of heterogeneity in systematic reviews of diagnostic tests. *Statistics in Medicine* 21: 1525–1537.

Macaskill, P. 2004. Empirical Bayes estimates generated in a hierarchical summary ROC analysis agreed closely with those of a full Bayesian analysis. *Journal of Clinical Epidemiology* 57: 925–932.

Mallett, S., J. J. Deeks, S. Halligan, S. Hopewell, V. Cornelius, and D. G. Altman. 2006. Systematic reviews of diagnostic tests in cancer: Review of methods and reporting. *British Medical Journal* 333: 413–416.

Moses, L. E., D. Shapiro, and B. Littenberg. 1993. Combining independent studies of a diagnostic test into a summary ROC curve: Data-analytic approaches and some additional considerations. *Statistics in Medicine* 12: 1293–1316.

Nordic Cochrane Centre. 2007. *Review Manager (RevMan): Version 5.* Software program. Copenhagen: The Nordic Cochrane Centre, The Cochrane Collaboration.

Rabe-Hesketh, S., A. Skrondal, and A. Pickles. 2004. GLLAMM manual. Working Paper 160, Division of Biostatistics, University of California–Berkeley. http://www.bepress.com/ucbbiostat/paper160/.

————. 2005. Maximum likelihood estimation of limited and discrete dependent variable models with nested random effects. *Journal of Econometrics* 128: 301–323.

Reitsma, J. B., A. S. Glas, A. W. S. Rutjes, R. J. P. M. Scholten, P. M. Bossuyt, and A. H. Zwinderman. 2005. Bivariate analysis of sensitivity and specificity produces informative summary measures in diagnostic reviews. *Journal of Clinical Epidemiology* 58: 982–990.

Riley, R. D., K. R. Abrams, A. J. Sutton, P. C. Lambert, and J. R. Thompson. 2007. Bivariate random-effects meta-analysis and the estimation of between-study correlation. *BMC Medical Research Methodology* 7: 3.

Rutter, C. M., and C. A. Gatsonis. 2001. A hierarchical regression approach to meta-analysis of diagnostic test accuracy evaluations. *Statistics in Medicine* 20: 2865–2884.

Scheidler, J., H. Hricak, K. K. Yu, L. Subak, and M. R. Segal. 1997. Radiological evaluation of lymph node metastases in patients with cervical cancer: A meta-analysis. *Journal of the American Medical Association* 278: 1096–1101.

Skrondal, A., and S. Rabe-Hesketh. 2004. *Generalized Latent Variable Modeling: Multilevel, Longitudinal, and Structural Equation Models*. Boca Raton, FL: Chapman & Hall/CRC.

Sterne, J. A. C., R. J. Harris, R. M. Harbord, and T. J. Steichen. 2007. What meta-analysis features are available in Stata? FAQ. Available at http://www.stata.com/support/faqs/stat/meta.html.

Tatsioni, A., D. A. Zarin, N. Aronson, D. J. Samson, C. R. Flamm, C. H. Schmid, and J. Lau. 2005. Challenges in systematic reviews of diagnostic technologies. *Annals of Internal Medicine* 142: 1048–1055.

Walter, S. D. 2002. Properties of the summary receiver operating characteric (SROC) curve for diagnostic test data. *Statistics in Medicine* 21: 1237–1256.

Zwinderman, A. H., and P. M. Bossuyt. 2008. We should not pool diagnostic likelihood ratios in systematic reviews. *Statistics in Medicine* 27: 687–697.

Multivariate random-effects meta-analysis

Ian R. White
MRC Biostatistics Unit
Cambridge, UK
ian.white@mrc-bsu.cam.ac.uk

Abstract. Multivariate meta-analysis combines estimates of several related parameters over several studies. These parameters can, for example, refer to multiple outcomes or comparisons between more than two groups. A new Stata command, `mvmeta`, performs maximum likelihood, restricted maximum likelihood, or method-of-moments estimation of random-effects multivariate meta-analysis models. A utility command, `mvmeta_make`, facilitates the preparation of summary datasets from more detailed data. The commands are illustrated with data from the Fibrinogen Studies Collaboration, a meta-analysis of observational studies; I estimate the shape of the association between a quantitative exposure and disease events by grouping the quantitative exposure into several categories.

Keywords: st0156, mvmeta, mvmeta_make, mvmeta_l, meta-analysis, multivariate meta-analysis, individual participant data, observational studies

1 Introduction

Standard meta-analysis combines estimates of one parameter over several studies (Normand 1999). Multivariate meta-analysis is an extension that can combine estimates of several related parameters (van Houwelingen, Arends, and Stijnen 2002). In such work, it is important to allow for heterogeneity between studies, usually by fitting a random-effects model (Thompson 1994).

Multivariate meta-analysis has a variety of applications in randomized controlled trials. The simplest is modeling the outcome separately in each arm of a clinical trial (van Houwelingen, Arends, and Stijnen 2002). Other published applications explore treatment effects simultaneously on two clinical outcomes (Berkey, Anderson, and Hoaglin 1996; Berkey et al. 1998; Riley et al. 2007a,b) or on cost and effectiveness (Pinto, Willan, and O'Brien 2005), and explore combining trials comparing more than one treatment (Hasselblad 1998; Lu and Ades 2004). Further applications have been reviewed by Riley et al. (2007b).

There are also possible applications of multivariate meta-analysis in observational studies. These applications include assessing the shape of the association between a quantitative exposure and a disease, which will be illustrated in this article.

One difficulty in random-effects meta-analysis is estimating the between-studies variance. In the univariate case, this is commonly performed by using the method of DerSimonian and Laird (1986). However, maximum likelihood (ML) and restricted maximum likelihood (REML) methods are alternatives (van Houwelingen, Arends, and

Stijnen 2002); in Stata, they are not available in `metan` but can be obtained from `metareg` (Sharp 1998). This article describes a new command, `mvmeta`, that performs REML and ML estimation in the multivariate case by using a Newton–Raphson procedure. `mvmeta` requires a dataset of study-specific point estimates and their variance–covariance matrix. I also describe a utility command, `mvmeta_make`, that facilitates forming this dataset.

2 Multivariate random-effects meta-analysis with mvmeta

2.1 Syntax

mvmeta *b V* [*if*] [*in*] [, reml ml mm <u>f</u>ixed <u>var</u>s(*varlist*) corr(*expression*)

    start(*matrix* | *matrix_expression* | mm) <u>showst</u>art <u>showch</u>ol

    <u>keep</u>mat(*bname Vname*) <u>nounc</u>ertainv eform(*name*) bscorr bscov

    missest(*#*) missvar(*#*) *maximize_options*]

where the data are arranged with one line per study, the point estimates are held in variables whose names start with *b* (excluding *b* itself), the variance of *bx* is held in variable *Vxx*, and the covariance of *bx* and *by* is held in variable *Vxy* or *Vyx* (or the corr() option is specified).

If the dataset includes variables whose names start with *b* that do not represent point estimates, then the vars() option must be used.

2.2 Options

reml, the default, specifies that REML be used for estimation. Specify only one of the reml, ml, mm, or fixed options.

ml specifies that ML be used for estimation. ML is likely to underestimate the variance, so REML is usually preferred. Specify only one of the reml, ml, mm, or fixed options.

mm specifies that the multivariate method-of-moments procedure (Jackson, White, and Thompson 2010) be used for estimation. This procedure is a multivariate generalization of the procedure of DerSimonian and Laird (1986) and is faster than the likelihood-based methods. Specify only one of the reml, ml, mm, or fixed options.

fixed specifies that the fixed-effects model be used for estimation. Specify only one of the reml, ml, mm, or fixed options.

vars(*varlist*) specifies which variables are to be used. By default, all variables *b** are used (excluding *b* itself). The order of variables in *varlist* does not affect the model itself but does affect the parameterization.

corr(*expression*) specifies that all within-study correlations take the given value. This means that covariance variable *Vxy* need not exist. (If it does exist, corr() is ignored.)

start(*matrix* | *matrix_expression* | mm) specifies a starting value for the between-studies variance, except start(mm) specifies that the starting value is computed by the mm method. If start() is not specified, the starting value is the weighted between-studies variance of the estimates, not allowing for the within-study variances; this ensures that the starting value is greater than zero (the iterative procedure never moves away from zero). start(0) uses a starting value of 0.001 times the default. The starting value for the between-studies mean is the fixed-effects estimate.

showstart reports the starting values used.

showchol reports the estimated values of the basic parameters underlying the between-studies variance matrix (the Cholesky decomposition).

keepmat(*bname Vname*) saves the vector of study-specific estimates and the vector of the variance–covariance matrix for study *i* as *bnamei* and *Vnamei*, respectively.

nouncertainv invokes alternative (smaller) standard errors that ignore the uncertainty in the estimated variance–covariance matrix and therefore agree with results produced by procedures such as SAS PROC MIXED (without the ddfm=kr option) and metareg. (Note, however, that the confidence intervals do not agree because mvmeta uses a normal approximation, whereas the other procedures approximate the degrees of freedom of a *t* distribution.)

eform(*name*) exponentiates the reported mean parameters, labeling them *name*.

bscorr reports the between-studies variance–covariance matrix as the standard deviations and reports the correlation matrix. This is the default if bscov is not specified.

bscov reports the between-studies variance–covariance matrix without transformation.

missest(#) specifies the value to be used for missing point estimates; the default is missest(0). This is of minor importance because the variance of these missing estimates is specified to be very large.

missvar(#) is used in imputing the variance of missing point estimates. For a specific variable, the variance used is the largest observed variance multiplied by the specified value. The default is missvar(1E4); this value is unlikely to need to be changed.

maximize_options are any options allowed by ml maximize.

3 Details of mvmeta

3.1 Notation

The data for mvmeta comprise the point estimate, y_i, and the within-study variance–covariance matrix, S_i, for each study $i = 1$ to n.

We assume the model

$$
\begin{aligned}
y_i &\sim N(\mu_i, S_i) \\
\mu_i &\sim N(\mu, \Sigma) \\
\Sigma &= \begin{pmatrix} \tau_1^2 & \kappa_{12}\tau_1\tau_2 & \cdot \\ \kappa_{12}\tau_1\tau_2 & \tau_2^2 & \cdot \\ \cdot & \cdot & \cdot \end{pmatrix}
\end{aligned}
$$

where y_i, μ_i, and μ are $p \times 1$ vectors, and S_i and Σ are $p \times p$ matrices. The within-study variance, S_i, is assumed to be known. Our aim is to estimate μ and Σ.

We set $W_i = (\Sigma + S_i)^{-1}$, noting that this depends on the unknown Σ. If Σ were known (or assumed to be the zero matrix, as in fixed-effects meta-analysis), then we would have

$$
\widehat{\mu} = \left(\sum_i W_i \right)^{-1} \left(\sum_i W_i y_i \right)
$$

3.2 Estimating Σ

Methods proposed for estimating Σ in the multivariate setting include extensions of Cochran's method (Berkey et al. 1998), of the DerSimonian and Laird method (Pinto, Willan, and O'Brien 2005) for diagonal W_i, and of likelihood-based methods (van Houwelingen, Arends, and Stijnen 2002). We use the latter because of their generality and optimality properties. Respectively, the likelihood and restricted likelihood are

$$
-2L = \sum_i \left\{ \log |\Sigma + S_i| + (y_i - \mu)' W_i (y_i - \mu) \right\} + np\log 2\pi
$$

$$
-2RL = -2L + \log \left| \sum_i W_i \right| - p\log 2\pi \tag{1}
$$

where W_i is a function of the unknown Σ, as noted above.

We maximize the (restricted) likelihood with a Newton–Raphson algorithm by using Stata's `ml` procedure. To ensure that Σ is nonnegative definite (for example, in the bivariate case, to ensure that the between-studies variances are nonnegative and that the between-studies correlation lies between -1 and 1), the basic model parameters are taken as the elements of a Cholesky decomposition of Σ (Riley et al. 2007b).

3.3 Saved results

As well as the usual `e()` information, `mvmeta` returns the estimated overall mean in `e(Mu)` and the between-studies variance–covariance matrix, the standard deviation vector, and the correlation matrix in `e(Sigma)`, `e(Sigma_SD)`, and `e(Sigma_corr)`, respectively.

3.4   Files required

mvmeta uses the likelihood program mvmeta_l.ado.

4   A utility command to produce data in the correct format: mvmeta_make

4.1   Syntax

mvmeta_make *regression_command* [*if*] [*in*] [*weight*], by(*by_variable*)
   <u>sav</u>ing(*savefile*) [replace append <u>names</u>(*bname Vname*) keepmat
   <u>usev</u>ars(*varlist*) <u>usec</u>onstant esave(*namelist*) <u>nodet</u>ails pause
   <u>ppf</u>ix(none | check | all) <u>augwt</u>(#) <u>noaug</u>list <u>ppcmd</u>(*regcmd*[, *options*])
   hard *regression_options*]

   mvmeta_make performs *regression_command* for each level of *by_variable* and stores the results in *savefile* in the format required by mvmeta. *weight* is any weight allowed by *regression_command*.

4.2   Options

by(*by_variable*) is required; it identifies the studies in which the regression command will be performed.

saving(*savefile*) is required; it specifies to save the regression results to *savefile*.

replace specifies to overwrite the existing file called *savefile*.

append specifies to append the current results to the existing file called *savefile*.

names(*bname Vname*) specifies that the estimated coefficients for variable x are to be stored in variable *bnamex* and that the estimated covariance between coefficients *bnamex* and *bnamey* is to be stored in variable *Vnamexy*. The default is names(y S).

keepmat specifies that the results are also to be stored as matrices. The estimate vector and the covariance matrix for study i are stored as matrices *bnamei* and *Vnamei*, respectively, where *bname* and *Vname* are specified with names().

usevars(*varlist*) identifies the variables whose regression coefficients are of interest. The default is all variables in the model, excluding the constant.

useconstant specifies that the constant is also of interest.

esave(*namelist*) adds the specified e() statistics to the saved data. For example, esave(N ll) saves e(N) and e(ll) as variables _e_N and _e_ll.

`nodetails` suppresses the results of running *regression_command* on each study.

`pause` pauses output after the analysis of each study, provided that `pause on` has been set.

`ppfix(none | check | all)` specifies whether perfect prediction should be fixed in no studies, only in studies where it is detected (the default), or in all studies.

`augwt(#)` specifies the total weight of augmented observations to be added in any study in which perfect prediction is detected (see section 7). `augwt(0)` turns off augmentation but is not recommended. The default is `augwt(0.01)`.

`noauglist` suppresses listing of the augmented observations.

`ppcmd(`*regcmd*[, *options*]`)` specifies that perfect prediction should be fixed by using regression command *regcmd* with options *options* instead of by using the default augmentation procedure.

`hard` is useful when convergence cannot be achieved in some studies. It captures the results of initial model fitting in each study and treats any nonzero return code as a symptom of perfect prediction.

regression_options are any options for *regression_command*.

5 Example 1: Telomerase data

Data from 10 studies of the value of telomerase measurements in the diagnosis of primary bladder cancer were reproduced by Riley et al. (2007b). In the table below, taken from that article, `y1` is logit sensitivity, `y2` is logit specificity, and `s1` and `s2` are their respective standard errors, all estimated from 2×2 tables of true status versus test status.

```
. use telomerase
(Riley's telomerase data)
. format y1 s1 y2 s2 %6.3f
. list, noobs clean
    study      y1       s1       y2       s2
        1    1.139    0.406    3.219    1.020
        2    1.447    0.556    1.299    0.651
        3    1.705    0.272    0.661    0.308
        4    0.470    0.403    3.283    0.588
        5    0.856    0.290    4.920    1.004
        6    1.440    0.371    1.386    0.456
        7    0.187    0.306    3.219    1.442
        8    1.504    0.451    2.197    0.745
        9    1.540    0.636    2.269    0.606
       10    1.665    0.412   -1.145    0.434
. generate S11=s1^2
. generate S22=s2^2
```

5.1   Univariate meta-analysis

We first analyze the data by two univariate meta-analyses:

```
. mvmeta y S, vars(y1) bscov
Note: using method reml
Note: using variable y1
Note: 10 observations on 1 variables
    (output omitted)
```

	Number of obs	=	10
	Wald chi2(1)	=	38.52
Log likelihood = -8.7276382	Prob > chi2	=	0.0000

	Coef.	Std. Err.	z	P>|z|	[95% Conf. Interval]	
Overall_mean						
y1	1.154606	.1860421	6.21	0.000	.7899701	1.519242

```
Estimated between-studies covariance matrix Sigma:
          y1
y1   .18579341
. mvmeta y S, vars(y2) bscov
Note: using method reml
Note: using variable y2
Note: 10 observations on 1 variables
    (output omitted)
```

	Number of obs	=	10
	Wald chi2(1)	=	12.93
Log likelihood = -18.728644	Prob > chi2	=	0.0003

	Coef.	Std. Err.	z	P>|z|	[95% Conf. Interval]	
Overall_mean						
y2	1.963801	.5460555	3.60	0.000	.8935515	3.03405

```
Estimated between-studies covariance matrix Sigma:
          y2
y2   2.386426
```

These results agree with SAS PROC MIXED as reported by Riley et al. (2007b), except that the standard errors for the overall means are slightly larger (0.5461 for y2, compared with 0.5414 from SAS). This is because SAS does not, by default, allow for uncertainty in the estimated between-studies variance (SAS Institute 1999). mvmeta's nouncertainv option inverts just the elements of the information matrix relating to the overall mean and agrees with SAS PROC MIXED:

```
. mvmeta y S, vars(y2) nouncertainv
Note: using method reml
Note: using variable y2
Note: 10 observations on 1 variables
```
   (*output omitted*)

Alternative standard errors, ignoring uncertainty in V:

	Coef.	Std. Err.	z	P>|z|	[95% Conf. Interval]	
Overall_mean						
y2	1.963801	.5413727	3.63	0.000	.9027297	3.024872

5.2 Multivariate analysis

Because sensitivity and specificity are estimated on separate groups of individuals, their within-study covariance is zero. We could generate a new variable, S12=0, but it is easier to use the corr(0) option:

```
. mvmeta y S, corr(0) bscov
Note: using method reml
Note: using variables y1 y2
Note: 10 observations on 2 variables
Note: corr(0) used for all covariances
```
   (*output omitted*)

```
                                     Number of obs   =         10
                                     Wald chi2(2)    =     159.58
Log likelihood = -24.415968          Prob > chi2     =     0.0000
```

	Coef.	Std. Err.	z	P>|z|	[95% Conf. Interval]	
Overall_mean						
y1	1.166187	.1863275	6.26	0.000	.8009913	1.531382
y2	2.057752	.5607259	3.67	0.000	.9587493	3.156755

```
Estimated between-studies covariance matrix Sigma:
          y1          y2
y1   .20219111
y2  -.7227506   2.5835381
```

Again these results agree with those of Riley et al. (2007b), except that our standard errors are slightly larger because they allow for uncertainty in the between-studies covariance, Σ.

6 Example 2: Fibrinogen Studies Collaboration data

Fibrinogen Studies Collaboration (FSC) is a meta-analysis of individual data on 154,012 adults from 31 prospective studies with information on plasma fibrinogen and major disease outcomes (Fibrinogen Studies Collaboration 2004). As part of the published analysis, the incidence of coronary heart disease was compared across 10 groups defined by baseline levels of fibrinogen (Fibrinogen Studies Collaboration 2005). That analysis used a fixed-effects model; here we allow for heterogeneity between studies by using a random-effects model, but we reduce the analysis to five groups to avoid presenting lengthy output.

In the first stage of analysis, we start with individual-level data including fibrinogen concentration, `fg`, in five levels. Following standard practice in the analysis of these data (Fibrinogen Studies Collaboration 2005), all analyses are stratified by sex and, for two studies that were randomized trials, by trial arm (variable `tr`). We adjust all analyses for age (variable `ages`), although in practice, more confounders would be adjusted for. We use the `esave(N)` option to record the sample size used in each study in variable `_e_N`.

```
. stset duration allchd
  (output omitted)
. xi: mvmeta_make stcox ages i.fg, strata(sex tr) nohr
> saving(FSCstage1) replace by(cohort) usevars(i.fg) names(b V) esave(N)
i.fg              _Ifg_1-5           (naturally coded; _Ifg_1 omitted)
Using coefficients: _Ifg_2 _Ifg_3 _Ifg_4 _Ifg_5

-> cohort==1

        failure _d:  allchd
   analysis time _t:  duration

Iteration 0:    log likelihood = -5223.9564
Iteration 1:    log likelihood = -5135.3888
Iteration 2:    log likelihood = -5129.5633
Iteration 3:    log likelihood =  -5129.551
Refining estimates:
Iteration 0:    log likelihood =  -5129.551

Stratified Cox regr. -- Breslow method for ties

No. of subjects =        14436               Number of obs   =      14436
No. of failures =          603
Time at risk    = 127969.6428
                                             LR chi2(5)      =     188.81
Log likelihood  =    -5129.551               Prob > chi2     =     0.0000
```

_t	Coef.	Std. Err.	z	P>|z|	[95% Conf. Interval]	
ages	.0501925	.0072871	6.89	0.000	.03591	.064475
_Ifg_2	.2523666	.1895222	1.33	0.183	-.11909	.6238233
_Ifg_3	.5317069	.1804709	2.95	0.003	.1779905	.8854233
_Ifg_4	.9464425	.1761563	5.37	0.000	.6011824	1.291703
_Ifg_5	1.400935	.1779354	7.87	0.000	1.052188	1.749682

```
                                                      Stratified by sex tr
-> cohort==2
  (output omitted)
```

Here are the data stored for the first 15 of the 31 studies; the data also include covariances `V_Ifg_2_Ifg_3`, etc., which are not displayed to save space. The first row of the data below reproduces the results from the `stcox` analysis given above.

```
. use FSCstage1, clear

. format b* V* %5.3f

. list cohort b_Ifg_2 b_Ifg_3 b_Ifg_4 b_Ifg_5 V_Ifg_2_Ifg_2 V_Ifg_3_Ifg_3,
> clean noobs
    cohort   b_Ifg_2   b_Ifg_3   b_Ifg_4   b_Ifg_5   V_Ifg_~2   ~3_Ifg_3
         1     0.252     0.532     0.946     1.401      0.036      0.033
         2    -0.184    -0.032     0.119     0.567      0.348      0.344
         3     0.001    -0.529    -0.339     0.416      0.375      0.323
         4     0.066     0.184     0.407     0.645      0.058      0.053
         5     0.078     0.406     0.544     1.088      0.101      0.083
         6    -0.113     0.456     0.456     0.875      0.065      0.054
         7    -2.149    -0.264    -0.494     0.169      1.336      0.421
         8    -0.039     0.170     0.420     1.053      0.042      0.038
         9     0.443     0.595     0.922     0.797      0.202      0.175
        10     0.356     1.312     0.628     2.133      1.500      1.170
        11     1.297     1.052     1.421     1.752      0.559      0.542
        12     0.323     0.545     0.681     0.540      0.132      0.122
        13    -0.042     0.509     0.560     0.998      0.088      0.072
        14    -2.667    -2.524    -2.010    -1.767      1.337      0.584
        15     5.946     5.420     6.088     7.057    189.088    189.271
```

(output omitted)

Note the large parameter estimates and very large variances in study 15, which occur because this study has no events in category 1 of fg. Details of how such *perfect prediction* is handled are described in section 7.

Now the second stage of analysis:

```
. mvmeta b V
Note: using method reml
Note: using variables b_Ifg_2 b_Ifg_3 b_Ifg_4 b_Ifg_5
Note: 31 observations on 4 variables
```

(output omitted)

```
                                         Wald chi2(4)    =      139.59
Log likelihood = -79.489126             Prob > chi2     =      0.0000
```

	Coef.	Std. Err.	z	P>|z|	[95% Conf. Interval]	
Overall_mean						
b_Ifg_2	.1615842	.0796996	2.03	0.043	.005376	.3177925
b_Ifg_3	.3926019	.0878114	4.47	0.000	.2204947	.5647091
b_Ifg_4	.5620076	.0905924	6.20	0.000	.3844497	.7395654
b_Ifg_5	.8973289	.0942603	9.52	0.000	.712582	1.082076

```
Estimated between-studies SDs and correlation matrix:
                   SD     b_Ifg_2     b_Ifg_3     b_Ifg_4     b_Ifg_5
b_Ifg_2    .22734097           1   .98953788   .97421937   .70621223
b_Ifg_3    .28611302   .98953788           1   .99657543   .80096928
b_Ifg_4    .30834247   .97421937   .99657543           1   .84773246
b_Ifg_5    .32742861   .70621223   .80096928   .84773246           1
```

It is interesting to compare the estimates with those obtained from four univariate meta-analyses, which can be run by mvmeta b V, vars(b_Ifg_2), etc., and are summarized in table 1.

Table 1. Summary of estimates from four univariate meta-analyses

Group	Univariate			Multivariate						
	$\widehat{\mu}_i$	$\mathrm{se}(\widehat{\mu}_i)$	$\widehat{\tau}_i$	$\widehat{\mu}_i$	$\mathrm{se}(\widehat{\mu}_i)$	$\widehat{\tau}_i$	Correlations $\widehat{\kappa}_{ij}$			
2 vs 1	0.200	0.066	0.134	0.162	0.080	0.227	1			
3 vs 1	0.430	0.073	0.196	0.393	0.088	0.286	0.990	1		
4 vs 1	0.568	0.084	0.263	0.562	0.091	0.308	0.974	0.997	1	
5 vs 1	0.840	0.101	0.363	0.897	0.094	0.327	0.706	0.801	0.848	1

The univariate and multivariate methods give broadly similar point estimates, $\widehat{\mu}_i$, but the multivariate method gives rather larger estimates of three between-studies standard deviations, $\widehat{\tau}_i$, and, consequently, larger standard errors for $\widehat{\mu}_i$. A different choice of reference category would yield the same multivariate results but different univariate results. Of course, the multivariate method also has the advantage of estimating the between-studies correlations.

7   Perfect prediction

7.1   The problem

One difficulty that can occur in regression models with a categorical or time-to-event outcome is *perfect prediction* or *separation* (Heinze and Schemper 2002). In logistic regression, for example, perfect prediction occurs if there is a level of a categorical explanatory variable for which the observed values of the outcome are all one (or all zero); in Cox regression, it occurs if there is a category in which no events are observed. Here, as one or more regression parameters go to plus or minus infinity, the log likelihood increases to a limit and the second derivative of the log likelihood tends to zero.

Stata handles this problem in two ways. Stata first attempts to detect perfect prediction. If successful, it drops the relevant observations and term from the model. However, sometimes (in particular, if perfect prediction is in the reference category of a variable with more than two levels) Stata fails to detect perfect prediction. Here Stata reports very large ML estimates, observes that the variance–covariance matrix is singular, and reports a generalized inverse.

In the meta-analysis context, perfect prediction is likely to occur in some studies and not in others. (In the FSC analysis, it occurred in four studies.) Unfortunately, neither of the above solutions is satisfactory. In the first case, the model fit to a study with perfect prediction differs from that fit to other studies and has fewer parameters, so combination across studies is not meaningful. In the second case, some extremely large coefficients have inappropriately moderate standard errors, so they can have an excessive influence on meta-analytic results.

As an example, we use data from FSC study 15, which has no events in the reference category fg==1:

```
. xi: stcox ages i.fg if cohort==15, nohr

  (output omitted)
No. of subjects =         3134                   Number of obs   =       3134
No. of failures =           17
Time at risk    =  9465.954814
                                                 LR chi2(5)      =      16.43
Log likelihood  =    -127.22742                  Prob > chi2     =     0.0057
```

_t	Coef.	Std. Err.	z	P>|z|	[95% Conf. Interval]
ages	.0357279	.0263705	1.35	0.175	-.0159573 .087413
_Ifg_2	21.36403	.9147602	23.35	0.000	19.57113 23.15692
_Ifg_3	20.84916	.	.	.	. .
_Ifg_4	21.50048	.8689028	24.74	0.000	19.79746 23.2035
_Ifg_5	22.47926	.7987255	28.14	0.000	20.91379 24.04473

Perfect prediction has not been detected, and the coefficients are appropriately large but with inappropriately small standard errors.

7.2   Solution: Augmentation

mvmeta_make checks for perfect prediction by checking that 1) all parameters are reported and 2) there are no zeros on the diagonal of the variance–covariance matrix of the parameter estimates. If perfect prediction is detected, mvmeta_make augments the data in such a way as to avoid perfect prediction but gives the added observations a tiny weight to minimize their impact on well-estimated parts of the model.

The augmentation is performed at two design points for each covariate x, defined by letting $x = \bar{x} \pm s_x$ (where $\bar{x}$ and s_x are the study-specific mean and standard deviation of x, respectively) and by fixing other covariates at their mean value. The records added at each design point depend on the form of regression model. For logistic regression, we add one event and one nonevent. For other regression models with discrete outcomes, we add one observation with each outcome level. For survival analyses, we add one event at time $t_{min}/2$ and one censoring at time $t_{max} + t_{min}/2$, where t_{min} and t_{max} are the first and last follow-up times in the study. For a stratified Cox model, the augmentation is performed for each stratum.

A total weight of wp is then shared equally between the added observations, where w is specified by the augwt() option (the default is augwt(0.01)), and p is the number of model parameters (treating the baseline hazard in a Cox model as one parameter). The regression model is then rerun including the weighted added observations. For study 15, this yields

```
No. of subjects =      3134.06                Number of obs   =       3134
No. of failures =        17.03
Time at risk    =   9466.077771
                                              LR chi2(5)      =      16.33
Log likelihood  =    -115.75111               Prob > chi2     =     0.0060
```

_t	Coef.	Std. Err.	z	P>|z|	[95% Conf. Interval]	
ages	.0353976	.0263231	1.34	0.179	-.0161948	.08699
_Ifg_2	5.946375	13.75093	0.43	0.665	-21.00495	32.89771
_Ifg_3	5.41975	13.75757	0.39	0.694	-21.54459	32.38409
_Ifg_4	6.088434	13.74965	0.44	0.658	-20.86039	33.03726
_Ifg_5	7.057288	13.74605	0.51	0.608	-19.88448	33.99905

```
                                              Stratified by sex tr
```

The coefficients for the _Ifg_* terms are reduced but still large, but their large standard errors now mean that they will not unduly influence the meta-analysis. The coefficient and standard error for ages are barely changed. It is useful to compare the variance–covariance matrix of the parameter estimates before augmentation,

```
ages        _Ifg_2       _Ifg_3      _Ifg_4       _Ifg_5
  ages    .00069444
_Ifg_2    .00156723    .83711768
_Ifg_3            0            0           0
_Ifg_4   -.00185585    .49628548           0    .75596628
_Ifg_5   -.00303957    .49370111           0    .50944939    .64022023
```

with that after augmentation:

```
ages        _Ifg_2       _Ifg_3      _Ifg_4       _Ifg_5
  ages    .00069291
_Ifg_2   -.00309014    189.08811
_Ifg_3   -.00465418    188.76205    189.27067
_Ifg_4   -.00650648    188.77085    188.78488    189.05294
_Ifg_5   -.00768805    188.77649    188.79309    188.81504    188.95394
```

Because the covariances in the latter matrix are large, contrasts between groups 2, 3, 4, and 5 will receive appropriately small standard errors. This study will therefore contribute information about contrasts between groups 2, 3, 4, and 5 to the meta-analysis, but it will contribute no information about contrasts between group 1 and other groups.

A related problem occurs if some study has no observations at all in a particular category. The augmentation algorithm is applied here, too, with the modification that the value s_x, used to define the added design points, is taken as the standard deviation across all studies, because the within-study standard deviation is zero.

8   Discussion

8.1   Difficulties and limitations

The main difficulty that might be encountered in fitting multivariate random-effects meta-analysis models is a nonpositive-definite Σ. However, the parameterization used here ensures that Σ is positive semidefinite and achieves a nonpositive-definite Σ if one or more elements of the Cholesky decomposition approach zero. I have encountered non-convergence of the Newton–Raphson algorithm only when the starting value is $\Sigma = 0$, which is avoided by a suitable nonzero choice of starting values, or when inappropriately handled perfect prediction has led to extreme parameter estimates with small standard errors.

The standard error provided for an REML analysis allows for uncertainty in estimating Σ by inverting the second derivative matrix of the restricted likelihood (3). This is not the standard approach (Kenward and Roger 1997), and its properties require further investigation. Confidence intervals based on a t distribution would be a useful enhancement.

At present, the augmentation routine in `mvmeta_make` effectively ignores any category in which perfect prediction occurs but allows information to be drawn from other categories from that study. A larger augmentation would allow information to be drawn from categories with perfect prediction. For example, if the data consist of 2×2 tables, then standard practice would add 0.5 observations to each cell (Sweeting, Sutton, and Lambert 2004). This amounts to assigning to the augmented observations a total weight equal to the number of parameters, and it is tempting to apply this rule more widely (by using `augment(1)`). However, larger augmentation weights have the undesirable property of not being invariant to reparameterization; for example, a different choice of reference category for the `fg` variable in section 6 would lead to somewhat different results. Larger augmentation is probably best implemented by the user.

There are alternate ways to handle perfect prediction, including various forms of penalized likelihood. The methods of Le Cessie and van Houwelingen (1992) and Verweij and van Houwelingen (1994) have been implemented in Stata by the `plogit` and `stpcox` commands, respectively, and both are currently being updated to allow for perfect prediction (G. Ambler, pers. comm.). The method of Firth (1993) is invariant to reparameterization and is being implemented by the author. When suitable routines become available in Stata, they can be called by the `ppcmd()` option in `mvmeta_make`.

8.2   Comparison to other procedures

All the models considered here can also be fit in SAS PROC MIXED, although some programming effort is required to specify the known within-study variances, S_i. The two approaches are very similar, but by default, SAS produces standard errors that ignore the uncertainty in Σ, and produces confidence intervals by using the t distribution on

$n - 1$ degrees of freedom. Further, SAS optionally provides a standard error adjusted to allow for uncertainty in estimating Σ and provides the approximate degrees of freedom of Kenward and Roger (1997), which has good small-sample properties.

Multivariate meta-analysis models cannot be fit by using existing Stata commands, but univariate models can. `metan` differs from `mvmeta` because it uses DerSimonian and Laird (1986) estimation of the random-effects variance. `metareg` offers the choice of DerSimonian and Laird, ML, or REML estimation, so if run without covariates, it can be compared to `mvmeta`. The original `metareg` (Sharp 1998) used the algorithm of Hardy and Thompson (1996) and did not always find the best solution. Version 2 of `metareg`, by Harbord and Higgins (2008), uses Newton–Raphson maximization via `ml`, and produces the same point estimates as `mvmeta` and the same standard errors as `mvmeta` with the `nouncertainv` option. `metareg` produces confidence intervals that allow for nonnormality of the sampling distributions by using the method of Knapp and Hartung (2003); its `z` option produces confidence intervals that agree with `mvmeta`. Of course, `metareg` also has the enormous advantage of handling meta-regression.

8.3   More than two outcomes

Although `mvmeta` handles several outcomes perfectly well, its computing time increases sharply as the number of outcomes increases. `mvmeta` can even computationally handle situations where there are more quantities of interest than studies ($p > n$); however, fitting such large models can be unwise and results can be untrustworthy.

9   Acknowledgments

I thank the FSC for providing access to their data for illustrative analyses: a full list of the FSC collaborators is given in Fibrinogen Studies Collaboration (2005). I also thank Li Su and Dan Jackson for helping me rediscover the Cholesky decomposition parameterization; Stephen Kaptoge and Sebhat Erquo for helpful comments on the programming; James Roger for help in understanding SAS PROC MIXED; and Patrick Royston and Gareth Ambler for discussions about augmentation and penalized likelihoods.

10   References

Berkey, C. S., J. J. Anderson, and D. C. Hoaglin. 1996. Multiple-outcome meta-analysis of clinical trials. *Statistics in Medicine* 15: 537–557.

Berkey, C. S., D. C. Hoaglin, A. Antczak-Bouckoms, F. Mosteller, and G. A. Colditz. 1998. Meta-analysis of multiple outcomes by regression with random effects. *Statistics in Medicine* 17: 2537–2550.

DerSimonian, R., and N. Laird. 1986. Meta-analysis in clinical trials. *Controlled Clinical Trials* 7: 177–188.

Fibrinogen Studies Collaboration. 2004. Collaborative meta-analysis of prospective studies of plasma fibrinogen and cardiovascular disease. *European Journal of Cardiovascular Prevention and Rehabilitation* 11: 9–17.

———. 2005. Plasma fibrinogen level and the risk of major cardiovascular diseases and nonvascular mortality: An individual participant meta-analysis. *Journal of the American Medical Association* 294: 1799–1809.

Firth, D. 1993. Bias reduction of maximum likelihood estimates. *Biometrika* 80: 27–38.

Harbord, R. M., and J. P. T. Higgins. 2008. Meta-regression in Stata. *Stata Journal* 8: 493–519. (Reprinted in this collection on pp. 85–111.)

Hardy, R. J., and S. G. Thompson. 1996. A likelihood approach to meta-analysis with random effects. *Statistics in Medicine* 15: 619–629.

Hasselblad, V. 1998. Meta-analysis of multitreatment studies. *Medical Decision Making* 18: 37–43.

Heinze, G., and M. Schemper. 2002. A solution to the problem of separation in logistic regression. *Statistics in Medicine* 21: 2409–2419.

Jackson, D., I. R. White, and S. G. Thompson. 2010. Extending DerSimonian and Laird's methodology to perform multivariate random effects meta-analyses. *Statistics in Medicine* 29: 1282–1297.

Kenward, M. G., and J. H. Roger. 1997. Small sample inference for fixed effects from restricted maximum likelihood. *Biometrics* 53: 983–997.

Knapp, G., and J. Hartung. 2003. Improved tests for a random-effects meta-regression with a single covariate. *Statistics in Medicine* 22: 2693–2710.

Le Cessie, S., and J. C. van Houwelingen. 1992. Ridge estimators in logistic regression. *Journal of the Royal Statistical Society, Series C* 41: 191–201.

Lu, G., and A. E. Ades. 2004. Combination of direct and indirect evidence in mixed treatment comparisons. *Statistics in Medicine* 23: 3105–3124.

Normand, S. L. T. 1999. Meta-analysis: Formulating, evaluating, combining and reporting. *Statistics in Medicine* 18: 213–259.

Pinto, E., A. Willan, and B. O'Brien. 2005. Cost-effectiveness analysis for multinational clinical trials. *Statistics in Medicine* 24: 1965–1982.

Riley, R. D., K. R. Abrams, P. C. Lambert, A. J. Sutton, and J. R. Thompson. 2007a. An evaluation of bivariate random-effects meta-analysis for the joint synthesis of two correlated outcomes. *Statistics in Medicine* 26: 78–97.

Riley, R. D., K. R. Abrams, A. J. Sutton, P. C. Lambert, and J. R. Thompson. 2007b. Bivariate random-effects meta-analysis and the estimation of between-study correlation. *BMC Medical Research Methodology* 7: 3.

SAS Institute. 1999. *SAS OnlineDoc Version Eight*. Cary, NC: SAS Institute. http://www.technion.ac.il/docs/sas/.

Sharp, S. 1998. sbe23: Meta-analysis regression. *Stata Technical Bulletin* 42: 16–22. Reprinted in *Stata Technical Bulletin Reprints*, vol. 7, pp. 148–155. College Station, TX: Stata Press. (Reprinted in this collection on pp. 112–120.)

Sweeting, M. J., A. J. Sutton, and P. C. Lambert. 2004. What to add to nothing? Use and avoidance of continuity corrections in meta-analysis of sparse data. *Statistics in Medicine* 23: 1351–1375.

Thompson, S. G. 1994. Systematic review: Why sources of heterogeneity in meta-analysis should be investigated. *British Medical Journal* 309: 1351–1355.

van Houwelingen, H. C., L. R. Arends, and T. Stijnen. 2002. Advanced methods in meta-analysis: Multivariate approach and meta-regression. *Statistics in Medicine* 21: 589–624.

Verweij, P. J. M., and H. C. van Houwelingen. 1994. Penalized likelihood in Cox regression. *Statistics in Medicine* 13: 2427–2436.

The Stata Journal (2011)
11, Number 2, st0156_1, pp. 255–270

249

Multivariate random-effects meta-regression: Updates to mvmeta

Ian R. White
MRC Biostatistics Unit
Cambridge, UK
ian.white@mrc-bsu.cam.ac.uk

Abstract. An extension of `mvmeta`, my program for multivariate random-effects meta-analysis, is described. The extension handles meta-regression. Estimation methods available are restricted maximum likelihood, maximum likelihood, method of moments, and fixed effects. The program also allows a wider range of models (Riley's overall correlation model and structured between-studies covariance); better estimation (using Mata for speed and correctly allowing for missing data); and new postestimation facilities (I-squared, standard errors and confidence intervals for between-studies standard deviations and correlations, and identification of the best intervention). The program is illustrated using a multiple-treatments meta-analysis.

Keywords: st0156_1, mvmeta, meta-analysis, meta-regression, I-squared

1 Introduction

Stata software for meta-analysis is well advanced and has been described in a recent collection of articles (Sterne 2009). Most software is designed for univariate meta-analysis, in which each study contributes an estimate of a single quantity; but there has been recent interest in multivariate meta-analysis, in which some studies contribute estimates of more than one quantity: for example, intervention effects on different outcomes or differences in one outcome among three or more groups (van Houwelingen, Arends, and Stijnen 2002; Jackson, Riley, and White Forthcoming). I have previously described a Stata routine, `mvmeta` (White 2009), that fits the multivariate random-effects meta-analysis model using restricted maximum likelihood (REML), maximum likelihood (ML), or the method of moments (MM).

This article presents various extensions to `mvmeta`. Covariates are allowed so that a multivariate meta-regression is performed. For the case in which within-study correlations are unknown, Riley's overall correlation model can be fit (Riley, Thompson, and Abrams 2008). The between-studies covariance matrix may now be structured. The I-squared statistic, which measures the impact of heterogeneity on a meta-analysis (Higgins and Thompson 2002), has been extended to the multivariate case and implemented. Confidence intervals are available both for variance components and for I-squared. Finally, in the case of comparisons of multiple interventions, the probability that each is the best intervention can be estimated.

The model considered is

$$
\begin{aligned}
y_i &\sim N(\mu_i, S_i) \\
\mu_i &\sim N(\beta X_i, \Sigma)
\end{aligned}
$$

where y_i is a vector of estimates from the ith study, S_i is their variance–covariance matrix, μ_i is the study-specific mean vector, and X_i is a matrix of study-specific covariates. In this model, the data are y_i, S_i, and X_i, and we aim to estimate the regression coefficients β and the between-studies variance–covariance matrix Σ.

I describe the new `mvmeta` options in section 3 and give technical details in section 3. The command is illustrated in a multiple-treatments meta-analysis in section 4, and limitations and possible extensions are discussed in section 5.

2   mvmeta: Multivariate random-effects meta-regression

2.1   Syntax

`mvmeta` $b\ V\ xvars$ [if] [in] [, $old\_options\ new\_options$]

where $old\_options$ are the options for `mvmeta` described in White (2009) and $new\_options$ are described below.

2.2   New model and estimation options

<u>wscorr</u>(`riley`) can be used when within-study correlations are unknown. It uses the alternative model of Riley, Thompson, and Abrams (2008) to estimate an overall correlation; see section 3.5. Riley (2009) discusses other ways to handle unknown within-study correlations.

<u>bscov</u>ariance(*string*) specifies the between-studies covariance structure; see section 2.4.

<u>eq</u>uations(*yvar1 : xvars1* [, *yvar2 : xvars2* [, ...]]) allows different outcomes to have different regression models. For example, for two-dimensional b, `mvmeta b V x` is the same as `mvmeta b V, eq(b1:x,b2:x)`.

<u>nocons</u>tant suppresses the constant in meta-regression.

`longparm` estimates the results as one regression model for each outcome. Without covariates, this is usually less convenient than the default (in which all outcomes form a single regression model) but is required if the `pbest()` option will be used. With covariates, `longparm` is the default and cannot be changed.

Other new estimation options, which are described in the help file, are `noposdef`, `psdcrit(#)`, *maximize_options*, `augment`, and `augquiet`.

2.3 New output options

For regression parameters

dof(*expression*) specifies the degrees of freedom for t tests and confidence intervals on the regression parameters. The expression may include n, the number of observations. The default is to use a normal distribution.

pbest(min|max [*if*] [*in*], [reps(#1) zero gen(*string*) seed(#2) format(%*fmt*) id(*varlist*)]) requests estimation of the probability that each linear predictor is the best—that is, the maximum or minimum, depending on the first argument of pbest(). The probability is estimated under a Bayesian model with flat priors, assuming that the posterior distribution of the parameter estimates is approximated by a normal distribution with mean and variance equal to the frequentist estimates and variance–covariance matrix. Rankings are constructed by drawing the coefficients #1 times (default is 100) from their approximate posterior density. For each draw, the linear predictor is evaluated for each study, and the largest linear predictor is noted. The zero option specifies that 0 be considered another linear predictor; its use is illustrated in the example in section 4. gen() specifies that the probabilities be saved in variables with the prefix *string*. seed() specifies the random-number seed, format() specifies the output format, and id() specifies identifiers for the output. Although the default behaviour is to rank linear predictors, the predict option ranks the true effects in a future study with the same covariates, thus allowing for heterogeneity as well as parameter uncertainty, as in the calculation of prediction intervals (Higgins, Thompson, and Spiegelhalter 2009). For models without covariates, pbest() is only available if longparm was specified when the model was fit.

For between-studies variance parameters

i2 reports the between-studies variance τ_j^2 and the I-squared statistic (Higgins and Thompson 2002) for each outcome, together with confidence intervals. See section 3.6 for details.

i2fmt(%*fmt*) specifies an output format for the I-squared statistics.

ncchi2 uses the option of heterogi in computing confidence intervals for τ_j^2 and I-squared. It is only relevant after MM estimation. See section 3.6 for details.

ciscale(sd|logsd|logh) determines the scale on which confidence intervals for τ_j^2 and I-squared are computed. The default is ciscale(sd). See section 3.6 for details.

testsigma is only allowed after ML or REML estimation. It performs a likelihood-ratio test or restricted likelihood-ratio test of $\Sigma = 0$. The latter is valid because the models compared have the same fixed part (Verbeke and Molenberghs 2000).

2.4   Covariance structures

bscovariance(<u>unstructured</u>) estimates an unstructured Σ and is the default. Starting
values for Σ may be specified explicitly by start(*matrix_expression*). start(mm)
(the default) specifies that the starting value be computed by the mm method.
start(0) uses a starting value of 0.001 times the default, because a starting value
of 0 leads to nonconvergence (White 2009). The starting value for β is derived from
Σ using (1) below.

bscovariance(<u>proportional</u> *matexp*) models $\Sigma = \tau^2\Sigma_0$, where τ is an unknown parameter and $\overline{\Sigma_0}$ is a known matrix expression (for example, a matrix name or I(2)).
start(#) then specifies the starting value for the scalar τ.

bscovariance(<u>equals</u> *matexp*) forces $\Sigma = \Sigma_0$, where Σ_0 is a known matrix expression
(for example, a matrix name or I(2)).

bscovariance(<u>correlation</u> *matexp*) models $\Sigma = \mathbf{D} \times matexp \times \mathbf{D}$, where *matexp*
is a known matrix expression containing the between-study correlations and $\mathbf{D}$ is
an unknown diagonal matrix containing the between-studies standard deviations.
start(*rowvector*) specifies the starting values for the diagonal of $\mathbf{D}$.

2.5   Other changes in version 2

The showchol option has been renamed showall, the corr() option has been renamed
wscorr(), and the bscorr and bscov options have been renamed print(bscorr) and
print(bscov), respectively. mvmeta typed without specifying b and V redisplays the
latest estimation results, and output options (including showall, eform, nouncertainv,
print(), level(), dof, i2, and pbest()) may be used.

3   Details

3.1   Notation

The data for mvmeta for the rth outcome from the ith study ($i = 1, \ldots, n$) are the point
estimate y_{ir} (a scalar) and the covariates x_{ir} (a $q_r \times 1$ vector). For standard meta-
analysis, $x_{ir} = (1)$, a vector of ones. The mean of y_{ir} is assumed to be $\beta_r x_{ir}$, where β_r
is a $1 \times q_r$ (row)vector. Thus we have the marginal models

$$\begin{aligned} y_{ir} &\sim N(\mu_{ir}, s_{ir}^2) \\ \mu_{ir} &\sim N(\beta_r x_{ir}, \tau_r^2) \end{aligned}$$

In matrix notation, we write the (row)vector outcome $y_i = (y_{i1}, y_{i2}, \ldots, y_{ip})$. We also
know the within-study variance–covariance matrix S_i (a $p \times p$ matrix). We assume the
following joint model:

$$y_i \sim N(\mu_i, S_i)$$
$$\mu_i \sim N(\beta X_i, \Sigma)$$

$$S_i = \begin{pmatrix} s_{i1}^2 & \rho_{i12}s_{i1}s_{i2} & \cdots & \rho_{i1p}s_{i1}s_{ip} \\ \rho_{i12}s_{i1}s_{i2} & s_{i2}^2 & \cdots & \rho_{i2p}s_{i2}s_{ip} \\ \vdots & \vdots & \ddots & \vdots \\ \rho_{i1p}s_{i1}s_{ip} & \rho_{i2p}s_{i2}s_{ip} & \cdots & s_{ip}^2 \end{pmatrix}$$

$$\beta = (\beta_1, \beta_2, \ldots, \beta_p)$$

$$X_i = \begin{pmatrix} x_{i1} & 0 & \cdots & 0 \\ 0 & x_{i2} & \cdots & 0 \\ \vdots & \vdots & \ddots & \vdots \\ 0 & 0 & \cdots & x_{ip} \end{pmatrix}$$

$$\Sigma = \begin{pmatrix} \tau_1^2 & \kappa_{12}\tau_1\tau_2 & \cdots & \kappa_{1p}\tau_1\tau_p \\ \kappa_{12}\tau_1\tau_2 & \tau_2^2 & \cdots & \kappa_{2p}\tau_2\tau_p \\ \vdots & \vdots & \ddots & \vdots \\ \kappa_{1p}\tau_1\tau_p & \kappa_{2p}\tau_2\tau_p & \cdots & \tau_p^2 \end{pmatrix}$$

where y_i and μ_i are $1 \times p$, S_i and Σ are $p \times p$, β is $1 \times q_+$, and X_i is $q_+ \times p$. Σ may be constrained as explained in section 2.4. Our aim is to estimate β and Σ.

3.2 Estimating β, knowing Σ

We set $W_i = (\Sigma + S_i)^{-1}$. Then

$$\hat{\beta} = \left(\sum_i y_i W_i X_i'\right)\left(\sum_i X_i W_i X_i'\right)^{-1} \tag{1}$$

3.3 Estimating Σ: likelihood-based methods

We still use the notation $W_i = (\Sigma + S_i)^{-1}$, noting that this now depends on the unknown Σ. The log likelihood and restricted log likelihood, respectively, are

$$-2L = \sum_i \{\log|\Sigma + S_i| + (y_i - X_i\beta)W_i(y_i - X_i\beta)' + p_i \log 2\pi\} \tag{2}$$

$$-2RL = -2L + \log\left|\sum_i X_i W_i X_i'\right| - q_+ \log 2\pi \tag{3}$$

where p_i is the number of observed outcomes in y_i. Where a study reports only a subset of outcomes, y_i and S_i are of reduced dimension, so Σ in (2) and (3) is replaced by its corresponding submatrix. This makes unnecessary the augmentation procedures in the previous version (White 2009), but they can still be implemented using the `augment` option.

The (restricted) log likelihood is maximized by a Newton–Raphson algorithm using Stata's `ml` procedure. The code has been speeded up by computing the log likelihood using Mata. For unstructured Σ, the basic model parameters are taken as the elements of a Cholesky decomposition of Σ, ensuring that Σ is nonnegative definite (Riley et al. 2007). For the model $\Sigma = \tau^2 \Sigma_0$, the basic parameter is τ.

3.4   Estimating Σ: method of moments

Jackson, White, and Thompson (2010) define a matrix generalization of the univariate Q statistic of DerSimonian and Laird (1986). With unstructured Σ, this satisfies $E(Q_{rs}) = A_{rs} + B_{rs}\Sigma_{rs}$ for $r, s = 1, \ldots, p$, where A and B are matrices that can be computed from the observed data. Estimation of Σ is therefore straightforward and fast. The MM has not yet been developed for meta-regression with structured Σ or for the overall correlation model described in section 3.5 below.

3.5   Unknown within-study correlations

When within-study correlations are unknown, various options are available, including sensitivity analysis over alternative values (Riley 2009). Alternatively, Riley, Thompson, and Abrams (2008) proposed an "overall correlation model" that does not involve the within-study correlations. Let $\mathrm{var}(y_i) = V_i$; the standard model of section 3.1 has $V_i = S_i + \Sigma$. The alternative model has the same diagonal elements, $V_{irr} = S_{irr} + \Sigma_{rr}$, but off-diagonal elements $V_{irs} = \rho_{rs}^O \sqrt{V_{irr}V_{iss}}$ for $r \neq s$. Here ρ_{rs}^O represents an overall correlation between outcomes r and s.

3.6   I-squared

I-squared measures the impact of heterogeneity on the meta-analysis (Higgins and Thompson 2002). In univariate meta-analysis, I-squared is computed as the ratio of a "between" variance (the appropriate element of Σ) to the sum of the "between" variance and a "within" variance given by (9) of Higgins and Thompson (2002). To generalize this, I propose computing I-squared separately for each outcome and handling covariates by defining I-squared for the rth outcome as

$$I_r^2 = \frac{\tau_r^2}{A_{rr}/B_{rr} + \tau_r^2} \tag{4}$$

where A_{rr}/B_{rr} is a "typical" squared standard error, and A_{rr} and B_{rr} are as defined in section 3.4. If there are no covariates, then (4) corresponds exactly to the definition of Higgins and Thompson (2002). If, further, τ^2 is estimated by MM, then (4) gives the standard quantity $I_r^2 = \max[0, \{Q_{rr} - (n_r - 1)/Q_{rr}\}]$ (for example, as output by `metan`), where n_r is the number of studies reporting outcome y_r. However, (4) applies equally well if τ^2 is estimated by REML or ML. This definition of I-squared does not account for "borrowing strength" between outcomes.

When estimation uses the MM, confidence intervals for I_r^2 are computed on the scale of $\log(H_r)$, where $H_r^2 = (1 - I_r^2)^{-1} = Q_{rr}/(n_r - 1)$, as suggested by Higgins and Thompson (2002; they also called them "uncertainty intervals") and implemented in Stata by `heterogi`. The noncentral chi-squared option of `heterogi` is available through the `ncchi2` option. Confidence intervals for τ_r^2 are derived from (4). Exact methods (Biggerstaff and Jackson 2008) are computationally intensive and have not been implemented in Stata.

When estimation uses REML or ML, confidence intervals are first estimated for τ_r^2 using the estimated standard errors. The confidence interval may be computed on the scale of τ (the default), $\log(\tau)$, or $\log(H_r)$. Confidence intervals for I_r^2 are then derived using (4). With unstructured Σ, confidence intervals for the between-studies correlations κ_{rs} are also available; they are computed on the scale of $\log\{(1 + \kappa_{rs})/(1 - \kappa_{rs})\}$. All standard errors are computed using Stata's `nlcom` command. When one or more basic variance parameters is estimated as zero, the corresponding zero term is dropped from the expression for τ_r^2 to avoid causing `nlcom` to fail. This fix can be checked by changing the order of the variables (using `mvmeta`'s `vars()` option), which often avoids causing `nlcom` to fail. In my experience, the two methods always give the same confidence intervals.

4 Example

4.1 Data

We use data from a multiple-treatments meta-analysis comparing four interventions to promote smoking cessation. These data have been previously presented and analyzed by Lu and Ades (2004). The interventions are coded A, B, C, and D, and the data for each trial arm are summarized as the number of individuals and the number who quit smoking. The original dataset is

```
. use smoking_raw, clear
(Smoking data from Lu & Ades (2006))
. list, noo clean
      study    design   dA    nA   dB    nB    dC     nC   dD    nD
          1       ACD    9   140    .     .    23    140   10   138
          2       BCD    .     .   11    78    12     85   29   170
          3        AB   79   702   77   694     .      .    .     .
          4        AB   18   671   21   535     .      .    .     .
          5        AB    8   116   19   146     .      .    .     .
          6        AC   75   731    .     .   363    714    .     .
          7        AC    2   106    .     .     9    205    .     .
          8        AC   58   549    .     .   237   1561    .     .
          9        AC    0    33    .     .     9     48    .     .
         10        AC    3   100    .     .    31     98    .     .
         11        AC    1    31    .     .    26     95    .     .
         12        AC    6    39    .     .    17     77    .     .
         13        AC   95  1107    .     .   134   1031    .     .
         14        AC   15   187    .     .    35    504    .     .
         15        AC   78   584    .     .    73    675    .     .
         16        AC   69  1177    .     .    54    888    .     .
         17        AC   64   642    .     .   107    761    .     .
         18        AC    5    62    .     .     8     90    .     .
         19        AC   20   234    .     .    34    237    .     .
         20        AD    0    20    .     .     .      .    9    20
         21        BC    .     .   20    49    16     43    .     .
         22        BD    .     .    7    66     .      .   32   127
         23        CD    .     .    .     .    12     76   20    74
         24        CD    .     .    .     .     9     55    3    26
```

The first stage of analysis constructs a dataset of estimated intervention effects and their variance–covariance matrices. We choose A as the reference category. Trials without an arm A (trials 2 and 21–24) are augmented with an arm A with 0.01 individuals and 0.001 successes. Trials containing zero cells (trials 9 and 20) have 1 individual with 0.5 successes added to each arm. This leads to the following augmented dataset:

```
. use smoking_aug
(Smoking data from Lu & Ades (2006))

. list, noo clean
    study    design      dA      nA     dB      nB      dC      nC     dD     nD
        1       ACD       9     140      .       .      23     140     10    138
        2       BCD    .001     .01     11      78      12      85     29    170
        3        AB      79     702     77     694       .       .      .      .
        4        AB      18     671     21     535       .       .      .      .
        5        AB       8     116     19     146       .       .      .      .
        6        AC      75     731      .       .     363     714      .      .
        7        AC       2     106      .       .       9     205      .      .
        8        AC      58     549      .       .     237    1561      .      .
        9        AC      .5      34      .       .     9.5      49      .      .
       10        AC       3     100      .       .      31      98      .      .
       11        AC       1      31      .       .      26      95      .      .
       12        AC       6      39      .       .      17      77      .      .
       13        AC      95    1107      .       .     134    1031      .      .
       14        AC      15     187      .       .      35     504      .      .
       15        AC      78     584      .       .      73     675      .      .
       16        AC      69    1177      .       .      54     888      .      .
       17        AC      64     642      .       .     107     761      .      .
       18        AC       5      62      .       .       8      90      .      .
       19        AC      20     234      .       .      34     237      .      .
       20        AD      .5      21      .       .       .       .    9.5     21
       21        BC    .001     .01     20      49      16      43      .      .
       22        BD    .001     .01      7      66       .       .     32    127
       23        CD    .001     .01      .       .      12      76     20     74
       24        CD    .001     .01      .       .       9      55      3     26
```

We now compute the log odds-ratios for arms B, C, and D relative to arm A, as well as the variance–covariance matrix of these three estimates. We could use mvmeta_make (White 2009), but it is easy to run the loop

```
foreach trt in A B C D {
    if "`trt'"=="A" continue
    gen y`trt' = log(d`trt'/(n`trt'-d`trt')) - log(dA/(nA-dA))
    gen S`trt'`trt' = 1/d`trt' + 1/(n`trt'-d`trt') + 1/dA + 1/(nA-dA)
    foreach trt2 in A B C D {
        if "`trt2'"=="A" continue
        if "`trt2'">"`trt'" gen S`trt'`trt2' = 1/dA + 1/(nA-dA) ///
            if !mi(d`trt') & !mi(d`trt2')
    }
}
format y* S* %6.2g
```

which yields the following data:

```
. list study design y* S*, noo clean
```

study	design	yB	yC	yD	SBB	SBC	SBD	SCC	SCD	SDD
1	ACD	.	1.1	.13	.	.	.	.17	.12	.23
2	BCD	.39	.39	.62	1111	1111	1111	1111	1111	1111
3	AB	-.016	.	.	.029	.	.	.	.	.
4	AB	.39	.	.	.11	.	.	.	.	.
5	AB	.7	.	.	.19	.	.	.	.	.
6	AC	.	2.2	.	.	.	.	.02	.	.
7	AC	.	.87	.	.	.	.	.63	.	.
8	AC	.	.42	.	.	.	.	.024	.	.
9	AC	.	2.8	.	.	.	.	2.2	.	.
10	AC	.	2.7	.	.	.	.	.39	.	.
11	AC	.	2.4	.	.	.	.	1.1	.	.
12	AC	.	.44	.	.	.	.	.27	.	.
13	AC	.	.46	.	.	.	.	.02	.	.
14	AC	.	-.16	.	.	.	.	.1	.	.
15	AC	.	-.24	.	.	.	.	.03	.	.
16	AC	.	.039	.	.	.	.	.035	.	.
17	AC	.	.39	.	.	.	.	.028	.	.
18	AC	.	.11	.	.	.	.	.35	.	.
19	AC	.	.58	.	.	.	.	.089	.	.
20	AD	.	.	3.5	.	.	.	.	.	2.2
21	BC	1.8	1.7	.	1111	1111	.	1111	.	.
22	BD	.066	.	1.1	1111	.	1111	.	.	1111
23	CD	.	.52	1.2	.	.	.	1111	1111	1111
24	CD	.	.57	.16	.	.	.	1111	1111	1111

The first stage of analysis is now complete. In the second stage of analysis, we use `mvmeta` to model the intervention effects across studies, using consistency and inconsistency models.

4.2 Consistency model

A consistency model (Lu and Ades 2004) allows the intervention effects to be heterogeneous between studies but assumes that there are no systematic differences between designs. It is easy to fit:

```
. mvmeta y S
Note: using method reml
Regressing yB on
Regressing yC on
Regressing yD on
Note: using variables yB yC yD
Note: 24 observations on 3 variables
Variance-covariance matrix: unstructured

   (output omitted)

Multivariate meta-analysis
Variance-covariance matrix = unstructured
Method = reml                                Number of dimensions    =    3
Restricted log likelihood = -53.826928       Number of observations  =   24
```

	Coef.	Std. Err.	z	P>|z|	[95% Conf. Interval]	
Overall_mean						
yB	.3326048	.3048747	1.09	0.275	-.2649385	.9301482
yC	.6810167	.218959	3.11	0.002	.2518649	1.110168
yD	.8357459	.3664475	2.28	0.023	.117522	1.55397

```
Estimated between-studies SDs and correlation matrix:
          SD          yB          yC          yD
yB  .31410047          1           .           .
yC  .7497773    .9362371          1           .
yD  .72247338   .85588029   .61958804          1
```

We see that, under the consistency assumption, interventions C and D are significantly
superior to A, and D appears to be the best. We could perform significance tests between
B, C, and D using lincom. The heterogeneity (between-studies variation) is larger for
C versus A and D versus A than for B versus A.

It is often of interest to find the best intervention. We can do this using

```
. mvmeta y S, longparm pbest(max in 1, zero reps(1000) seed(478))
   (output omitted)
Estimated probabilities (%) of being the maximum
(allows for parameter uncertainty):
```

	zero	yB	yC	yD
1.	0.0	3.1	31.9	65.0

Let the intervention effects be μ_B, μ_C, and μ_D, all representing contrasts from the
reference intervention A. Positive values indicate better interventions in this dataset, so
if μ_B, μ_C, and μ_D are all negative, then A is best; otherwise, the intervention with the
largest μ is best. Thus we want to find the largest member of the set $\{0, \mu_B, \mu_C, \mu_D\}$,
which is coded using max to find the largest and zero to include 0 in the set. We specify
in 1 to output results for the first study only; because there are no covariates, the
results for all studies are the same.

In the output, the columns headed `zero`, `yB`, `yC`, and `yD` each indicate the posterior probability that intervention A, B, C, or D is the best, respectively. The best intervention is probably D and is very likely to be either C or D.

4.3   Estimating I^2

We can estimate the contribution of between-studies heterogeneity to the meta-analyses:

```
. mvmeta, i2

  (output omitted )
```

Approximate confidence intervals for between-studies SDs and I^2:

Variable	SD	[95% Conf. Interval]		I^2	[95% Conf. Interval]	
yB	.31410107	0	.90877776	31	0	79
yC	.74977167	.39813052	1.1014128	88	68	94
yD	.72246304	0	1.8312487	8	0	35

```
Note: I^2 computed from estimated between-studies and typical within-studies
> variances
Note: CI computed on SD scale
Note: one or more CIs for I^2 were computed by dropping zero terms
```

Between-study correlations:

Variables	Correl.	[95% Conf. Interval]	
yB & yC	.93624729	-.99999763	1
yB & yD	.85586496	-.99994596	.99999967
yC & yD	.61958754	-.83474606	.99011097

```
Note: CI computed on log((1+corr)/(1-corr)) scale
```

The main contribution of between-studies heterogeneity appears to arise from the A–C contrast. Note that the between-studies correlations are very poorly estimated. In fact, the unstructured Σ matrix is barely identified in this problem. We next consider a structured Σ matrix.

4.4   Structured Σ

In sparser problems, it may be useful to assume that the heterogeneity variance is the same for each intervention contrast. This can be done by forcing Σ to be proportional to the matrix P defined below (Salanti et al. 2008):

```
. mat P = I(3) + J(3,3,1)
. mat l P

symmetric P[3,3]
    c1  c2  c3
r1   2
r2   1   2
r3   1   1   2
```

```
. mvmeta y S, bscov(prop P)
Note: using method reml
Regressing yB on
Regressing yC on
Regressing yD on
Note: using variables yB yC yD
Note: 24 observations on 3 variables
Variance-covariance matrix: proportional to P
```

  (*output omitted*)

```
Multivariate meta-analysis
Variance-covariance matrix = proportional P
Method = reml                                 Number of dimensions    =     3
Restricted log likelihood = -54.946189        Number of observations  =    24
```

	Coef.	Std. Err.	z	P>|z|	[95% Conf. Interval]	
Overall_mean						
yB	.3984951	.3310639	1.20	0.229	-.2503782	1.047368
yC	.7023595	.1990896	3.53	0.000	.312151	1.092568
yD	.8658847	.3762281	2.30	0.021	.1284912	1.603278

```
Estimated between-studies SDs and correlation matrix:
          SD        yB        yC        yD
yB   .6744175        1         .         .
yC   .6744175       .5         1         .
yD   .6744175       .5        .5         1
```

4.5   Inconsistency model

An inconsistency model allows intervention effects to differ between designs (to a greater extent than can be explained by the heterogeneity). It therefore requires a multivariate meta-regression, with particular dummy variables for design as covariates. There are many ways to parameterize this model: we choose the two-arm designs involving A as basic contrasts, and we introduce one extra effect for each two-arm design that does not include A and two extra effects for each three-arm design.

```
. tab design, gen(des)
```

design	Freq.	Percent	Cum.
ACD	1	4.17	4.17
BCD	1	4.17	8.33
AB	3	12.50	20.83
AC	14	58.33	79.17
AD	1	4.17	83.33
BC	1	4.17	87.50
BD	1	4.17	91.67
CD	2	8.33	100.00
Total	24	100.00	

```
. mvmeta y S, bscov(prop P) eq(yC: des1 des2 des6, yD: des1 des2 des7 des8)
Note: using method reml
Regressing yB on
Regressing yC on des1 des2 des6
Regressing yD on des1 des2 des7 des8
Note: using variables yB yC yD
Note: 24 observations on 3 variables
Variance-covariance matrix: proportional to P

  (output omitted )

Multivariate meta-analysis
Variance-covariance matrix = proportional P
Method = reml                              Number of dimensions    =     3
Restricted log likelihood = -45.783933     Number of observations  =    24
```

	Coef.	Std. Err.	z	P>|z|	[95% Conf. Interval]	
yB						
_cons	.3303086	.4673829	0.71	0.480	-.5857451	1.246362
yC						
des1	.3468573	.882037	0.39	0.694	-1.381903	2.075618
des2	-.3728619	1.013567	-0.37	0.713	-2.359417	1.613693
des6	-.5253268	1.004197	-0.52	0.601	-2.493516	1.442862
_cons	.7044357	.2347562	3.00	0.003	.2443219	1.164549
yD						
des1	-3.393989	1.889914	-1.80	0.073	-7.098153	.3101744
des2	-2.966854	1.926324	-1.54	0.124	-6.742379	.8086707
des7	-2.148826	1.940325	-1.11	0.268	-5.951792	1.654141
des8	-2.576181	1.80985	-1.42	0.155	-6.123422	.9710605
_cons	3.522517	1.67126	2.11	0.035	.2469077	6.798126

```
Estimated between-studies SDs and correlation matrix:
          SD        yB        yC        yD
yB   .7430402        1         .         .
yC   .7430402       .5         1         .
yD   .7430402       .5        .5         1
```

We can now test for inconsistency by jointly testing the seven inconsistency parameters:

```
. test ([yC]: des1 des2 des6) ([yD]: des1 des2 des7 des8)
 ( 1)  [yC]des1 = 0
 ( 2)  [yC]des2 = 0
 ( 3)  [yC]des6 = 0
 ( 4)  [yD]des1 = 0
 ( 5)  [yD]des2 = 0
 ( 6)  [yD]des7 = 0
 ( 7)  [yD]des8 = 0

           chi2( 7) =      5.11
         Prob > chi2 =    0.6464
```

There is no evidence of inconsistency here. It is not valid to test consistency by comparing restricted likelihoods between models, because the models' fixed parts differ—but we could instead reestimate the models by maximum likelihood and perform a likelihood-ratio test.

5   Difficulties and limitations

`mvmeta` implements a two-stage meta-analysis procedure. This is common practice, but it does involve a quadratic approximation to the log likelihood, which may perform poorly with sparse data. One-stage procedures are possible with individual participant data (Smith, Williamson, and Marson 2005). They are implemented for Stata by `metandi` for diagnostic test data (Harbord and Whiting 2009), but they are not implemented more generally.

The MM is a fast alternative to REML, but further research is required to extend it to new situations, including structured Σ and Riley's overall correlation model.

6   Acknowledgments

I was supported by MRC grant U1052.00.006. I would like to thank Stephen Kaptoge, Dan Jackson, Richard Riley, Julian Higgins, Jessica Barrett, and Antonio Gasparrini for their help and encouragement.

7   References

Biggerstaff, B. J., and D. Jackson. 2008. The exact distribution of Cochran's heterogeneity statistic in one-way random effects meta-analysis. *Statistics in Medicine* 27: 6093–6110.

DerSimonian, R., and N. Laird. 1986. Meta-analysis in clinical trials. *Controlled Clinical Trials* 7: 177–188.

Harbord, R. M., and P. Whiting. 2009. metandi: Meta-analysis of diagnostic accuracy using hierarchical logistic regression. *Stata Journal* 9: 211–229. (Reprinted in this collection on pp. 213–231.)

Higgins, J. P. T., and S. G. Thompson. 2002. Quantifying heterogeneity in a meta-analysis. *Statistics in Medicine* 21: 1539–1558.

Higgins, J. P. T., S. G. Thompson, and D. J. Spiegelhalter. 2009. A re-evaluation of random-effects meta-analysis. *Journal of the Royal Statistical Society, Series A* 172: 137–159.

Jackson, D., R. D. Riley, and I. R. White. 2011. Multivariate meta-analysis: Potential and promise. *Statistics in Medicine* 30: 2481–2498.

Jackson, D., I. R. White, and S. G. Thompson. 2010. Extending DerSimonian and Laird's methodology to perform multivariate random effects meta-analyses. *Statistics in Medicine* 29: 1282–1297.

Lu, G., and A. E. Ades. 2004. Combination of direct and indirect evidence in mixed treatment comparisons. *Statistics in Medicine* 23: 3105–3124.

Riley, R. D. 2009. Multivariate meta-analysis: The effect of ignoring within-study correlation. *Journal of the Royal Statistical Society, Series A* 172: 789–811.

Riley, R. D., K. R. Abrams, A. J. Sutton, P. C. Lambert, and J. R. Thompson. 2007. Bivariate random-effects meta-analysis and the estimation of between-study correlation. *BMC Medical Research Methodology* 7: 3.

Riley, R. D., J. R. Thompson, and K. R. Abrams. 2008. An alternative model for bivariate random-effects meta-analysis when the within-study correlations are unknown. *Biostatistics* 9: 172–186.

Salanti, G., J. P. T. Higgins, A. E. Ades, and J. P. A. Ioannidis. 2008. Evaluation of networks of randomized trials. *Statistical Methods in Medical Research* 17: 279–301.

Smith, C. T., P. R. Williamson, and A. G. Marson. 2005. Investigating heterogeneity in an individual patient data meta-analysis of time to event outcomes. *Statistics in Medicine* 24: 1307–1319.

Sterne, J. A. C., ed. 2009. *Meta-Analysis in Stata: An Updated Collection from the Stata Journal.* College Station, TX: Stata Press.

van Houwelingen, H. C., L. R. Arends, and T. Stijnen. 2002. Advanced methods in meta-analysis: Multivariate approach and meta-regression. *Statistics in Medicine* 21: 589–624.

Verbeke, G., and G. Molenberghs. 2000. *Linear Mixed Models for Longitudinal Data.* New York: Springer.

White, I. R. 2009. Multivariate random-effects meta-analysis. *Stata Journal* 9: 40–56. (Reprinted in this collection on pp. 232–248.)

Part 5

Individual patient data meta-analysis: ipdforest and ipdmetan

Some systematic reviews assemble data on each participant in each study, in which case an individual participant data (IPD) meta-analysis is possible (Stewart 1995; Riley, Lambert, and Abo-Zaid 2010). IPD meta-analysis can be performed for studies reporting either numerical or binary outcomes, respectively using random-effects linear regression fitted with `mixed` or logistic regression fitted with `meqrlogit`. This is automated in the `ipdforest` command (Kontopantelis and Reeves 2013). The command can also produce a forest plot of the IPD meta-analysis.

Two-stage IPD meta-analysis methods are implemented in the `ipdmetan` command (Fisher 2015). The command, which is very flexible, can incorporate covariates and interactions and can also incorporate studies reporting aggregate data into the IPD meta-analysis. `ipdmetan` also has a `forestplot` option that can change the appearance of certain aspects of complex forest plots.

1   References

Fisher, D. J. 2015. Two-stage individual participant data meta-analysis and generalized forest plots. *Stata Journal* 15: 369–396. (Reprinted in this collection on pp. 280–307.)

Kontopantelis, E., and D. Reeves. 2013. A short guide and a forest plot command (ipdforest) for one-stage meta-analysis. *Stata Journal* 13: 574–587. (Reprinted in this collection on pp. 321–354.)

Riley, R. D., P. C. Lambert, and G. Abo-Zaid. 2010. Meta-analysis of individual participant data: Rationale, conduct, and reporting. *British Medical Journal* 340: c221.

Stewart, L. A. 1995. Practical methodology of meta-analyses (overviews) using updated individual patient data. *Statistics in Medicine* 14: 2057–2079.

The Stata Journal (2013)
13, Number 3, st0309, pp. 574–587

A short guide and a forest plot command (ipdforest) for one-stage meta-analysis

Evangelos Kontopantelis
NIHR School for Primary Care Research
Institute of Population Health
University of Manchester
Manchester, UK
e.kontopantelis@manchester.ac.uk

David Reeves
Institute of Population Health
University of Manchester
Manchester, UK
david.reeves@manchester.ac.uk

Abstract. In this article, we describe a new individual patient data meta-analysis postestimation command, `ipdforest`. The command produces a forest plot following a one-stage meta-analysis with `xtmixed` or `xtmelogit`. (These commands have been renamed in Stata 13 to `mixed` and `meqrlogit`, respectively; `ipdforest` is currently not compatible with the new names.) The overall effect is obtained from the preceding mixed-effects regression and the study effects from linear or logistic regressions on each study, which are executed within `ipdforest`. Individual patient data meta-analysis models with Stata are discussed.

Keywords: st0309, ipdforest, meta-analysis, forest plot, individual patient data, IPD, one-stage

1 Introduction

Meta-analysis, the methodology that allows results from independent studies to be combined, is usually a two-stage process. First, the relevant summary effect statistics are extracted from published articles on the included studies. These are then combined into an overall effect estimate using a suitable meta-analysis model (Harris et al. 2008; Kontopantelis and Reeves 2010). However, problems often arise when an article does not report all the statistical information required as input for the meta-analysis (for example, it fails to provide a variance estimate for the outcome measure); reports a statistic other than the effect size (such as a t-value or p-value) that needs to be transformed with a loss of precision; or provides a sample too clinically heterogeneous for the study to be included in the meta-analysis (Kontopantelis and Reeves 2009).

When individual patient data (IPD) from each study are available, meta-analysts can avoid these problems when estimating study effects; outcomes can be easily standardized, while clinical heterogeneity can be addressed, at least partially, with subgroup analyses and patient-level covariate control. Furthermore, when IPD data are available, meta-analysts can use a mixed-effects regression model to combine information across studies in a single stage. This is recognized as the best approach for performing an IPD meta-analysis, with the two-stage method being at best equivalent in certain scenarios (Mathew and Nordström 2010).

Despite these advantages of the one-stage approach, one obvious advantage of two-stage meta-analysis is the ability to convey information graphically through a forest plot. Because study effects have been calculated or extracted in the first stage of the process, they and their respective confidence intervals can be used to demonstrate the relative strength of the intervention in each study and across all the studies. Forest plots are informative, easy to follow, and particularly useful for readers with little or no experience in meta-analysis methods. It is not surprising, then, that they have become a key feature of meta-analysis and are always presented when two-stage meta-analyses are performed.

However, under a one-stage meta-analysis model, only the overall effect is calculated, not individual study effects; thus creating a forest plot is not straightforward. A search by the authors failed to identify one-stage meta-analysis forest-plot modules in any general or meta-analysis specialist statistical package. We attempt to address this gap in Stata with the `ipdforest` command.

This article is divided into two sections. In the first section, we describe IPD meta-analysis models and their implementation in Stata with available mixed-effects models. In the second section, we describe the `ipdforest` command in detail and provide an example.

2 Individual patient data meta-analysis

A description of IPD meta-analysis methods for continuous and binary outcomes has been provided by Higgins et al. (2003) and Turner et al. (2000), respectively. Although we will only explore a representative selection of linear random-effects models in Stata (using the `xtmixed` command), application to the logistic case using `xtmelogit` should be straightforward. Let us assume IPD from a group of studies. For each trial, we have the exposure variable, which is continuous or binary (for example, control or intervention group membership), and baseline and follow-up data for the continuous outcome and covariates. We will also assume that both the outcome measure and any covariates have been measured in the same way across studies and that, therefore, standardization is not required. In the models that follow, in general, we denote a fixed effect by γ and a random effect by β.

Possibly the simplest approach is to assume that there is a common intercept across studies and that baseline is a fixed effect but to allow the treatment effect to vary at random across studies. Thus we have

$$
\begin{aligned}
\acute{Y}_{ij} &= \gamma_0 + \beta_{1j}\mathrm{group}_{ij} + \gamma_2 Y_{ij} + \epsilon_{ij} \\
\beta_{1j} &= \gamma_1 + u_{1j}
\end{aligned}
\tag{1a}
$$

and

$$
\begin{aligned}
\epsilon_{ij} &\sim N(0, \sigma_j^2) \\
u_{1j} &\sim N(0, \tau_1{}^2)
\end{aligned}
\tag{1b}
$$

where i is the patient; j is the study; $\acute{Y}_{ij}$ is the outcome for patient i in study j; γ_0 is the fixed common intercept; β_{1j} is the random treatment effect for study j; γ_1 is the mean treatment effect; group_{ij} is the exposure for patient i in study j; γ_2 is the fixed baseline effect; Y_{ij} is the baseline score for patient i in study j; u_{1j} is the random treatment effect for study j (shifting the regression line up or down by study); $\tau_1{}^2$ is the between-study variance; ϵ_{ij} is the error term for patient i in study j; and σ_j^2 is the within-study variance for study j.

However, the common intercept and fixed baseline assumptions are difficult to justify, and such a model should be approached with caution—if at all. A more accepted model allows for different fixed intercepts and fixed baseline effects for each study:

$$
\begin{aligned}
\acute{Y}_{ij} &= \gamma_{0j} + \beta_{1j}\text{group}_{ij} + \gamma_{2j}Y_{ij} + \epsilon_{ij} \\
\beta_{1j} &= \gamma_1 + u_{1j}
\end{aligned}
\tag{2}
$$

where γ_{0j} is the fixed intercept for study j and γ_{2j} is the fixed baseline effect for study j.

Another possibility, although contentious (Whitehead 2002), is to assume that study intercepts are random, as in a multicenter study; for example,

$$
\begin{aligned}
\acute{Y}_{ij} &= \beta_{0j} + \beta_{1j}\text{group}_{ij} + \gamma_{2j}Y_{ij} + \epsilon_{ij} \\
\beta_{0j} &= \gamma_0 + u_{0j} \\
\beta_{1j} &= \gamma_1 + u_{1j}
\end{aligned}
\tag{3a}
$$

In this case, it is probably wiser to assume a nonzero correlation ρ between the random effects:

$$
\begin{aligned}
\epsilon_{ij} &\sim N(0, \sigma_j^2) \\
u_{0j} &\sim N(0, \tau_0^2) \\
u_{1j} &\sim N(0, \tau_1^2) \\
\text{cov}(u_{0j}, u_{1j}) &= \rho\tau_0\tau_1
\end{aligned}
\tag{3b}
$$

The baseline could also have been modeled as a random effect, and we could have allowed for nonzero correlations between the three random effects, thus complicating (3) further:

$$
\begin{aligned}
\acute{Y}_{ij} &= \beta_{0j} + \beta_{1j}\text{group}_{ij} + \beta_{2j}Y_{ij} + \epsilon_{ij} \\
\beta_{0j} &= \gamma_0 + u_{0j} \\
\beta_{1j} &= \gamma_1 + u_{1j} \\
\beta_{2j} &= \gamma_2 + u_{2j}
\end{aligned}
\tag{4a}
$$

with effects

$$\epsilon_{ij} \sim N(0, \sigma_j^2)$$
$$u_{0j} \sim N(0, \tau_0^2)$$
$$u_{1j} \sim N(0, \tau_1^2)$$
$$u_{2j} \sim N(0, \tau_2^2) \tag{4b}$$
$$\mathrm{cov}(u_{0j}, u_{1j}) = \rho_1 \tau_0 \tau_1$$
$$\mathrm{cov}(u_{0j}, u_{2j}) = \rho_2 \tau_0 \tau_2$$
$$\mathrm{cov}(u_{1j}, u_{2j}) = \rho_3 \tau_1 \tau_2$$

In some cases, the focus might be on interactions. For example, if we assume a continuous and standardized variable X, we can expand (2) to include fixed effects, in this instance, for both X and its interaction with the treatment:

$$\acute{Y}_{ij} = \gamma_{0j} + \beta_{1j}\mathrm{group}_{ij} + \gamma_{2j}Y_{ij} + \gamma_3 X_{ij} + \gamma_4\mathrm{group}_{ij}X_{ij} + \epsilon_{ij}$$
$$\beta_{1j} = \gamma_1 + u_{1j} \tag{5}$$

If we consider `Yfin` and `Ybas` as representing the outcome and baseline, respectively, the exposure variable `group`, and the study identifier `studyid` for four studies, we can implement the models described above by using `xtmixed`.

Model (1): Fixed common intercept; random treatment effect; fixed effect for baseline.

```
. xtmixed Yfin i.group Ybas || studyid:group, nocons
```

The `nocons` option suppresses estimation of the intercept as a random effect.

Model (2): Fixed study-specific intercepts; random treatment effect; fixed study-specific effects for baseline (where `Ybas'i'=Ybas` if `studyid='i'` and equals 0 otherwise).

```
. xtmixed Yfin i.group i.studyid Ybas1 Ybas2 Ybas3 Ybas4 || studyid:group,
> nocons
```

Model (3): Random study intercept; random treatment effect; fixed study-specific effects for baseline.

```
. xtmixed Yfin i.group Ybas1 Ybas2 Ybas3 Ybas4 || studyid:group, cov(uns)
```

Model (4): Random study intercept; random treatment effect; random effect for baseline.

```
. xtmixed Yfin i.group Ybas || studyid:group Ybas, cov(uns)
```

In general, a covariate (or an interaction term) can be modeled as a fixed or random effect, but in the latter case, the complexity of the model increases and nonconvergence issues are more likely to be encountered. If we also consider patient covariate `age` and its interaction with the treatment effect, then (5) will be

```
. xtmixed Yfin i.group i.studyid Ybas1 Ybas2 Ybas3 Ybas4 age i.group#c.age
> || studyid: group, nocons
```

Or alternatively, `age` can be modeled as a random effect:

```
. xtmixed Yfin i.group i.studyid Ybas1 Ybas2 Ybas3 Ybas4 age i.group#c.age
> || studyid: group age, nocons
```

3   The ipdforest command

3.1   Syntax

ipdforest *varname* [, re(*varlist*) fe(*varlist*) fets(*namelist*) ia(*varname*) auto
    label(*varlist*) or gsavedir(*string*) gsavename(*string*) eps gph
    export(*string*)]

where *varname* is the exposure variable, continuous or binary (for example, intervention or control).

3.2   Options

re(*varlist*) specifies covariates to be included as random factors. For each covariate specified, a different regression coefficient is estimated for each study.

fe(*varlist*) specifies covariates to be included as fixed factors. For each covariate specified, the respective coefficient in the study-specific regressions is fixed to the value returned by the multilevel regression.

fets(*namelist*) specifies covariates to be included as study-specific fixed factors (that is, by using the estimated study fixed effects from the main regression in all individual study regressions). Only baseline scores and study identifiers can be included. For each covariate specified, the respective coefficient in the study-specific regressions is fixed to the value returned by the multilevel regression for the specific study. For study-specific intercepts, the study identifier (not in factor-variable format, for example, `studyid`) or the *stub* of the dummy variables (for example, `studyid_` when dummy study identifiers are `studyid_1 studyid_3`, etc.) would be included. For study-specific baseline scores, only the *stub* of the dummy variables is accepted (for example, `dept0s_` when dummy study baseline scores are `dept0s_1 dept0s_3`, etc.).

ia(*varname*) specifies covariates for which the interaction with the exposure variable will be calculated and displayed. The covariate should also be specified as a fixed, random, or study-specific fixed effect. If binary, the command will provide two sets of results, one for each group. If categorical, it will provide as many sets of results as there are categories. If continuous, it will provide one set of results for the main effect and one for the interaction. Although the command will allow a variable to be interacted with the exposure variable as a fixed or study-specific fixed effect, the variable necessarily will be included as a random effect in the individual regressions (it will not run a regression with the interaction term only; the main effects must be included as well). Therefore, although the overall effect will differ between a model with a fixed-effects interacted variable and a random-effects one, the individual study effects will be identical across the two approaches.

auto allows ipdforest to automatically detect the specification of the preceding model. This option cannot be issued along with options re(), fe(), fets(), or ia(). The auto option will work in most situations, but it comes with certain limitations. It uses the returned command string of the preceding command, which is effectively constrained to 244 characters; therefore, the auto option will return an error if ipdforest follows a very wide regression model—in such a situation, only the manual specification can be used. In addition, the variable names used in the preceding model must follow certain rules: 1) fixed-effects covariates (manually with option fe()) must not contain underscores; 2) for study-specific intercepts (manually with option fets()), factor-variable format is allowed or a *varlist* (for example, cons_2–cons_16), but each variable must contain a single underscore followed by the study number (not necessarily sequential); and 3) for study-specific baseline scores (manually with option fets()), each variable must contain a single underscore followed by the study number (again, not necessarily sequential). There are no restrictions for random-effects covariates (manually with option re()). For interactions (manually with option ia()), the factor-variable notation should be preferred (for example, i.group#c.age) and, alternatively, the older xi: notation. Interactions expanded to dummy variables cannot be identified with the auto option, and only the manual specification should be used in this case. Variables whose names start with an _I and contain a capital X will be assumed to be expanded interaction terms, and if detected in the last model, ipdforest will terminate with a syntax error.

label(*varlist*) specifies labels for the studies. Up to two variables can be specified and converted to strings. If two variables are specified, they will be separated by a comma. Usually, the author names and the year of study are selected as labels. If label() is not specified, the command automatically uses the value labels of the numeric cluster variable, if any, to label the forest plot. Either way, the final string is truncated to 30 characters.

or reports odds ratios instead of coefficients. It can only be used following the execution of xtmelogit.

gsavedir(*string*) specifies the directory in which to save the graph, if different from the active directory.

gsavename(*string*) specifies the optional name prefix for the graph. Graphs are saved as *gsavename_graphname*.gph or *gsavename_graphname*.eps, where *graphname* includes a description of the summary effect (for example, main_group for the main effect if group is the exposure variable).

eps saves the graph in .eps format instead of the default .gph.

gph saves the graph in .gph format. gph is the default. Use it to save in both formats: including only the eps option will save the graph in .eps format only.

export(*string*) exports the study identifiers, weights, effects, and standard errors in a Stata dataset (named after *string*). It is provided for users who wish to use other commands or software to draw the forest plots.

3.3 Stored results

ipdforest stores the following in r():

Scalars

r(Isq)	heterogeneity measure I^2	r(eff1pe_ov)	overall effect estimate
r(Hsq)	heterogeneity measure H_M^2	r(eff1se_ov)	standard error of the overall
r(tausq)	$\hat{\tau}^2$, between-study variance		effect
	estimate	r(eff1pe_st$i$)	effect estimate for study i
r(tausqlo)	$\hat{\tau}^2$, lower 95% confidence	r(eff1se_st$i$)	standard error of the effect
	interval		for study i
r(tausqup)	$\hat{\tau}^2$, upper 95% confidence		
	interval		

If an interaction with a continuous variable is included in the model, it also stores the following:

Scalars

r(eff2pe_ov)	overall interaction effect	r(eff2pe_st$i$)	interaction effect estimate for
	estimate		study i
r(eff2se_ov)	standard error of the overall	r(eff2se_st$i$)	interaction effect standard
	interaction effect		error for study i

If the interaction variable is binary, the first set of results corresponds to the effects for the first category of the binary (for example, sex = 0) and the second set for the second category (for example, sex = 1). If the variable is categorical, the command returns as many sets of effect results as there are categories (with each set corresponding to one category). Estimation results from xtmixed or xtmelogit in e() are restored after the execution of ipdforest.

3.4 Methods

The ipdforest command is issued following a random-effects IPD meta-analysis conducted using a linear (xtmixed) or logistic (xtmelogit) two-level regression with patients nested within studies. The command provides a meta-analysis summary table and a forest plot. Study effects are calculated within `ipdforest`, while the overall effect and variance estimates are extracted from the preceding regression. The default estimation methods for `xtmixed` and `xtmelogit` are restricted maximum likelihood and maximum likelihood, respectively. A description of these methods is beyond the scope of this article.

`ipdforest` estimates individual study effects and their standard errors by using one-level linear or logistic regression analyses. Following `xtmixed`, `regress` is used, and following `xtmelogit`, `logit` is used for each study in the meta-analysis. The `ipdforest` command controls these regressions for fixed- or random-effects covariates that were specified in the preceding two-level regression. The user has full control over the covariates to be included in the `ipdforest` command, including their specification as fixed or random effects. However, we strongly recommend using the same specification as in the preceding `xtmixed` or `xtmelogit` command because the reported overall effect and its confidence interval is taken from that model.

In the estimation of individual study effects, `ipdforest` controls for a random-effects covariate (that is, allowing the regression coefficient to vary by study) by including the covariate as an independent variable in each regression. Control for a fixed-effects covariate (where the regression coefficient is assumed constant across studies and is given by the coefficient estimated under `xtmixed` or `xtmelogit`) is a little more complex. Because it is not possible to specify a fixed value for a regression coefficient under `regress`, the continuous outcome variable is adjusted by subtracting the contribution of the fixed covariates to its values prior to analysis. For a binary outcome, the equivalent is achieved using the `offset` option in `logit`. Patient weights are uniform; therefore, each study's weight is the ratio of its participants over the total number of participants across all studies.

Between-study variability in the treatment effect, known as heterogeneity, arises from differences in study design, quality, outcomes, or populations and needs to be accounted for in the meta-analysis model when present. Heterogeneity is usually reported in the form of measures or tests that compare the between- and within-study variance estimates. For continuous outcomes, `ipdforest` reports two heterogeneity measures, I^2 and H_M^2, based on the `xtmixed` output. I^2 values of 25%, 50%, and 75% correspond to low, moderate, and high heterogeneity, respectively (Higgins et al. 2003), while H_M^2 takes values in the $[0, +\infty)$ range with 0 indicating perfect homogeneity (Mittlböck and Heinzl 2006). We have not attempted to calculate an IPD version of Cochran's Q, the orthodox χ_{k-1}^2 homogeneity test, considering its poor performance when the number of studies k is small (Hardy and Thompson 1998). For binary outcomes, an estimate of the within-study variance is not reported under `xtmelogit`, and hence, heterogeneity measures cannot be computed. The between-study variance estimate $\hat{\tau}^2$ and its confidence interval are reported under both models.

Fixed-effects meta-analysis models are widely used when heterogeneity is very low or 0. However, a more conservative approach is to take account of even low levels of between-study variability by adopting a random-effects model (Hunter and Schmidt 2000). When between-study variance is estimated to be close to 0, results with the two approaches converge. Therefore, although `ipdforest` is a postestimation command for random-effects IPD meta-analysis, output is close to that for a fixed-effects model when $\widehat{\tau}^2 \approx 0$.

3.5   Example

As an example, we apply the `ipdforest` command to a dataset of four depression intervention studies. Data were provided by the authors of the studies, and we had complete information in terms of age, gender, exposure (control and intervention group membership), continuous outcome baseline, and endpoint values for 518 patients. Because the findings of the IPD meta-analysis had not been published when this article was being prepared, we used fake author names and generated random continuous and binary outcome variables for the purposes of this example while keeping the covariates at their actual values. We introduced correlation between baseline and endpoint scores and between-study variability, although the exact specification of the data generation is unimportant.

Using the semiartificial dataset, we perform a logistic IPD meta-analysis, followed by the `ipdforest` command.

```
. use ipdforest_example

. describe

Contains data from ipdforest_example.dta
  obs:           518
  vars:           17                              6 Feb 2012 11:35
  size:        20,202
```

variable name	storage type	display format	value label	variable label
studyid	byte	%22.0g	stid	Study identifier
patid	int	%8.0g		Patient identifier
group	byte	%20.0g	grplbl	Intervention/control group
sex	byte	%10.0g	sexlbl	Gender
age	float	%10.0g		Age in years
depB	byte	%9.0g		Binary outcome, endpoint
depBbas	byte	%9.0g		Binary outcome, baseline
depBbas1	byte	%9.0g		Bin outcome baseline, trial 1
depBbas2	byte	%9.0g		Bin outcome baseline, trial 2
depBbas5	byte	%9.0g		Bin outcome baseline, trial 5
depBbas9	byte	%9.0g		Bin outcome baseline, trial 9
depC	float	%9.0g		Continuous outcome, endpoint
depCbas	float	%9.0g		Continuous outcome, baseline
depCbas1	float	%9.0g		Cont outcome baseline, trial 1
depCbas2	float	%9.0g		Cont outcome baseline, trial 2
depCbas5	float	%9.0g		Cont outcome baseline, trial 5
depCbas9	float	%9.0g		Cont outcome baseline, trial 9

```
Sorted by:  studyid  patid
```

We generate a centered age variable, interacted with the exposure variable in a mixed-effects logistic regression model. The model includes fixed study-specific intercepts and fixed study-specific effects for baseline and random treatment and age effects. The ipdforest command follows the regression model, requesting outcomes for both the main effect and the interaction.

```
. quietly summarize age

. quietly generate agec = age-r(mean)

. xtmelogit depB group agec sex i.studyid depBbas1 depBbas2 depBbas5 depBbas9
> i.group#c.agec || studyid:group agec, var nocons or

Refining starting values:
Iteration 0:   log likelihood = -347.40378  (not concave)
Iteration 1:   log likelihood = -336.07882  (not concave)
Iteration 2:   log likelihood = -329.28268

Performing gradient-based optimization:

Iteration 0:   log likelihood = -329.28268  (not concave)
Iteration 1:   log likelihood = -326.79754
Iteration 2:   log likelihood =  -326.5689
Iteration 3:   log likelihood = -326.55747
Iteration 4:   log likelihood = -326.55747

Mixed-effects logistic regression          Number of obs      =        518
Group variable: studyid                    Number of groups   =          4

                                           Obs per group: min =         42
                                                          avg =      129.5
                                                          max =        214

Integration points =    7                  Wald chi2(11)      =      42.06
Log likelihood = -326.55747                Prob > chi2        =     0.0000
```

depB	Odds Ratio	Std. Err.	z	P>|z|	[95% Conf. Interval]	
group	1.840804	.3666167	3.06	0.002	1.245894	2.71978
agec	.9867902	.0119059	-1.10	0.270	.9637288	1.010403
sex	.7117592	.1540753	-1.57	0.116	.4656639	1.087912
studyid						
2	1.050007	.5725515	0.09	0.929	.3606168	3.057302
5	.8014552	.5894511	-0.30	0.763	.1896011	3.387799
9	1.281413	.6886055	0.46	0.644	.4469621	3.673734
depBbas1	3.152909	1.49528	2.42	0.015	1.244587	7.987251
depBbas2	4.480302	1.863908	3.60	0.000	1.982385	10.12573
depBbas5	2.387336	1.722993	1.21	0.228	.5802064	9.823007
depBbas9	1.881203	.7086506	1.68	0.093	.8990571	3.936261
group#c.agec						
1	1.011776	.0163748	0.72	0.469	.9801858	1.044385
_cons	.5533714	.2398341	-1.37	0.172	.2366473	1.293993

Random-effects Parameters	Estimate	Std. Err.	[95% Conf. Interval]	
studyid: Independent				
var(group)	6.15e-21	2.03e-11	0	.
var(agec)	6.03e-18	4.41e-11	0	.

LR test vs. logistic regression: chi2(2) =   0.00   Prob > chi2 = 1.0000
Note: LR test is conservative and provided only for reference.

. ipdforest group, fe(sex) re(agec) ia(agec) or

One-stage meta-analysis results using xtmelogit (ML method) and ipdforest
Main effect (group)

Study	Effect	[95% Conf. Interval]		% Weight
Hart 2005	2.118	0.942	4.765	19.88
Richards 2004	2.722	1.336	5.545	30.69
Silva 2008	2.690	0.748	9.676	8.11
Kompany 2009	1.895	0.969	3.707	41.31
Overall effect	1.841	1.246	2.720	100.00

One-stage meta-analysis results using xtmelogit (ML method) and ipdforest
Interaction effect (group x agec)

Study	Effect	[95% Conf. Interval]		% Weight
Hart 2005	0.972	0.901	1.049	19.88
Richards 2004	0.995	0.937	1.055	30.69
Silva 2008	0.987	0.888	1.098	8.11
Kompany 2009	1.077	1.015	1.144	41.31
Overall effect	1.012	0.980	1.044	100.00

Heterogeneity Measures

	value	[95% Conf. Interval]	
I (%)	.		
H	.		
tau est	0.000	0.000	.

Maximum likelihood converged successfully in this example, and the between-study variance estimate $\hat{\tau}^2$ was practically 0. Note that the intercept for the reference study (studyid = 1) was estimated in _cons. The reported coefficients under studyid are the differences in intercept compared with the first study. I^2 and H_M^2 could not be estimated because residual variability is not reported under xtmelogit. The overall treatment effect was significant at the 95% level, but the overall effect for the interaction of treatment and age was not. The forest plots created by ipdforest are displayed in figures 1 and 2.

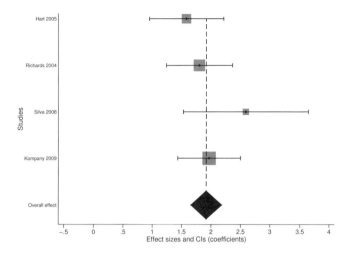

Figure 1. Main-effect IPD forest plot reporting odds ratios

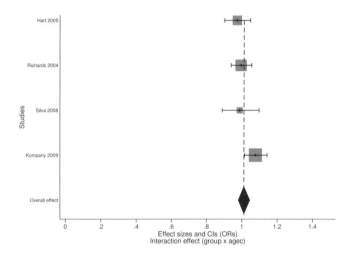

Figure 2. Interaction-effect IPD forest plot reporting odds ratios

4 Discussion

The aim of this article was to provide a practical guide for conducting one-stage IPD meta-analysis and to present ipdforest, a new forest-plot command. ipdforest aims to help meta-analysts better communicate their results through the familiar and distinctive forest plot—a graphical output not previously available in one-stage IPD meta-analysis software routines.

Although only binary or continuous exposure variables can be modeled, categorical exposures can also be investigated with the use of dummy variables and a focus on the comparison of interest through one of these. In addition, `ipdforest` is fully compatible with the estimates produced by the multiple-imputation estimation command `mi estimate: xtmixed` or `mi estimate: xtmelogit`.

Note that these commands were renamed in Stata 13: `xtmixed` to `mixed` and `xtmelogit` to `meqrlogit`. `ipdforest` is not yet compatible with the new commands, but users of version 13 can still use the older commands before calling `ipdforest`.

5   Acknowledgments

We would like to thank Isabel Canette, senior statistician at StataCorp, for her help with advanced aspects of `xtmixed` and `xtmelogit` and the anonymous reviewer whose comments and suggestions improved the command significantly. Evan Kontopantelis is on an NIHR School for Primary Care fellowship.

6   References

Hardy, R. J., and S. G. Thompson. 1998. Detecting and describing heterogeneity in meta-analysis. *Statistics in Medicine* 17: 841–856.

Harris, R. J., M. J. Bradburn, J. J. Deeks, R. M. Harbord, D. G. Altman, and J. A. C. Sterne. 2008. metan: Fixed- and random-effects meta-analysis. *Stata Journal* 8: 3–28. (Reprinted in this collection on pp. 29–54.)

Higgins, J. P. T., S. G. Thompson, J. J. Deeks, and D. G. Altman. 2003. Measuring inconsistency in meta-analyses. *British Medical Journal* 327: 557–560.

Hunter, J. E., and F. L. Schmidt. 2000. Fixed effects vs. random effects meta-analysis models: Implications for cumulative research knowledge. *International Journal of Selection and Assessment* 8: 275–292.

Kontopantelis, E., and D. Reeves. 2009. MetaEasy: A meta-analysis add-in for Microsoft Excel. *Journal of Statistical Software* 30: 1–25.

———. 2010. metaan: Random-effects meta-analysis. *Stata Journal* 10: 395–407. (Reprinted in this collection on pp. 55–67.)

Mathew, T., and K. Nordström. 2010. Comparison of one-step and two-step meta-analysis models using individual patient data. *Biometrical Journal* 52: 271–287.

Mittlböck, M., and H. Heinzl. 2006. A simulation study comparing properties of heterogeneity measures in meta-analyses. *Statistics in Medicine* 25: 4321–4333.

Turner, R. M., R. Z. Omar, M. Yang, H. Goldstein, and S. G. Thompson. 2000. A multilevel model framework for meta-analysis of clinical trials with binary outcomes. *Statistics in Medicine* 19: 3417–3432.

Whitehead, A. 2002. *Meta-Analysis of Controlled Clinical Trials.* Chichester, UK: Wiley.

The Stata Journal (2015)
15, Number 2, st0384, pp. 369–396

Two-stage individual participant data meta-analysis and generalized forest plots

David J. Fisher
MRC Clinical Trials Unit at University College London
London, UK
d.fisher@ucl.ac.uk

Abstract. In this article, I describe a command, `ipdmetan`, that facilitates two-stage individual participant data meta-analysis of any measure of effect and its standard error by fitting a specified model to data from each study. The command can estimate random effects and heterogeneity statistics and include additional covariates and interactions. If individual participant data are available for certain studies and aggregate data for others, `ipdmetan` allows them to be combined in one analysis. This command can produce detailed and flexible forest plots, including ones outside the context of formal meta-analysis.

Keywords: st0384, ipdmetan, forestplot, admetan, ipdover, meta-analysis, forest plot, individual participant data, survival, two-stage, time-to-event, prefix command, subgroup

1   Introduction

Meta-analysis is a statistical technique for combining results from multiple independent studies, usually to estimate one overall effect with increased precision (Deeks, Altman, and Bradburn 2001). Stata has various tools for performing meta-analysis, many of which are described in a previous collection of articles (Sterne 2009). These have focused on the pooling of study results in the form of aggregate (published summary) data (AD)—traditionally the most common form of meta-analysis. However, individual participant (or patient) data (IPD) are increasingly used and permit a greater range of possible analyses while minimizing bias from heterogeneity of analysis protocols between studies (for example, patient exclusions or the precise model used) and maximizing data completeness and follow-up (Stewart and Tierney 2002).

Both "one-stage" and "two-stage" approaches exist for analyzing IPD; both are commonly used and are consistent estimators of the pooled effect. Although a one-stage estimator is always at least as efficient as its equivalent two-stage estimator for linear models where the full covariance matrix is directly observed (Mathew and Nordström 2010), it remains unclear whether this holds for nonlinear outcomes (for example, time-to-event) and situations where the covariate matrix is only an estimate. Moreover, it is unclear whether such differences are great enough to be statistically or clinically relevant. For instance, simulation studies (Tudur Smith and Williamson 2007; Bowden et al. 2011) and case studies (Stewart et al. 2012) have shown little or no differences between one-stage and two-stage models under specific conditions. Furthermore, for situations such as random-effects analysis of time-to-event data, one-stage solutions are just becoming available (Bowden et al. 2011; Crowther, Look, and Riley 2014). Partly for this reason, the most common approach used with time-to-event data is the two-stage approach (Simmonds et al. 2005).

In practice, the two-stage approach is appealing in its simplicity and because it can call on extensive AD meta-analysis literature, including the *Cochrane Handbook* (Higgins and Green 2011). Two-stage models can offer quick and unbiased answers, especially for reviews of randomized controlled trials (RCTs), where studies are wholly independent and confounding is minimized. Also, for some situations, a one-stage model would require innovative techniques. Furthermore, studies without IPD available may be included as a sensitivity analysis (Riley et al. 2008) by simply including AD from publications and by using secondary requests for additional data (for example, an effect estimate adjusted for extra covariates) after an initial request for IPD has failed. Nevertheless, two-stage IPD analysis in Stata has not previously been straightforward. Now, `ipdmetan` provides a comprehensive two-stage package.

One motivation for collecting IPD is the possibility of assessing differential responses to treatment—in other words, of analyzing treatment-covariate interactions. In previous work (Fisher et al. 2011), we discussed the fundamentally different approaches to the analysis of such data and concluded that a two-stage approach is again suitable in many situations, especially compared with older methods. In future work, we hope to expand on those suggestions and discuss how trialists might facilitate meta-analysis of interactions, even if they are unwilling or unable to provide IPD, by presenting subgroup analyses clearly and consistently and thus allowing aggregate estimates to be calculated.

The standard method of presenting meta-analysis results is a forest plot, in which the results from each study are plotted together with the overall result. Previous Stata AD meta-analysis commands produce forest plots as standard, and such plots transfer naturally to two-stage IPD analysis. Results from one-stage models may also be plotted in this way, but with greater complications. The `ipdforest` command (Kontopantelis and Reeves 2013) is one approach to this; another might be to present two-stage study results alongside a one-stage pooled result. Forest plots for `ipdmetan` are created using code based heavily on that written for the `metan` command (Harris et al. 2008) but modified and updated for increased flexibility. They are produced by the program `forestplot`, which is called by `ipdmetan` but can also be run by itself, giving the user more control.

Finally, the `ipdmetan` package also includes the following "wrapper" programs, which I do not describe in detail here: `admetan`, an AD routine similar to `metan`; `petometan`, a routine for performing IPD meta-analysis of time-to-event data using the Peto log-rank approach (Yusuf et al. 1985); and `ipdover`, a routine that creates forest plots of subgroups within one trial.

2   Two-stage IPD meta-analysis

2.1   Basic principles

Two-stage meta-analysis methodology involves first fitting the desired model to the data from each study in turn. This gives study-effect estimates and variances, which are denoted here by θ_i and σ_i^2, respectively. These are assumed to be observed without error in the second (pooling) stage. Let θ denote the true pooled effect to be estimated by $\widehat{\theta}$. A suitable reference for most of what follows is Deeks, Altman, and Bradburn (2001), with extra references given where appropriate.

`ipdmetan` uses inverse-variance weighting, with weights denoted by $w_i = 1/\sigma_i^2$. The Peto (or log-rank) method for time-to-event outcomes is equivalent to an inverse-variance analysis, and it can be performed using `ipdmetan` as described in section 4.5. The built-in commands `cc` or `tabodds` can be used to perform IPD meta-analysis with Mantel–Haenszel weighting for binary outcomes.

The fixed-effects inverse-variance weighted pooled effect and its variance are calculated as

$$\widehat{\theta} = \frac{\sum\limits_i w_i \theta_i}{\sum\limits_i w_i}, \ \mathrm{Var}\left(\widehat{\theta}\right) = \frac{1}{\sum\limits_i w_i}$$

The standard Cochran heterogeneity statistic Q is defined as

$$Q = \sum_i w_i \left(\theta_i - \widehat{\theta}\right)^2$$

and is distributed as chi-squared with $k - 1$ degrees of freedom (d.f.), where k is the number of included studies.

Given an estimate of between-study variance τ^2, the random-effects inverse-variance weighted pooled effect (and variance) is calculated by replacing the fixed-effects weights w_i with random-effects weights $w_i^* = 1/(\sigma_i^2 + \tau^2)$, which incorporate the between-study variance τ^2. The most common estimator of τ^2 is the DerSimonian–Laird (DL) estimator τ_{DL}^2 (see section 2.2).

In addition to Q and τ^2, two further heterogeneity measures are commonly used: I^2 and H_M^2 (introduced by Higgins and Thompson [2002] and Mittlböck and Heinzl [2006]). We first define the "typical" or "average" within-study variance (Higgins and Thompson 2002) as

$$s^2 = \frac{\sum\limits_i w_i (k - 1)}{\left(\sum\limits_i w_i\right)^2 - \sum\limits_i w_i^2}$$

I^2 and H_M^2 can then be defined as follows:

$$I^2 = \frac{\tau^2}{\tau^2 + s^2}; \ H_M^2 = \frac{\tau^2}{s^2}$$

If $\tau^2 = \tau^2_{\mathrm{DL}}$, then the algebra may be rearranged to form the alternative formula

$$I^2 = \frac{Q - \mathrm{d.f.}}{Q}; \ H^2_M = \frac{Q - \mathrm{d.f.}}{\mathrm{d.f.}}$$

where $\mathrm{d.f.} = k - 1$ is the number of d.f. associated with Q (as defined in section 2.1).

Therefore, I^2 and H^2_M are measures of the relative magnitude of the between-study variance τ^2 and within-study variance s^2. Alternatively, if $\tau^2 = \tau^2_{\mathrm{DL}}$, then I^2 and H^2_M are measures of how much the point estimate of Q exceeds (or otherwise) its d.f. I^2 is the more well known of the two and is readily interpretable on the percentage scale. However, it is a nonlinear function of τ^2, such that an increase from, say, 80% to 90% implies a far greater increase in τ^2 than an increase from 20% to 30%. H^2_M does not have this limitation and is also less affected by the value of k, but it is perhaps less readily interpretable.

2.2 Random-effects methods

In this section, I describe the random-effects models currently available within the `ipdmetan` command.

Noniterative τ^2 estimators

The noniterative formulas for τ^2 presented in table 1 are based on moment-based approximations to the expectation of Q (DerSimonian–Laird; Hedges), a reparameterization of the total variance of θ_i (Sidik–Jonkman), or Bayesian considerations (Rukhin). Estimates of τ^2 are typically truncated at zero, but the Sidik–Jonkman and Rukhin BP estimators are always positive.

Table 1. Noniterative τ^2 estimators

Name	Reference	Definition
DerSimonian–Laird	DerSimonian and Laird (1986)	$\tau^2_{\mathrm{DL}} = \dfrac{Q - (k-1)}{\sum\limits_i w_i - \dfrac{\sum\limits_i w_i^2}{\sum\limits_i w_i}}$
Hedges "variance component" Cochran	Sidik and Jonkman (2007) DerSimonian and Kacker (2007)	$\tau^2_{\mathrm{VC}} = \dfrac{\sum\limits_i \left(\theta_i - \overline{\theta}\right)^2}{k-1} - \dfrac{\sum\limits_i \sigma_i^2}{k}$ where $\overline{\theta} = \sum_i \theta_i / k$
Sidik–Jonkman	Sidik and Jonkman (2007)	$\tau^2_{\mathrm{SJ}} = \dfrac{\sum\limits_i \widetilde{w_i}\left(\theta_i - \widetilde{\theta}\right)}{k-1}$ where $\widetilde{w_i} = \{(\sigma_i^2)/(\tau_0^2) + 1\}^{-1}$ are inverse-variance weights where the variance estimator is based on a parameterization involving ratios of the within-study variances σ_i^2 to a crude initial estimate $\tau_0^2 > 0$ of the true heterogeneity variance, and $\widetilde{\theta}$ is the pooled estimate based on weights $\widetilde{w_i}$
Rukhin Bayes estimators	Rukhin (2013)	$\tau^2_{\mathrm{B0}} = \dfrac{\sum\limits_i \left(\theta_i - \overline{\theta}\right)^2}{k+1}$ $- \dfrac{(N-k)(k-1)\sum\limits_i \sigma_i^2}{k(k+1)(N-k+2)}$ where N is the total number of participants within the k included studies $\tau^2_{\mathrm{BP}} = \dfrac{\sum\limits_i \left(\theta_i - \overline{\theta}\right)^2}{k+1}$

The Sidik–Jonkman estimator requires a nonzero initial estimate τ_0^2, for which τ_{VC}^2 was suggested by Sidik and Jonkman (2007), with an arbitrary small value of 0.01 being used in case $\tau_{\text{VC}}^2 = 0$. However, τ^2 is not scale invariant, so it is arguably more appropriate to assume an arbitrarily small I^2 instead, such as 1% (see the `isq()` suboption to `re()` in section 3.2).

Iterative τ^2 estimators

Iterative methods can be used to estimate τ^2 with confidence limits and, in the case of profile likelihood (PL), iterative confidence limits for θ. In table 2, w_i^* and $\widehat{\theta}^*$ are defined in terms of the current iteration of τ^2.

Table 2. Iterative τ^2 estimators

Name	Reference	Definition
Approximate gamma	Biggerstaff and Tweedie (1997)	Point estimate $\tau^2 = \tau_{\mathrm{DL}}^2$ Confidence limits for τ^2 obtained by profiling an approximate gamma distribution for Q Point estimate and standard error for θ obtained from modified weights generated by numerical integration
Mandel–Paule Generalized Q Empirical Bayes	DerSimonian and Kacker (2007) Viechtbauer (2007) Sidik and Jonkman (2007)	$$\tau_{\mathrm{GQ}}^2 = \frac{k}{C}\frac{\sum\limits_i w_i^*\left(\theta_i - \widehat{\theta}^*\right)^2}{\sum\limits_i w_i^*} - \frac{\sum\limits_i w_i^*\sigma_i^2}{\sum\limits_i w_i^*}$$ where C is a critical value, as follows: Point estimate $\quad\quad\quad C = k - 1$ Lower confidence limit $\quad C = \chi_{k-1;1-\alpha/2}^2$ Upper confidence limit $\quad C = \chi_{k-1;\alpha/2}^2$
Maximum likelihood (ML)	Viechtbauer (2007)	$$\tau_{\mathrm{ML}}^2 = \frac{\sum\limits_i w_i^{*2}\left(\theta_i - \widehat{\theta}^*\right)^2}{\sum\limits_i w_i^{*2}} - \frac{\sum\limits_i w_i^{*2}\sigma_i^2}{\sum\limits_i w_i^{*2}}$$ Confidence limits for τ^2 obtained by likelihood profiling
PL	Hardy and Thompson (1996)	τ^2 same as for ML. Confidence limits for θ obtained by nested likelihood profiling
Restricted maximum-likelihood (REML)	Viechtbauer (2007)	$$\tau_{\mathrm{REML}}^2 = \frac{\sum\limits_i w_i^{*2}\left(\theta_i - \widehat{\theta}^*\right)^2}{\sum\limits_i w_i^{*2}}$$ $$-\frac{\sum\limits_i w_i^{*2}\sigma_i^2}{\sum\limits_i w_i^{*2}} + \frac{1}{\sum\limits_i w_i^*}$$ Confidence limits for τ^2 obtained by likelihood profiling

Extensions to the DerSimonian–Laird model

a. Hartung–Knapp variance estimator

This approach, proposed by Hartung and Knapp (2001) and independently by Sidik and Jonkman (2002), uses an alternative estimator of $\text{Var}(\widehat{\theta})$ to give improved coverage probability in the presence of between-study heterogeneity rather than relying on accurate estimation of τ^2. Given standard DerSimonian–Laird random-effects weights $w_i^* = 1/(\sigma_i^2 + \tau_{\mathrm{DL}}^2)$ and corresponding pooled effect $\widehat{\theta}^*$, the Hartung–Knapp variance estimator is defined as

$$\text{Var}\left(\widehat{\theta}^*\right) = \frac{\sum_i w_i^* \left(\theta_i - \widehat{\theta}^*\right)^2}{(k-1)\sum_i w_i^*}$$

This is used to construct the following t-based $(1-\alpha)$ confidence interval for $\widehat{\theta}^*$:

$$\widehat{\theta}^* \pm t_{k-1;1-\alpha/2}\sqrt{\text{Var}\left(\widehat{\theta}^*\right)}$$

b. Nonparametric bootstrap

Kontopantelis, Springate, and Reeves (2013) more recently suggested a simple nonparametric bootstrap version of DerSimonian–Laird by sampling studies with replacement and taking the mean of the truncated τ_{DL}^2 estimates. This is intended to increase the chance of detecting heterogeneity when the true value of τ^2 is small.

Various simulation studies have been performed to assess the properties of the estimators of θ and τ^2 described above. Kontopantelis and Reeves (2012) concluded that the PL and Hartung–Knapp methods gave the most accurate coverage probabilities for θ, even when the true study effects were extremely nonnormally distributed. However, confidence intervals may be too wide for small numbers of studies k; in this situation, they recommend the standard DerSimonian–Laird method. Rukhin (2013) and Kontopantelis, Springate, and Reeves (2013) found that the Bayes estimator τ_{BP}^2 gave reliable coverage and accurate confidence intervals for small k, except when the true heterogeneity was also small, in which case the bootstrapped DerSimonian–Laird estimator was preferable.

In terms of τ^2, Sidik and Jonkman (2007) concluded that τ_{GQ}^2 and τ_{SJ}^2 were the least biased point estimates of τ^2. Viechtbauer (2007) found that the generalized Q and approximate gamma models gave the most accurate confidence intervals for τ^2.

Kontopantelis, Springate, and Reeves (2013) examined real-life meta-analysis data from the Cochrane Collaboration and concluded that heterogeneity is often underestimated, especially for small k. They therefore propose that sensitivity analyses be performed at specified levels of heterogeneity. Such a level might be moderate to high to test the sensitivity of conclusions regarding θ, or it may be minimal to simply force nonzero heterogeneity (see the `isq()` suboption to `re()` in section 3.2).

2.3   Interactions

What initially motivated the development of `ipdmetan` was unbiased estimation of
patient-level treatment-covariate interactions in IPD meta-analysis. As discussed in
Fisher et al. (2011), additional complexities exist for IPD that do not occur with AD.
With AD, there is just one observation per study; hence, only "study-level" characteristics can be recorded—data such as year of publication, specifics of treatment, or
study-average patient characteristics. Such data can be used only to investigate how
characteristics of studies may affect study-aggregate treatment effects via techniques
such as meta-regression and study-subgroup analysis. However, to investigate how treatment effects differ at the patient level requires IPD. An AD analysis of study-average
patient characteristics (for example, mean age or proportion male) is at risk of bias as
an estimate of the true patient-level interaction effect.

Patient-level characteristics are often categorical—for example, they may be categorized by sex, age subgroups, or disease stage. It may be tempting to compare by-
subgroup treatment effects to estimate an interaction. However, this is again biased if
interpreted as a measure of treatment-effect difference between patients (Fisher et al.
2011). Instead, treatment-covariate interactions must be estimated within each study
separately and pooled thereafter in the same way as overall treatment effects. (Of
course, continuous patient-level characteristics do not present such issues.)

Assume a series of linear regressions (extension to generalized linear models or time-
to-event regressions is straightforward) on patients j within studies i as follows (the
parentheses around i highlight the two-stage nature of the analysis):

$$y_{(i)j} = \alpha_{(i)} + \beta_{(i)}x_{(i)j} + \gamma_{(i)}z_{(i)j} + \delta_{(i)}x_{(i)j}z_{(i)j}$$

y is the outcome; x the treatment; z the covariate; and α, β, γ, and δ represent study,
treatment, covariate, and interaction effect coefficients, respectively. We equate the
interaction-effect estimates $\delta_{(i)}$ with the study estimates θ_i in section 2.1 and proceed
accordingly (Simmonds and Higgins 2007; Fisher et al. 2011, *Appendix B*). Such models may be extended to include adjustment for (within-study) covariates.

Interaction models are often complex, and the two-stage approach is suitable for only
simple cases—principally meta-analyses of RCTs where other factors may be assumed
to be balanced. Furthermore, only one coefficient may be pooled, placing limitations
on variable transformation (for example, use of fractional polynomials [Royston and
Altman 1994]) and precluding the recovery of absolute effects. However, they are certain to be free of bias from between-study effects (Simmonds et al. 2005; Fisher et al.
2011, *Appendix B*) and to provide a quick and accurate estimate of the within-study
interaction effect size. Although one-stage models permit a far greater variety of random effects, which may increase efficiency, the resulting coefficients must be interpreted
carefully to ensure the intended question is answered.

3   The ipdmetan command

3.1   Syntax

ipdmetan is a prefix command; see [U] **11.1.10 Prefix commands**. Its basic syntax is

ipdmetan $\big[\,exp\_list\,\big]$, <u>s</u>tudy(*study_id*$\big[$, <u>m</u>issing$\big]$) $\big[$by(*subgroup_id*$\big[$, <u>m</u>issing$\big]$)

   eform_option <u>eff</u>ect(*string*) <u>inter</u>action keepall <u>m</u>essages <u>nogra</u>ph

   notable nohet nooverall <u>nosu</u>bgroup <u>notot</u>al ovwt sgwt

   <u>poolvar</u>(*model_coefficient*) random re$\big[$(*re_model*)$\big]$ sortby(*varname*|_n)

   <u>lcols</u>(*cols_info*) <u>rcols</u>(*cols_info*) plotid(*varname*|_BYAD $\big[$, list <u>nogra</u>ph$\big]$)

   ovstat(q) <u>sa</u>ving(*filename*$\big[$, replace <u>stack</u>label$\big]$)

   <u>forest</u>plot(*forestplot_options*) ad(...)$\big]$:   *estimation_command* ...

where *estimation_command* is the model to be fit within each study. Typically, this will be a built-in command that changes the contents of e(b), but it could be any command that returns an effect size and standard error (or statistics that may be combined to form such). if, in, and weights should be supplied to *estimation_command* rather than to ipdmetan (that is, after the colon rather than before it). Relevant declarations of data structure (for example, stset, tsset, and xtset) must be done before running ipdmetan.

   When *estimation_command* does not change the contents of e(b), expressions that evaluate the effect size and standard error to be collected and pooled must be specified manually by using *exp_list*. (An optional third statistic is the number of observations.) Otherwise, *exp_list* defaults to _b[*varname*] _se[*varname*], where *varname* is the first independent variable within *estimation_command*.

3.2   Options

The options in this section (supplied directly to ipdmetan) control aspects of model fitting, computation, creation of the results set, on-screen tabulation of results, and saved statistics. Some options included are relevant to the forest plot, but they are used within the main routine of ipdmetan. The only suboptions that should be supplied to the forestplot() option are those controlling graphical presentation of the results set outputted by ipdmetan.

study(*study_id*$\big[$, missing$\big]$) specifies the variable that contains the study identifier, which must be either integer valued or string. study() is required.

   missing requests that missing values be treated as potential study identifiers (the default is to exclude them).

by(*subgroup_id*[, missing]) specifies a variable identifying subgroups of studies (and must therefore be constant within studies), which must be integer valued or string. If an AD dataset is specified and contains a variable named *subgroup_id*, subgrouping will be extended to the AD observations.

> missing requests that missing values be treated as potential subgroup identifiers (the default is to exclude them).

eform_option (see [R] *eform_option*) specifies that effect sizes and confidence limits be exponentiated in the output (table and forest plot) and generates a heading for the effect-size column.

> Note that ipdmetan does not check the validity of the particular *eform_option*; for example, it does not check whether *estimation_command* is a survival model if hr is supplied.

effect(*string*) specifies a heading for the effect-size column, overriding any heading generated by *eform_option*.

interaction specifies that *estimation_command* contain one or more interaction effects expressed using factor-variable syntax (see [U] **11.4.3 Factor variables**) and that the first valid interaction effect be pooled across studies. This is a helpful shortcut for simple interaction analyses, but it is not foolproof or comprehensive. The output should be checked and, if necessary, the analysis rerun with the poolvar() option.

keepall specifies that all values of *study_id* be visible in the output (table and forest plot), even if no effect could be estimated (for example, because of insufficient observations or missing data). For such studies, (Insufficient data) will appear in place of effect estimates and weights.

messages requests that information be printed to screen whether the effect size and standard-error statistics have been successfully obtained from each study and, if applicable, whether the iterative random-effects calculations converged successfully.

nograph and notable suppress construction of the forest plot and the table of effect sizes, respectively.

nohet suppresses all heterogeneity statistics.

nooverall suppresses the overall pooled effect so that, for instance, subgroups are considered independently. It also suppresses between-subgroup heterogeneity statistics (if applicable).

nosubgroup suppresses the within-subgroup pooled effects (if applicable) so that subgroups are displayed separately but with a single overall pooled effect with associated heterogeneity statistics.

nototal requests that *estimation_command* not be fit within the entire dataset, for example, for time-saving reasons. By default, such fitting is done to check for problems in convergence and in the validity of requested coefficients and returned expressions. If nototal is specified, either poolvar() or *exp_list* must be supplied, and a message appears above the table of results warning that estimates should be double checked by the user.

`ovwt` and `sgwt` override the default choice of whether to display overall weights or within-subgroup weights in the screen output and forest plot. Note that because weights are normalized, these options do not affect estimation of pooled effects or heterogeneity statistics.

`poolvar`(*model_coefficient*) specifies the coefficient from `e(b)` that is to be pooled when the default behavior of `ipdmetan` fails or is incorrect. *model_coefficient* should be a variable name, a level indicator, an interaction indicator, or an interaction involving continuous variables (c.f. syntax of [R] **test**). Equation names can be specified using the format `poolvar`(*eqname*:*varname*). This option is appropriate only when *estimation_command* changes the contents of `e(b)`; otherwise, see *exp_list*.

`random` or `re` specifies DerSimonian and Laird random effects.

`re`(*re_model*) specifies other possible random-effects models. The default is the DerSimonian and Laird random-effects model. Other currently supported models (see section 2.2) are the following:

 `dl` specifies the DerSimonian–Laird estimator (equivalent to specifying `re` alone, with no suboption); this is the default.

 `dlt` or `hk` specifies the DerSimonian–Laird with Hartung–Knapp *t*-based variance estimator.

 `bdl` or `dlb` specifies the bootstrapped DerSimonian–Laird estimator.

 `ca`, `he`, or `vc` specifies the Hedges variance-component estimator, also known as the Cochran estimator.

 `sj` specifies the Sidik–Jonkman two-step estimator.

 `b0` and `bp` specify Rukhin's $B0$ and BP estimators, respectively.

 `bs`, `bt`, or `gamma` specifies the Biggerstaff–Tweedie approximate gamma model.

 `eb`, `gq`, `genq`, `mp`, or `q` specifies the empirical Bayes estimator, also known as the generalized Q estimator and the Mandel–Paule estimator.

 `ml` specifies the ML estimator.

 `pl` specifies the PL model.

 `reml` specifies the REML estimator.

 `sa`[, `isq`(*real*)] specifies the sensitivity analysis with a fixed user-specified value of I^2 (between 0 and 1, with default `isq(0.8)`). See Kontopantelis, Springate, and Reeves (2013).

Note that the approximate gamma, generalized Q, ML, PL, and REML models require the `mm_root()` function; and the bootstrapped DerSimonian–Laird model requires the `mm_bs()` and `mm_jk()` functions from the `moremata` package (`ssc install moremata`) (Jann 2005). The approximate gamma model also requires the `integrate` command (`ssc install integrate`) (Mander 2012).

`sortby(`*varname*`| _n)` allows user-specified ordering of studies in the table and forest plot. The default ordering is by *study_id*. Note that `sortby()` does not alter the data in memory.

Specify `sortby(_n)` to order the studies by their first appearance in the data by using the current sort order.

`lcols(`*cols_info*`)` and `rcols(`*cols_info*`)` define columns of additional data to be presented to the left or right of the forest plot. These options are carried over from `metan`; however, for IPD, they must first be generated from the existing dataset and thus require more complex syntax.

cols_info has the following syntax, which is based on that of `collapse` (see [D] **collapse**),

$$\big[(stat) \big] \big[newname= \big] item \big[\%fmt \text{ "}label\text{" } \big] \big[\big[newname= \big] item \big[\%fmt \text{ "}label\text{" } \big] \big]$$
$$\ldots \big[\big[(stat) \big] \ldots \big]$$

where *stat* is as defined for `collapse`; *newname* is an optional user-specified variable name; *item* is either a numeric expression (in parentheses) involving returned quantities from *estimation_command* or a variable currently in memory; *%fmt* is an optional format; and "*label*" is an optional variable label. An example using these options is in section 4.5.

`plotid(`*varname*`| _BYAD [, list nograph]`) specifies one or more categorical variables to form a series of groups of observations in which specific aspects of plot rendition may be affected using *plot*[#]`opts` (see *Forest plot suboptions*). The groups of observations will automatically be assigned ordinal labels $(1, 2, \ldots)$ on the basis of the ordering of *varlist*. Note that `plotid()` does not alter the data in memory.

`ovstat(q)` displays Q statistics instead of I-squared statistics.

`saving(`*filename*`[, replace stacklabel]`) saves the forest plot "results set" (see section 4.6) created by `ipdmetan` in a dataset for further use or manipulation.

`replace` overwrites *filename*.

`stacklabel` takes the variable label from the leftmost column variable (usually *study_id*), which would usually appear outside the plot region as the column heading, and copies it into a new first row in *filename*. This allows multiple such datasets to be `append`ed without this information being overwritten.

`forestplot(`*forestplot_options*`)` specifies other options to pass to `forestplot`. See *Forest plot suboptions* below.

The AD option

ipdmetan is designed so that IPD and AD can be analyzed together with a single command. Because of their differing structures (IPD is one observation per patient, so there are multiple observations per study; AD is one observation per study), AD must be stored in a separate Stata dataset.

The syntax for the ad() option is as follows:

ad(*filename* [*if*] [*in*], vars(*namelist*) *ad_options*)

where *filename* is an existing Stata dataset containing the AD.

vars(*namelist*) contains the names of variables (within *filename*) containing the effect
size and either a standard error or lower and upper 95% confidence limits on the
linear scale. If confidence limits are supplied, they must be derived from a normal
distribution, or the pooled result will be incorrect (see help admetan).

It is assumed that a study identifier variable exists with the same name as in the data currently in memory (that is, the IPD). If such a variable cannot be found, studies in the AD dataset are numbered sequentially, following the largest study identifier value used for the IPD. Subgroup variables may also be present in the AD dataset and may be referenced using the main by() option.

ad_options can be the following:

byad specifies that IPD and AD be treated as subgroups rather than as a single set
of estimates.

npts(*varname*) allows participant numbers (stored in *varname* within *filename*) to
be displayed in tables and forest plots.

Forest plot suboptions

Most of the forest plot options for metan can be supplied to ipdmetan via the option
forestplot(), together with most appropriate [G-3] *twoway_options*. The most significant extension that ipdmetan brings is the ability to apply specific plot rendition options (color, line patterns, etc.) to specific groups of observations via the plotid() option (see above).

The syntax for these options is

plot[*#*]opts(*plot_options*)

where *plot* refers to the plot feature, which can be box, ci, diam, oline, point, pci, or ppoint; *#* is the (optional) observation group identifier assigned by plotid(); and *plot_options* are [G-3] *twoway_options* appropriate to the plot type used to construct the feature, as described in Harris et al. (2008).

The extra plot features `pci` and `ppoint`, not included in `metan`, are used when pooled estimates must be represented by (differently rendered) point estimates and confidence intervals instead of by diamonds. See the `forestplot` documentation for more details.

3.3   Stored results

`ipdmetan` stores the following in `r()` (with some variation):

Scalars

`r(k)`	number of included studies k	`r(tausq)`	between-study variance tau-squared
`r(n)`	number of included participants	`r(sigmasq)`	average within-study variance
`r(eff)`	overall pooled effect size	`r(Isq)`	heterogeneity measure I^2
`r(se_eff)`	standard error of pooled effect size	`r(HsqM)`	heterogeneity measure H_M^2
`r(Q)`	Cochran Q heterogeneity statistic (on $k - 1$ d.f.)		

Macros

`r(command)`	full estimation command line	`r(estvar)`	name of pooled coefficient
`r(cmdname)`	estimation command name	`r(re_model)`	random-effects model used

Matrices

`r(coeffs)`	matrix of study and subgroup identifiers, effect coefficients, numbers of participants, and weights

Certain iterative random-effects models can save the following additional results (see `help mf_mm_root` for interpretations of convergence success values):

Scalars

`r(tsq_var)`	estimated variance of tau-squared
`r(tsq_lci)`	lower confidence limit for tau-squared
`r(tsq_uci)`	upper confidence limit for tau-squared
`r(rc_tausq)`	convergence of tau-squared point estimate
`r(rc_tsq_lci)`	convergence of tau-squared lower confidence limit
`r(rc_tsq_uci)`	convergence of tau-squared upper confidence limit
`r(rc_eff_lci)`	convergence of effect-size lower confidence limit
`r(rc_eff_uci)`	convergence of effect-size upper confidence limit

4   Example

The motivating example is an IPD meta-analysis of RCTs evaluating the effectiveness of postoperative radiotherapy in patients with completely resected nonsmall cell lung cancer (Burdett, Stewart, and PORT Meta-analysis Group 2005; Burdett et al. 2013). The outcome measure in each trial is overall survival, that is, censored time to death from any cause. A number of covariate measurements were also available but not necessarily for all trials or patients.

For an example dataset to be made publicly available, a simulated dataset was generated with similar characteristics to the postoperative radiotherapy data. It is this simulated dataset that will be discussed henceforth. There are 10 trials of varying size

in 2 trial subgroups with arbitrary names. Sex, age, and disease-stage covariates have
also been generated.

4.1   Basic use

Our first example aims to cover various aspects of inverse-variance meta-analysis simultaneously, because none of this functionality is particularly complex. We perform a
meta-analysis of main treatment effects fit from Cox models stratified by sex, pooling
the log hazard-ratios using the fixed-effects inverse-variance method.

```
. use ipdmetan_example

. quietly stset tcens, fail(fail)

. ipdmetan, study(trialid) hr by(region)
> forest(favours(Favours treatment # Favours control)): stcox trt, strata(sex)

Studies included: 10
Patients included: 1642

Meta-analysis pooling of main (treatment) effect estimate trt
using Fixed-effects
```

Region and Trial name	Haz. Ratio	[95% Conf. Interval]		% Weight
Europe				
London	1.475	1.015	2.143	10.93
Paris	1.197	0.789	1.816	8.78
Amsterdam	1.742	1.080	2.810	6.68
Stockholm	1.440	0.816	2.542	4.73
Madrid	2.242	1.501	3.350	9.47
Subgroup effect	1.593	1.312	1.934	40.59
North America				
New York	0.740	0.465	1.176	7.10
Chicago	0.774	0.558	1.074	14.29
Los Angeles	1.058	0.707	1.584	9.38
Toronto	1.082	0.816	1.434	19.21
College Station, TX	0.689	0.461	1.031	9.42
Subgroup effect	0.885	0.754	1.039	59.41
Overall effect	1.123	0.993	1.271	100.00

```
Tests of effect size = 1:
Europe            z =   4.707  p =    0.000
North America     z =  -1.495  p =    0.135
Overall           z =   1.846  p =    0.065

Q statistics for heterogeneity (calculated using Inverse Variance weights)
```

	Value	df	p-value
Europe	5.01	4	0.287
North America	5.40	4	0.249
Overall	31.39	9	0.000
Between	20.98	1	0.000
Between:Within (F)	6.02	1, 9	0.037

The command `stcox trt, strata(sex)` was first fit consecutively to each level of `trialid`. The resulting log hazard-ratios and standard errors `_b[trt]` and `_se[trt]` are displayed in the table and forest plot (figure 1). Specifying the *eform_option* `hr` exponentiated the estimates into hazard ratios, while the `by()` option gave us within-subgroup pooled estimates and tests of heterogeneity within and between subgroups and overall. The form of the rest of the output and the forest plot should be familiar to users of `metan` (Harris et al. 2008).

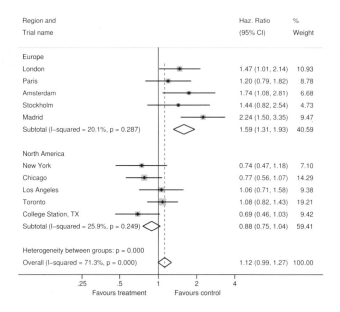

Figure 1. Basic forest plot

4.2   Treatment-covariate interactions

The syntax of `ipdmetan` allows any (single) estimated quantity and its standard error to be pooled across studies. This includes simple interaction models of the form given in section 2.3. Specifying the `interaction` option tells `ipdmetan` to identify and pool the treatment-covariate interaction coefficient and to use different symbols and effect-size column headings on the forest plot to differentiate such analyses from those of main effects. (Note that all of this behavior can be reproduced manually using other options.)

Studies often have insufficient covariate data, and such studies will be excluded from the corresponding interaction analysis. The default is to remove excluded studies from the output. However, this can be overridden with the `keepall` option, which will instead output "`(Insufficient data)`" for such studies in place of the effect size, the standard error, and the weighting in both table and forest plot. (This option applies generally. I mention it here because it is likely to have greater relevance.)

The screen output has only minor differences from that of a main-effects analysis, so we show only the forest plot here (figure 2). The default for interaction analyses is to use solid circles for the study effects and a clear circle for the pooled effect.

```
. ipdmetan, study(trialid) interaction hr keepall forest(favours("Favours
> greater treatment effect" "with higher disease stage"
> # "Favours greater treatment effect" "with lower disease stage")
> boxsca(200) fp(1)): stcox trt##c.stage
```

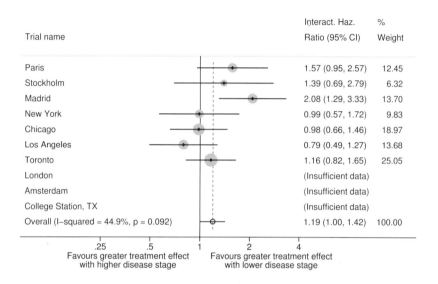

Figure 2. Forest plot of interactions

4.3   Random effects

We now show the syntax and output of the random-effects options. DerSimonian–Laird, the most common random-effects model, is specified with "`re`".

▷ **Example 1: DerSimonian–Laird random effects**

```
. ipdmetan, study(trialid) nograph re: stcox trt
  (output omitted)
Heterogeneity Measures
```

	Value	df	p-value
Cochran Q	31.73	9	0.000
I^2 (%)	71.6%		
Modified H^2	2.525		
tau^2	0.0985		

```
I² = between-study variance (tau²) as a percentage of total variance
Modified H² = ratio of tau² to typical within-study variance
```

◁

Other random-effects models are specified as options to "re", as shown in the following example (see section 2.2 for details of the Generalized Q model). Note that iterative models output confidence limits for heterogeneity in addition to point estimates.

▷ **Example 2: Generalized Q random effects**

```
. ipdmetan, study(trialid) nograph re(q): stcox trt
  (output omitted)
Heterogeneity Measures
```

	Value	df	p-value
Cochran Q	31.73	9	0.000

	Value	[% Conf. Interval]	
I^2 (%)	82.7%	73.8%	93.1%
Modified H^2	4.779	2.823	13.390
tau^2	0.1864	0.1101	0.5223

```
I² = between-study variance (tau²) as a percentage of total variance
Modified H² = ratio of tau² to typical within-study variance
```

◁

4.4   Aggregate data

Often IPD cannot be obtained for all eligible studies. In such cases, it may help to do a sensitivity analysis comparing "extra" AD estimates with the main body of IPD. Some authors (Riley et al. 2008) have even suggested pooling both IPD and AD together. Either of these approaches is straightforward with **ipdmetan**. Here we demonstrate the former by first artificially constructing an aggregate dataset from one of the trial subgroups we defined earlier by using the **saving()** option. Note the following (see also section 4.6):

- Saved datasets have a standard format where the study variable is named _STUDY. Therefore, we change our `trialid` variable to match (using `clonevar`) and so obtain correct labels throughout.

- We restrict the `ad()` option to _USE==1 so that only individual study estimates are included and not the pooled estimate.

```
. quietly ipdmetan, study(trialid) hr nograph saving(region2.dta): stcox
> trt if region==2, strata(sex)
. clonevar _STUDY = trialid
. ipdmetan, study(_STUDY) hr ad(region2.dta if _USE==1, vars(_ES _seES)
> npts(_NN) byad) nooverall: stcox trt if region==1, strata(sex)
Studies included from IPD: 5
Patients included: 656
Studies included from aggregate data: 5
Patients included: 986
Meta-analysis pooling of main (treatment) effect estimate trt
using Fixed-effects
```

Data source and Trial name	Haz. Ratio	[95% Conf. Interval]		% Weight
IPD				
London	1.475	1.015	2.143	26.93
Paris	1.197	0.789	1.816	21.63
Amsterdam	1.742	1.080	2.810	16.46
Stockholm	1.440	0.816	2.542	11.64
Madrid	2.242	1.501	3.350	23.34
Subgroup effect	1.593	1.312	1.934	100.00
Aggregate				
New York	0.740	0.465	1.176	11.96
Chicago	0.774	0.558	1.074	24.06
Los Angeles	1.058	0.707	1.584	15.79
Toronto	1.082	0.816	1.434	32.33
College Station, TX	0.689	0.461	1.031	15.86
Subgroup effect	0.885	0.754	1.039	100.00

```
Tests of effect size = 1:
IPD            z =   4.707  p =   0.000
Aggregate      z =  -1.495  p =   0.135
Q statistics for heterogeneity (calculated using Inverse Variance weights)
```

	Value	df	p-value
IPD	5.01	4	0.287
Aggregate	5.40	4	0.249

The numerical output is the same as in our first example, as expected. Note that to treat the IPD and AD as two subgroups, we specified the `byad` option; this also labels the subgroups appropriately in both the table and forest plot (not shown). The `nooverall` option requests that the table and forest plot present IPD and AD separately and not present an "overall" pooled result. This will also present weights and effect lines by subgroup rather than overall. Above the table, the number of trials and patients are also given separately for IPD and for AD. If the number of patients is missing for one or more observations in the AD dataset, `ipdmetan` will output a missing value here.

4.5   Further functionality

In this section, I discuss a few more useful features of `ipdmetan` and `forestplot`. For this next example, we return to the original analysis in section 4.1, but instead of Cox regression, we will use the noniterative Peto log-rank method. This is based on the log-rank test done by `sts test`, but it requires values stored in the vector **u** and matrix **V** (see documentation of the `mat()` option in [ST] **sts test**) to estimate the effect size (log hazard-ratio) and standard error for each trial. Therefore, we must supply `ipdmetan` with appropriate expressions for these quantities as an *exp_list*. Using our current example, we also demonstrate the use of `lcols()` and `rcols()` by presenting the $O - E$ and V statistics (Yusuf et al. 1985), and we demonstrate the use of *plot*[*#*]*opts* by rendering the weighted boxes and confidence interval lines in a light shade for the European trials (with capped confidence intervals) and in a darker shade for the North American trials. We proceed as follows (see figure 3):

```
. ipdmetan (u[1,1]/V[1,1]) (1/sqrt(V[1,1])), study(trialid) hr
> rcols((u[1,1]) %5.2f "o-E(o)" (V[1,1]) %5.1f "V(o)")
> by(region) plotid(region)
> forest(nooverall nostats nowt
> favours(Favours treatment # Favours control)
> box1opts(mcolor(gs10)) ci1opts(lcolor(gs10) rcap)
> box2opts(mcolor(gs6)) ci2opts(lcolor(gs6))): sts test trt, mat(u V)
```

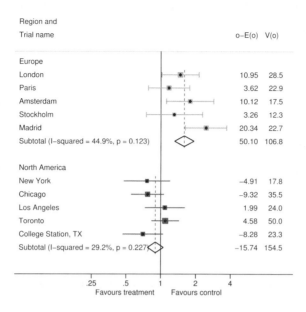

Figure 3. Use of advanced forest plot options

For the right-hand columns, we have requested that the usual statistics and weights be suppressed (options `nostats` and `nowt`) and replaced with the observed minus-expected event count and variance of observed event count (stored by `sts test` in `u[1,1]` and `V[1,1]`, respectively). Following each statistic is an (optional) format and column heading. (Note that this plot may be re-created with a Peto log-rank meta-analysis by using `petometan`, which is part of the `ipdmetan` package.)

Forest plots, with their columns of text and numbers, are unnatural to Stata, and sometimes, plots produced by `ipdmetan` or `forestplot` may be deficient in appearance (text too small or overlapping, too much white space, etc.). There are many options available that can correct such deficiencies (see the help file).

4.6   Use of results sets

Within a two-stage framework, IPD and AD may be considered a "long form" and "short form" of the same data. It may be useful to convert IPD to AD to perform certain tasks. This can be accomplished by using the `saving()` option with `ipdmetan` (see section 3.2), which saves an AD results set in a format suitable for `forestplot`. Hence, running `forestplot` on the results set will give the same plot as running `ipdmetan` on the original data.

As an example, the results set outputted by the analysis in section 4.1 is shown in the table below (value labels not displayed). This format includes not only the study estimates but also the pooled estimates with extra lines for subgroup headings and heterogeneity descriptions and blank lines for spacing. The values of the variable _USE specify the function of each observation, while text (study names, headings, etc., excluding extra data specified in `lcols()` and `rcols()`) is stored in the variable _LABELS.

Possible uses of such results sets include the following:

- Restricting the results set to observations for which _USE==1 (that is, the individual study estimates) creates a standard AD dataset that can be analyzed with any AD meta-analysis command. Hence, features not yet supported by `ipdmetan` can be used, such as prediction intervals with `metan` (Harris et al. 2008) or the "permutations" model of `metaan` (Kontopantelis and Reeves 2010), as well as related analyses such as meta-regression on study-level covariates (`metareg`, Harbord and Higgins [2008]) or network meta-analysis (`mvmeta`, White [2009]).

- If the initial stage in `ipdmetan` of fitting the model to each study proves time consuming, it need be performed only once. Subsequent tasks, such as investigating random-effects models (see previous bullet) or tweaking the forest plot (see following bullet), can be done with `admetan` or `forestplot` using the results set.

- The results set contains all the information that `forestplot` needs to construct the plot. Therefore, the layout of the forest plot can be customized by altering the results set; for instance, it can be altered to amend labels or titles, add or remove lines, or add or format extra columns of data (to be specified with `lcols()` and `rcols()`). For example, strings of the form "n/N" can be presented, where n is the number of events and N the total number of patients.

- More complex forest plots may not be possible to create with a single call to `ipdmetan`. Instead, `ipdmetan` might be run on a series of different datasets and results saved. The standard formatting of results sets allows straightforward appending to a single dataset using `append` (see [D] **append**), potentially with the `generate` suboption to create a variable to pass to the `plotid()` suboption of `forestplot()` (see section 4.5). It may then be necessary to alter labels or titles (see previous bullet).

```
. list, clean noobs nolabel
```

_USE	_BY	_STUDY	_LABELS	_ES	_seES	_LCI	_UCI	_WT	_NN
0	1	.	Europe	.	.	.	.	.	.
1	1	1	London	0.389	0.191	0.015	0.762	0.109	176
1	1	2	Paris	0.180	0.213	-0.237	0.597	0.088	141
1	1	3	Amsterdam	0.555	0.244	0.077	1.033	0.067	110
1	1	4	Stockholm	0.365	0.290	-0.204	0.933	0.047	69
1	1	5	Madrid	0.807	0.205	0.406	1.209	0.095	160
3	1	.	Subtotal (I-squared = 20.1%, p = 0.287)	0.466	0.099	0.272	0.660	0.406	656
6	1	.		.	.	.	.	.	.
0	2	.	North America	.	.	.	.	.	.
1	2	6	New York	-0.301	0.237	-0.765	0.162	0.071	115
1	2	7	Chicago	-0.256	0.167	-0.583	0.071	0.143	238
1	2	8	Los Angeles	0.056	0.206	-0.347	0.460	0.094	148
1	2	9	Toronto	0.079	0.144	-0.203	0.361	0.192	316
1	2	10	College Station, TX	-0.372	0.205	-0.775	0.030	0.094	169
3	2	.	Subtotal (I-squared = 25.9%, p = 0.249)	-0.122	0.082	-0.283	0.038	0.594	986
6	2	.		.	.	.	.	.	.
4	.	.	Heterogeneity between groups: p = 0.000	.	.	.	.	.	.
5	.	.	Overall (I-squared = 71.3%, p = 0.000)	0.116	0.063	-0.007	0.240	1.000	1642

5   Discussion

Meta-analysis is a widely used and constantly evolving technique that has made an enormous contribution to evidence-based policy and practice, particularly in health care. Although one-stage hierarchical models are increasingly seen and may have statistical advantages in some situations, the two-stage approach remains commonplace when IPD are available.

The `ipdmetan` command is a tool for performing two-stage IPD analysis simply and straightforwardly; it gives a standardized output and publication-standard forest plots (based on the original forest plot subroutine within `metan`). The syntax is designed to encompass varying levels of complexity, and the dedicated `forestplot` command allows great flexibility of presentation. The package also provides useful capabilities such as trial subgroup plots (via `ipdover`) and inclusion of AD.

Future work might include the ability to present prognostic effects in a forest plot with respect to a reference, as well as the ability to use multiple imputation for missing covariates using `mi`.

6   Acknowledgments

I thank the authors of `metan`, whose forest plot routine forms the basis of `forestplot`; I particularly thank Ross Harris for his comments and good wishes.

I also thank Tim Morris, James Carpenter, and Patrick Royston of the MRC Clinical Trials Unit at University College London, as well as the anonymous reviewer, for their comments and suggestions.

7   References

Biggerstaff, B. J., and R. L. Tweedie. 1997. Incorporating variability in estimates of heterogeneity in the random effects model in meta-analysis. *Statistics in Medicine* 16: 753–768.

Bowden, J., J. F. Tierney, M. C. Simmonds, A. J. Copas, and J. P. T. Higgins. 2011. Individual patient data meta-analysis of time-to-event outcomes: One-stage versus two-stage approaches for estimating the hazard ratio under a random effects model. *Research Synthesis Methods* 2: 150–162.

Burdett, S., L. Rydzewska, J. F. Tierney, D. J. Fisher, and PORT Meta-analysis Trialist Group. 2013. A closer look at the effects of postoperative radiotherapy by stage and nodal status: Updated results of an individual participant data meta-analysis in non-small-cell lung cancer. *Lung Cancer* 80: 350–352.

Burdett, S., L. A. Stewart, and PORT Meta-analysis Group. 2005. Postoperative radiotherapy in non-small-cell lung cancer: Update of an individual patient data meta-analysis. *Lung Cancer* 47: 81–83.

Crowther, M. J., M. P. Look, and R. D. Riley. 2014. Multilevel mixed effects parametric survival models using adaptive Gauss–Hermite quadrature with application to recurrent events and individual participant data meta-analysis. *Statistics in Medicine* 33: 3844–3858.

Deeks, J. J., D. G. Altman, and M. J. Bradburn. 2001. Statistical methods for examining heterogeneity and combining results from several studies in meta-analysis. In *Systematic Reviews in Health Care: Meta-analysis in Context*, ed. M. Egger, G. Davey Smith, and D. G. Altman, 2nd ed., 285–312. London: BMJ Books.

DerSimonian, R., and R. Kacker. 2007. Random-effects model for meta-analysis of clinical trials: An update. *Contemporary Clinical Trials* 28: 105–114.

DerSimonian, R., and N. Laird. 1986. Meta-analysis in clinical trials. *Controlled Clinical Trials* 7: 177–188.

Fisher, D. J., A. J. Copas, J. F. Tierney, and M. K. B. Parmar. 2011. A critical review of methods for the assessment of patient-level interactions in individual participant data meta-analysis of randomized trials, and guidance for practitioners. *Journal of Clinical Epidemiology* 64: 949–967.

Harbord, R. M., and J. P. T. Higgins. 2008. Meta-regression in Stata. *Stata Journal* 8: 493–519. (Reprinted in this collection on pp. 85–111.)

Hardy, R. J., and S. G. Thompson. 1996. A likelihood approach to meta-analysis with random effects. *Statistics in Medicine* 15: 619–629.

Harris, R. J., M. J. Bradburn, J. J. Deeks, R. M. Harbord, D. G. Altman, and J. A. C. Sterne. 2008. metan: Fixed- and random-effects meta-analysis. *Stata Journal* 8: 3–28. (Reprinted in this collection on pp. 29–54.)

Hartung, J., and G. Knapp. 2001. A refined method for the meta-analysis of controlled clinical trials with binary outcome. *Statistics in Medicine* 20: 3875–3889.

Higgins, J. P. T., and S. Green. 2011. *Cochrane Handbook for Systematic Reviews of Interventions Version 5.1.0*. http://www.cochrane-handbook.org.

Higgins, J. P. T., and S. G. Thompson. 2002. Quantifying heterogeneity in a meta-analysis. *Statistics in Medicine* 21: 1539–1558.

Jann, B. 2005. moremata: Stata module (Mata) to provide various functions. Statistical Software Components S455001, Department of Economics, Boston College. https://ideas.repec.org/c/boc/bocode/s455001.html.

Kontopantelis, E., and D. Reeves. 2010. metaan: Random-effects meta-analysis. *Stata Journal* 10: 395–407. (Reprinted in this collection on pp. 55–67.)

———. 2012. Performance of statistical methods for meta-analysis when true study effects are non-normally distributed: A simulation study. *Statistical Methods in Medical Research* 21: 409–426.

———. 2013. A short guide and a forest plot command (ipdforest) for one-stage meta-analysis. *Stata Journal* 13: 574–587. (Reprinted in this collection on pp. 321–354.)

Kontopantelis, E., D. A. Springate, and D. Reeves. 2013. A re-analysis of the Cochrane Library data: The dangers of unobserved heterogeneity in meta-analyses. *PLOS ONE* 8: e69930.

Mander, A. 2012. integrate: Stata module to perform one-dimensional integration. Statistical Software Components S457429, Department of Economics, Boston College. https://ideas.repec.org/c/boc/bocode/s457429.html.

Mathew, T., and K. Nordström. 2010. Comparison of one-step and two-step meta-analysis models using individual patient data. *Biometrical Journal* 52: 271–287.

Mittlböck, M., and H. Heinzl. 2006. A simulation study comparing properties of heterogeneity measures in meta-analyses. *Statistics in Medicine* 25: 4321–4333.

Riley, R. D., P. C. Lambert, J. A. Staessen, J. Wang, F. Gueyffier, L. Thijs, and F. Boutitie. 2008. Meta-analysis of continuous outcomes combining individual patient data and aggregate data. *Statistics in Medicine* 27: 1870–1893.

Royston, P., and D. G. Altman. 1994. Regression using fractional polynomials of continuous covariates: Parsimonious parametric modelling (with discussion). *Journal of the Royal Statistical Society, Series C* 43: 429–467.

Rukhin, A. L. 2013. Estimating heterogeneity variance in meta-analysis. *Journal of the Royal Statistical Society, Series B* 75: 451–469.

Sidik, K., and J. N. Jonkman. 2002. A simple confidence interval for meta-analysis. *Statistics in Medicine* 21: 3153–3159.

———. 2007. A comparison of heterogeneity variance estimators in combining results of studies. *Statistics in Medicine* 26: 1964–1981.

Simmonds, M. C., and J. P. T. Higgins. 2007. Covariate heterogeneity in meta-analysis: Criteria for deciding between meta-regression and individual patient data. *Statistics in Medicine* 26: 2982–2999.

Simmonds, M. C., J. P. T. Higgins, L. A. Stewart, J. F. Tierney, M. J. Clarke, and S. G. Thompson. 2005. Meta-analysis of individual patient data from randomized trials: A review of methods used in practice. *Clinical Trials* 2: 209–217.

Sterne, J. A. C., ed. 2009. *Meta-Analysis in Stata: An Updated Collection from the Stata Journal.* College Station, TX: Stata Press.

Stewart, G. B., D. G. Altman, L. M. Askie, L. Duley, M. C. Simmonds, and L. A. Stewart. 2012. Statistical analysis of individual participant data meta-analyses: A comparison of methods and recommendations for practice. *PLOS ONE* 7: e46042.

Stewart, L. A., and J. F. Tierney. 2002. To IPD or not to IPD? Advantages and disadvantages of systematic reviews using individual patient data. *Evaluation and the
Health Professions* 25: 76–97.

Tudur Smith, C., and P. R. Williamson. 2007. A comparison of methods for fixed effects
meta-analysis of individual patient data with time to event outcomes. *Clinical Trials*
4: 621–630.

Viechtbauer, W. 2007. Confidence intervals for the amount of heterogeneity in meta-
analysis. *Statistics in Medicine* 26: 37–52.

White, I. R. 2009. Multivariate random-effects meta-analysis. *Stata Journal* 9: 40–56.
(Reprinted in this collection on pp. 232–248.)

Yusuf, S., R. Peto, J. Lewis, R. Collins, and P. Sleight. 1985. Beta blockade during
and after myocardial infarction: An overview of the randomized trials. *Progress in
Cardiovascular Diseases* 27: 335–371.

Part 6

Network meta-analysis: indirect, network package, network_graphs package

Some of the most important developments in meta-analysis since publication of the first edition of this book have been in network meta-analysis (also referred to as "multiple treatments meta-analysis" and "mixed treatment comparisons"). Network meta-analysis is a method of synthesizing information from a collection of studies by combining evidence from all intervention comparisons that have been made among the studies. The results it produces for each pairwise comparison combine all the "direct evidence" (evidence based on head-to-head comparisons between interventions made within individual studies) with all the "indirect evidence" (comparisons between interventions inferred from the network via common comparator interventions). An important issue in these analyses is whether the direct evidence is consistent with the indirect evidence (Lu and Ades 2006). Three articles on network meta-analysis have appeared in the *Stata Journal*; they provide a comprehensive set of commands for performing these analyses.

Miladinovic et al. (2014) describe the `indirect` command, which estimates indirect treatment comparisons between multiple treatments.

In companion articles, Higgins et al. (2012) and White et al. (2012) described how to use `mvmeta` to estimate network meta-analysis models as special cases of multivariate meta-analysis. The new article by White (Forthcoming) incorporates and supersedes this material by providing the powerful `network` command. This implements the previously described methodology for fitting the models using `mvmeta` and implements many helpful procedures, such as completing data management tasks and producing graphical displays, as subcommands. These subcommands are particularly welcome because they provide a structure for completing complex programming tasks when performing a network meta-analysis. The graphical commands in the package can produce forest plots, rankograms, and maps of the network.

The data formats provided by White (Forthcoming) and used by the `network` package of commands have been adopted by Chaimani et al. (2013) within the package `network_graphs` of graphical commands for network meta-analysis. This article has been updated for the *Stata Journal* by Chaimani and Salanti (Forthcoming). The package includes commands for assessing the assumptions of the network meta-analysis models including plotting maps of the network (`networkplot`), plotting the contribution each direct treatment comparison to the network summary estimates (`netweight`), evaluating inconsistency in each closed loop of the network (`ifplot`), and plotting comparison-adjusted funnel plots (`netfunnel`). It also includes commands for viewing the results of network meta-analysis models, including plots of confidence and prediction intervals about summary estimates (`intervalplot`), plots of ranking probabilities for each treatment (`sucra`), plots of league tables of all possible pairwise comparisons (`netleague`, new in the updated article), and additional plots ranking the pairwise treatment comparisons (`mdsrank` and `clusterank`).

1   References

Chaimani, A., J. P. T. Higgins, D. Mavridis, P. Spyridonos, and G. Salanti. 2013. Graphical tools for network meta-analysis in Stata. *PLOS ONE* 8: e76654.

Chaimani, A., and G. Salanti. Forthcoming. Visualizing assumptions and results in network meta-analysis: The network graphs package. *Stata Journal* (Reprinted in this collection on pp. 355–400.)

Higgins, J. P. T., D. Jackson, J. K. Barrett, G. Lu, A. E. Ades, and I. R. White. 2012. Consistency and inconsistency in network meta-analysis: Concepts and models for multi-arm studies. *Research Synthesis Methods* 3: 98–110.

Lu, G., and A. E. Ades. 2006. Assessing evidence inconsistency in mixed treatment comparisons. *Journal of the American Statistical Association* 101: 447–459.

Miladinovic, B., A. Chaimani, I. Hozo, and B. Djulbegovic. 2014. Indirect treatment comparison. *Stata Journal* 14: 76–86. (Reprinted in this collection on pp. 311–320.)

White, I. R. Forthcoming. Network meta-analysis. *Stata Journal* (Reprinted in this collection on pp. 321–354.)

White, I. R., J. K. Barrett, D. Jackson, and J. P. T. Higgins. 2012. Consistency and inconsistency in network meta-analysis: Model estimation using multivariate meta-regression. *Research Synthesis Methods* 3: 111–125.

The Stata Journal (2014)
14, Number 1, st0325, pp. 76–86

Indirect treatment comparison

Branko Miladinovic
Center for Evidence-Based Medicine and Health Outcomes Research
University of South Florida
Tampa, FL
bmiladin@health.usf.edu

Anna Chaimani
Department of Hygiene and Epidemiology
University of Ioannina School of Medicine
Ioannina, Greece
achaiman@cc.uoi.gr

Iztok Hozo
Department of Mathematics
Indiana University Northwest
Gary, IN
ihozo@iun.edu

Benjamin Djulbegovic
Center for Evidence-Based Medicine and Health Outcomes Research
University of South Florida
Tampa, FL
bdjulbeg@health.usf.edu

Abstract. This article presents a command, `indirect`, for the estimation of effects of multiple treatments in the absence of randomized controlled trials for direct comparisons of interventions.

Keywords: st0325, indirect, Bucher, network meta-analysis

1 Introduction

Traditional meta-analyses that combine treatment effects across trials comparing the same interventions have been used in clinical medicine since the 1980s. In the absence of direct comparisons between two interventions, under certain conditions, a network of evidence can be constructed so that interventions may be compared indirectly (Glenny et al. 2005). The methods for indirect treatment comparison can be broadly categorized as frequentist or Bayesian. The frequentist methods are those described by Bucher et al. (1997), Lumley (2002), and White et al. (2012). The main difference between the two is that the former, also known as the adjusted indirect treatment comparison (AITC) method, is intended for situations where there is no direct evidence and comparisons are made pairwise. The Lumley method, like the Bayesian one, combines both direct and indirect comparisons within a total network of evidence. The Bayesian methods are statistically more flexible but computationally intensive and complex. They revolve around the choice of a prior estimate and depend on multiple-chain Monte Carlo simulations for the posterior estimates of treatment effects (Lu and Ades 2004; Caldwell, Ades, and Higgins 2005; Jansen et al. 2008). Interested readers are directed toward a special issue of *Research Synthesis Methods* for further information

(Salanti and Schmid 2012), especially about how network meta-analysis can accommodate more complicated networks in Stata (White et al. 2012; Chaimani et al. 2013). Motivated by AITC's desirability for simple networks, we implemented it as the Stata command `indirect`.

2   Adjusted indirect treatment comparison

The adjusted indirect method allows for the comparison of two treatments by using information from randomized controlled trials comparing each of the interventions with a common comparator. It assumes that the treatment effectiveness is the same across all trials used in the comparison. Formally and following notation by Wells et al. (2009), given k number of treatments $T_1, T_2, \ldots, T_k$ such that all consecutive pairs have been compared (T_1 versus T_2, T_2 versus T_3, ..., T_{k-1} versus T_k), the indirect $100(1 - \alpha/2)\%$ confidence interval (CI) estimator of the measure of association $\widehat{A}$ for a pair of treatments (T_i, T_{i+1}) is given by

$$\sum_{i=1}^{k-1} \widehat{A}_{T_i T_{i+1}} \pm Z_{\frac{\alpha}{2}} \sqrt{\sum_{i=1}^{k-1} \text{Var}\left(\widehat{A}_{T_i T_{i+1}}\right)}$$

where $\sum_{i=1}^{k-1} \widehat{A}_{T_i T_{i+1}}$ is the indirect estimator of treatments T_1 and T_k. The measure of association $\widehat{A}$ can be in the form of an odds ratio, a risk ratio, a hazard ratio (HR), a risk difference, or a mean difference. The test statistic for testing the indirect association between treatments T_1 and T_k for n number of studies used is

$$\chi^2_{df=n} = \frac{\displaystyle\sum_{i=1}^{k-2} \sum_{j=i+1}^{k-1} \left(\sum_{j=1}^{n} W_{T_i T_{i+1}, j}\right) \left(\sum_{j=1}^{n} W_{T_j T_{j+1}, j}\right) \left(\widehat{A}_{T_i T_{i+1}} - \widehat{A}_{T_j T_{j+1}}\right)^2}{\displaystyle\sum_{i=1}^{k-1} \sum_{j=1}^{n} W_{T_i T_{i+1}, j}}$$

where the weight assigned for the jth study evaluating treatments (T_i, T_{i+1}) is defined as

$$W_{T_i T_{i+1}, j} = \left\{\text{Var}\left(\widehat{A}_{T_i T_{i+1}, j}\right)\right\}^{-1}$$

AITC can calculate indirect treatment estimates for the networks given in figure 1 (star, ladder, and single loop) as long as the comparisons are made pairwise.

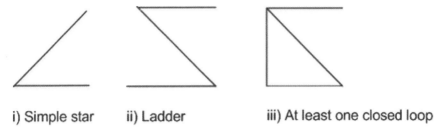

i) Simple star   ii) Ladder   iii) At least one closed loop

Figure 1. Examples of network patterns for the AITC

2.1 Syntax for indirect

Our command `indirect` assumes that Stata's `metan` command (Harris et al. 2008) has been installed. Because of the complexity of the syntax and to facilitate the ease of its implementation, we have included a dialog-box file, `indirect.dlg` (figure 2).

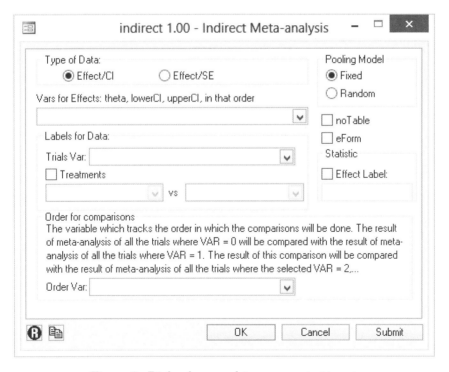

Figure 2. Dialog box used to process `indirect`

```
indirect varlist [if] [in], [random fixed eff(strvar) eform tabl
    trta(strvar) trtb(strvar)]
```

varlist contains a summary statistic (relative risk, odds ratio, HR) on the log scale and its standard error (SE) or a summary statistic on the log scale and its 95% CI limits; a variable that specifies the studies; and a variable that tracks the order in which the comparisons are done (trials comparing the same interventions will have the same order number).

2.2   Options

`random` specifies that a random-effects model should be used (the default).

`fixed` specifies that a fixed-effects model should be used.

`eff(strvar)` specifies the effect size (hazard ratio, relative risk, ..., etc.).

`eform` specifies that `eformat` should be used.

`tabl` specifies that the table of studies used should be displayed.

`trta(strvar)` specifies the experimental treatment.

`trtb(strvar)` specifies the standard treatment.

3   Example: Zoledronate versus Pamidronate in multiple myeloma

We illustrate the `indirect` command by using data from a systematic review of 13 studies on the effects of bisphosphonates on overall survival in patients with multiple myeloma (Mhaskar et al. 2012). The network is given in figure 3, while the trials, logHRs, and their SEs are presented in table 1. The dashed lines represent indirect comparisons. Suppose we wish to indirectly compare Zoledronate with Pamidronate (30mg) and Pamidronate (90mg) under the random-effects model. For this comparison, we discard Clodronate, Etidronate, and Ibandronate. Because there may be many trials comparing the same interventions, a variable, `order`, is introduced to keep track of comparisons being made (table 2).

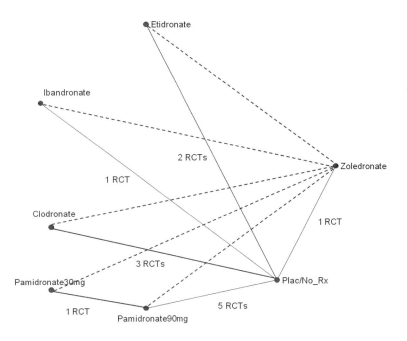

Figure 3. Evidence network of the reported 13 randomized controlled trials of bisphosphonates for overall survival in patients with multiple myeloma

Table 1. Effects of bisphosphonates on overall survival in multiple myeloma patients

Study	Active	Control	ln(HR), SE{ln(HR)}
Avilés et al. (2007)	Zoledronate	Plac/NoRx	−0.859, 0.333
Delmas et al. (1982)	Clodronate	Plac/NoRx	1.288, 0.894
Lahtinen et al. (1992)	Clodronate	Plac/NoRx	−0.287, 0.181
McCloskey et al. (2001)	Clodronate	Plac/NoRx	−0.016, 0.095
Belch et al. (1991)	Etidronate	Plac/NoRx	0.461, 0.198
Daragon et al. (1993)	Etidronate	Plac/NoRx	0.071, 0.034
Menssen et al. (2002)	Ibandronate	Plac/NoRx	0.063, 0.221
Brincker et al. (1998)	Pamidronate90mg	Plac/NoRx	−0.107, 0.945
Kraj et al. (2000)	Pamidronate90mg	Plac/NoRx	0.1168, 0.4
Terpos et al. (2000)	Pamidronate90mg	Plac/NoRx	−2.08, 1.41
Berenson et al. (1998)	Pamidronate90mg	Plac/NoRx	−0.29, 0.167
Musto et al. (2003)	Pamidronate90mg	Plac/NoRx	−0.02, 0.203
Gimsing et al. (2010)	Pamidronate30mg	Pamidronate90mg	−0.050, 0.120

Table 2. Network branches used to compare Zoledronate with Pamidronate

Study	Active	Control	ln(HR), SE{ln(HR)}	Order
Avilés et al. (2007)	Zoledronate	Plac/NoRx	−0.859, 0.333	0
Brincker et al. (1998)	Pamidronate90mg	Plac/NoRx	−0.107, 0.945	1
Kraj et al. (2000)	Pamidronate90mg	Plac/NoRx	0.1168, 0.4	1
Terpos et al. (2000)	Pamidronate90mg	Plac/NoRx	−2.08, 1.41	1
Berenson et al. (1998)	Pamidronate90mg	Plac/NoRx	−0.29, 0.167	1
Musto et al. (2003)	Pamidronate90mg	Plac/NoRx	−0.02, 0.203	1
Gimsing et al. (2010)	Pamidronate30mg	Pamidronate90mg	−0.050, 0.120	2

```
. use example

. indirect ln_hr se_ln_hr study order,  random eff(HR) eform trta(inn_rx)
> trtb(std_rx)
Meta-Analysis: comparing treatments  Zoledronate  and  Plac/No_Rx
Exponential Statistic HR = .424
Log statistic ln(HR) = -.859 and standard error = .333(var = .111)
----------------------------------------
Meta-Analysis: comparing treatments Pamidronate90  and  Plac/No_Rx
Exponential Statistic HR= .846
Log statistic ln(HR) = -.167 and standard error = .121 (var = .015)
----------------------------------------
----------------------------------------
Indirect comparison: Zoledronate vs Pamidronate90
Exponential Statistic HR =.5 with CI [ .25, 1.003]
Log statistic ln(HR) = -.692 and standard error =  .355 (var = .126)
Confidence Interval: [-1.387, .003]
Heterogeneity statistic ChiSquared: 3.81, p-value: = .051
----------------------------------------
Meta-Analysis: comparing treatments Pamidronate30  and  Pamidronate90
Exponential Statistic HR= .951
Log statistic ln(HR) = -.05 and standard error = .12 (var = .014)
----------------------------------------
----------------------------------------
Indirect comparison: Zoledronate vs Pamidronate30
Exponential Statistic HR =.526 with CI [ .253, 1.096]
Log statistic ln(HR) = -.642 and standard error =  .374 (var = .14)
Confidence Interval: [-1.376, .092]
Heterogeneity statistic ChiSquared: =2.942, p-value: = .086
```

In the network in figure 3, the indirect estimates favor Zoledronate over Pamidronate 30mg (HR = 0.526, 95% CI: [0.253, 1.096]) and Pamidronate 90mg (HR = 0.5, 95% CI: [0.25, 1.003]); however, they are both statistically nonsignificant. Both heterogeneity statistics are nonsignificant at $P = 0.05$. The remaining AITC estimates are given in table 3, and all significantly favor Zoledronate over Clodronate, Etidronate, and Ibandronate.

Table 3. Indirect comparison of Zoledronate versus other bisphosphonates under random effects

Comparison	Branches used: HR [95% CI]	HR [95% CI]
Zol versus Clo	Zol versus Plac: 0.42 [0.22, 0.81]	
	Clo versus Plac: 0.93 [0.77, 1.51]	0.46 [0.22, 0.95]
Zol versus Etid	Zol versus Plac: 0.42 [0.22, 0.81]	
	Etid versus Plac: 1.24 [0.66, 1.29]	0.34 [0.16, 0.72]
Zol versus Iban	Zol versus Plac: 0.42 [0.22, 0.81]	
	Iban versus Plac: 1.07 [0.69, 1.64]	0.39 [0.18, 0.87]
Zol versus Pam30	Zol versus Plac: 0.42 [0.22, 0.81]	
	Pam90 versus Plac: 0.85 [0.67, 1.07]	
	Pam30 versus Pam90: 0.95 [0.75, 1.2]	0.526 [0.253, 1.096]
Zol versus Pam90	Zol versus Plac: 0.42 [0.22, 0.81]	
	Pam versus Plac: 0.85 [0.67, 1.07]	0.5 [0.25, 1.003]

4 Conclusion

The application of indirect methods has grown in journal publications, and the issues related to the bias and power of indirect meta-analysis are well documented (Song et al. 2009; Mills et al. 2011). In the absence of direct treatment comparisons and in less complex networks, the AITC method we implemented has been found more favorable to Bayesian mixed-treatment comparisons because of its simplicity (O'Regan et al. 2009). Also, in the absence of systematic bias in primary studies, both methods are on average unbiased (Song et al. 2012). The major limitation of AITC is its inability to satisfactorily handle correlations that may exist between treatment effects in multiarm trials, which is a major advantage of the Bayesian approach. Bayesian methods can also be used to analyze more complex networks of evidence and can include study-level covariates. As we have pointed out, however, these advantages have to be weighed against their complexities. Because of the complexity and diversity of methods involved, the tools for the critical appraisal of methods do not yet exist, though there has been work in recent years to establish guidelines for conducting and interpreting indirect treatment comparisons (Jansen et al. 2011; Hoaglin et al. 2011). Direct meta-analyses estimate average treatment effects across trials and are fairly straightforward to interpret. The results obtained from indirect treatment comparisons rarely are, and readers are cautioned to interpret them with skepticism.

5   References

Avilés, A., M. J. Nambo, N. Neri, C. Castañeda, S. Cleto, and J. Huerta-Guzmán. 2007. Antitumor effect of zoledronic acid in previously untreated patients with multiple myeloma. *Medical Oncology* 24: 227–230.

Belch, A. R., D. E. Bergsagel, K. Wilson, S. O'Reilly, J. Wilson, D. Sutton, J. Pater, D. Johnston, and B. Zee. 1991. Effect of daily etidronate on the osteolysis of multiple myeloma. *Journal of Clinical Oncology* 9: 1397–1402.

Berenson, J. R., A. Lichtenstein, L. Porter, M. A. Dimopoulos, R. Bordoni, S. George, A. Lipton, A. Keller, O. Ballester, M. Kovacs, H. Blacklock, R. Bell, J. F. Simeone, D. J. Reitsma, M. Heffernan, J. Seaman, and R. D. Knight. 1998. Long-term pamidronate treatment of advanced multiple myeloma patients reduces skeletal events. Myeloma Aredia Study Group. *Journal of Clinical Oncology* 16: 593–602.

Brincker, H., J. Westin, N. Abildgaard, P. Gimsing, I. Turesson, M. Hedenus, J. Ford, and A. Kandra. 1998. Failure of oral pamidronate to reduce skeletal morbidity in multiple myeloma: A double-blind placebo-controlled trial. *British Journal of Haematology* 101: 280–286.

Bucher, H. C., G. H. Guyatt, L. E. Griffith, and S. D. Walter. 1997. The results of direct and indirect treatment comparisons in meta-analysis of randomized controlled trials. *Journal of Clinical Epidemiology* 50: 638–691.

Caldwell, D. M., A. E. Ades, and J. P. T. Higgins. 2005. Simultaneous comparison of multiple treatments: Combining direct and indirect evidence. *British Medical Journal* 331: 897–900.

Chaimani, A., H. S. Vasiliadis, N. Pandis, C. H. Schmid, N. J. Welton, and G. Salanti. 2013. Effects of study precision and risk of bias in networks of interventions: A network meta-epidemiological study. *International Journal of Epidemiology* 42: 1120–1131.

Daragon, A., C. Humez, C. Michot, X. Le Loet, B. Grosbois, F. Pouyol, L. Euller-Ziegler, I. Azais, J. F. Bernard, and J. F. Menard. 1993. Treatment of multiple myeloma with etidronate: Results of a multicentre double-blind study. Groupe d'Etudes et de Recherches sur le Myélome (GERM). *European Journal of Medicine* 2: 449–452.

Delmas, P. D., S. Charhon, M. C. Chapuy, E. Vignon, D. Briancon, C. Edouard, and P. J. Meunier. 1982. Long-term effects of dichloromethylene diphosphonate (CI2MDP) on skeletal lesions in multiple myeloma. *Metabolic Bone Disease and Related Research* 4: 163–168.

Gimsing, P., K. Carlson, I. Turesson, P. Fayers, A. Waage, A. Vangsted, A. Mylin, C. Gluud, G. Juliusson, H. Gregersen, H. Hjorth-Hansen, I. Nesthus, I. M. S. Dahl, J. Westin, J. L. Nielsen, L. M. Knudsen, L. Ahlberg, M. Hjorth, N. Abildgaard, N. F. Andersen, O. Linder, and F. Wisløff. 2010. Effect of pamidronate 30 mg versus 90 mg on physical function in patients with newly diagnosed multiple myeloma

(Nordic Myeloma Study Group): A double-blind, randomised controlled trial. *Lancet Oncology* 11: 973–982.

Glenny, A.-M., D. G. Altman, F. Song, C. Sakarovitch, J. J. Deeks, R. D'Amico, M. J. Bradburn, and A. J. Eastwood. 2005. Indirect comparisons of competing interventions. *Health Technology Assessment* 9: 1–134.

Harris, R. J., M. J. Bradburn, J. J. Deeks, R. M. Harbord, D. G. Altman, and J. A. C. Sterne. 2008. metan: Fixed- and random-effects meta-analysis. *Stata Journal* 8: 3–28. (Reprinted in this collection on pp. 29–54.)

Hoaglin, D. C., N. Hawkins, J. P. Jansen, D. A. Scott, R. Itzler, J. C. Cappelleri, C. Boersma, D. Thompson, K. M. Larholt, M. Diaz, and A. Barrett. 2011. Conducting indirect-treatment-comparison and network-meta-analysis studies: Report of the ISPOR Task Force on Indirect Treatment Comparisons Good Research Practices: Part 2. *Value in Health* 14: 429–437.

Jansen, J. P., B. Crawford, G. Bergman, and W. Stam. 2008. Bayesian meta-analysis of multiple treatment comparisons: An introduction to mixed treatment comparisons. *Value in Health* 11: 956–964.

Jansen, J. P., R. Fleurence, B. Devine, R. Itzler, A. Barrett, N. Hawkins, K. Lee, C. Boersma, L. Annemans, and J. C. Cappelleri. 2011. Interpreting indirect treatment comparisons and network meta-analysis for health-care decision making: Report of the ISPOR Task Force on Indirect Treatment Comparisons Good Research Practices: Part 1. *Value in Health* 14: 417–428.

Kraj, M., R. Pogłód, J. Pawlikowsky, and S. Maj. 2000. The effect of long-term pamidronate treatment on skeletal morbidity in advanced multiple myeloma. *Acta Haematologica Polonica* 31: 379–389.

Lahtinen, R., M. Laakso, I. Palva, I. Elomaa, and P. Virkkunen. 1992. Randomised, placebo-controlled multicentre trial of clodronate in multiple myeloma. *Lancet* 340: 1049–1052.

Lu, G., and A. E. Ades. 2004. Combination of direct and indirect evidence in mixed treatment comparisons. *Statistics in Medicine* 23: 3105–3124.

Lumley, T. 2002. Network meta-analysis for indirect treatment comparisons. *Statistics in Medicine* 21: 2313–2324.

McCloskey, E. V., J. A. Dunn, J. A. Kanis, I. C. MacLennan, and M. T. Drayson. 2001. Long-term follow-up of a prospective, double-blind, placebo-controlled randomized trial of clodronate in multiple myeloma. *British Journal of Haematology* 113: 1035–1043.

Menssen, H. D., A. Sakalová, A. Fontana, Z. Herrmann, C. Boewer, T. Facon, M. R. Lichinitser, C. R. J. Singer, L. Euller-Ziegler, M. Wetterwald, D. Fiere, M. Hrubisko, E. Thiel, and P. D. Delmas. 2002. Effects of long-term intravenous ibandronate therapy on skeletal-related events, survival, and bone resorption markers in patients with advanced multiple myeloma. *Journal of Clinical Oncology* 20: 2353–2359.

Mhaskar, R., J. Redzepovic, K. Wheatley, A. Clark, B. Miladinovic, A. Glasmacher, A. Kumar, and B. Djulbegovic. 2012. Bisphosphonates in multiple myeloma: A network meta-analysis. *Cochrane Database of Systematic Reviews* 5: CD003188.

Mills, E. J., I. Ghement, C. O'Regan, and K. Thorlund. 2011. Estimating the power of indirect comparisons: A simulation study. *PLOS ONE* 6: e16237.

Musto, P., A. Falcone, G. Sanpaolo, C. Bodenizza, N. Cascavilla, L. Melillo, P. R. Scalzulli, M. Dell'Olio, A. La Sala, S. Mantuano, M. Nobile, and A. M. Carella. 2003. Pamidronate reduces skeletal events but does not improve progression-free survival in early-stage untreated myeloma: Results of a randomized trial. *Journal of Leukemia and Lymphoma* 44: 1545–1548.

O'Regan, C., I. Ghement, O. Eyawo, G. H. Guyatt, and E. J. Mills. 2009. Incorporating multiple interventions in meta-analysis: An evaluation of the mixed treatment comparison with the adjusted indirect comparison. *Trials* 10: 86.

Salanti, G., and C. H. Schmid. 2012. Research Synthesis Methods special issue on network meta-analysis: Introduction from the editors. *Research Synthesis Methods* 3: 69–70.

Song, F., A. Clark, M. O. Bachmann, and J. Maas. 2012. Simulation evaluation of statistical properties of methods for indirect and mixed treatment comparisons. *BMC Medical Research Methodology* 12: 138.

Song, F., Y. K. Loke, T. Walsh, A.-M. Glenny, A. J. Eastwood, and D. G. Altman. 2009. Methodological problems in the use of indirect comparisons for evaluating healthcare interventions: Survey of published systematic reviews. *British Medical Journal* 338: b1147.

Terpos, E., J. Palermos, K. Tsionos, K. Anargyrou, N. Viniou, P. Papassavas, J. Meletis, and X. Yataganas. 2000. Effect of pamidronate administration on markers of bone turnover and disease activity in multiple myeloma. *European Journal of Haematology* 65: 331–336.

Wells, G. A., S. A. Sultan, L. Chen, M. Khan, and D. Coyle. 2009. Indirect evidence: Indirect treatment comparisons in meta-analysis. Ottawa: Canadian Agency for Drugs and Technologies in Health. http://www.cadth.ca/en/products/health-technology-assessment/publication/884.

White, I. R., J. K. Barrett, D. Jackson, and J. P. T. Higgins. 2012. Consistency and inconsistency in network meta-analysis: Model estimation using multivariate meta-regression. *Research Synthesis Methods* 3: 111–125.

The Stata Journal (2015)
15, Number 4, st0410, forthcoming

Network meta-analysis

Ian R. White
MRC Biostatistics Unit
Cambridge Institute of Public Health
Cambridge, UK
ian.white@mrc-bsu.cam.ac.uk

Abstract. Network meta-analysis is a popular way to combine results from several studies (usually randomized trials) comparing several treatments or interventions. It has usually been performed in a Bayesian setting, but recently it has become possible in a frequentist setting using multivariate meta-analysis and meta-regression, implemented in Stata with `mvmeta`. I describe a suite of Stata programs for network meta-analysis that perform the necessary data manipulation, fit consistency and inconsistency models using `mvmeta`, and produce various graphics.

Keywords: st0410, network components, network convert, network forest, network import, network map, network meta, network pattern, network query, network rank, network setup, network sidesplit, network table, network unset, network meta-analysis, multiple treatments meta-analysis, mixed-treatment comparisons

1 Introduction

Network meta-analysis, also called multiple treatments meta-analysis or mixed-treatment comparisons, is a popular way to combine evidence on multiple studies (usually randomized trials) comparing multiple treatments (or other interventions). Its key feature is the ability to combine direct and indirect evidence; for example, the comparison of treatments A and B is performed both using studies that directly compare A with B (direct evidence) and using studies that compare A with C and B with C (indirect evidence). Good general introductions are given by Mills, Thorlund, and Ioannidis (2013) for the concepts and Salanti et al. (2008) for the statistical methods.

A key issue in network meta-analysis is whether the network is consistent—that is, whether the direct evidence agrees with the indirect evidence (and, if there are multiple sources of indirect evidence, whether they agree with each other). Statistical models for inconsistency have been proposed and can be used to assess consistency (Lu and Ades 2006; Higgins et al. 2012).

Estimation of network meta-analysis models has usually been done in a Bayesian framework, with fitting in WinBUGS (Lu and Ades 2004). Frequentist estimation is possible by expressing the consistency and inconsistency models as multivariate random-effects meta-analysis or meta-regression (White et al. 2012). Graphical methods are well developed for presenting the results of network meta-analysis (Salanti, Ades, and Ioannidis 2011), although the individual study data are often not displayed.

My `mvmeta` package performs multivariate meta-analysis and meta-regression (White 2009, 2011). It can therefore be used to perform frequentist estimation of network meta-analysis models (White et al. 2012). However, difficulties remain in getting the data into the correct format and in specifying the `mvmeta` models. Many graphical tools for displaying the evidence base and the results of network meta-analysis have been written by Chaimani et al. (2013). Estimation of indirect treatment comparisons along a single path can also be done using the `indirect` command (Miladinovic et al. 2014).

In this paper, I introduce a suite of programs for performing network meta-analysis in Stata. The main aim of these programs is to provide a convenient tool that 1) performs the necessary data manipulation, allowing different data formats, 2) fits consistency and inconsistency models using `mvmeta`, and 3) produces various graphics, including a display of the individual study data. I also point out some methodological advances.

2   Model for network meta-analysis

I briefly describe the general model here; more details are given by White et al. (2012). The general model is a model for treatment contrasts (the "contrast-based" model of Salanti et al. [2008]). It allows for both heterogeneity (variation in the true treatment effect between studies) and inconsistency (additional variation in the true treatment effect between designs), where a design is the set of treatments compared in a study.

Consider a network including a total of T treatments: A, B, C, etc. Any treatment can be chosen as a reference treatment; for simplicity, let's choose A. Initially, assume that treatment A is included in every study. Let $d = 1, \ldots, D$ index the designs. Let y_{di}^{AJ} be the observed contrast of treatment J ($J = $ B, C, $\ldots$) with treatment A in the ith study in the dth design. y_{di}^{AJ} may represent any measure, such as a mean difference, a standardized mean difference, a log risk-ratio, or a log odds-ratio.

The model for the observed data is

$$y_{di}^{AJ} = \delta^{AJ} + \beta_{di}^{AJ} + \omega_d^{AJ} + \varepsilon_{di}^{AJ}, \quad J = \text{B}, \text{C}, \ldots \tag{1}$$

where the meaning of each term is described below. Equivalently, in vector notation with $\mathbf{y}_{di} = (y_{di}^{AB}, y_{di}^{AC}, \ldots)'$, the model is

$$\mathbf{y}_{di} = \boldsymbol{\delta} + \boldsymbol{\beta}_{di} + \boldsymbol{\omega}_d + \boldsymbol{\varepsilon}_{di} \tag{2}$$

where $\boldsymbol{\delta} = (\delta^{AB}, \delta^{AC}, \ldots)'$, $\boldsymbol{\beta}_{di} = (\beta_{di}^{AB}, \beta_{di}^{AC}, \ldots)'$, $\boldsymbol{\omega}_d = (\omega_d^{AB}, \omega_d^{AC}, \ldots)'$, and $\boldsymbol{\varepsilon}_{di} = (\varepsilon_{di}^{AB}, \varepsilon_{di}^{AC}, \ldots)'$ are described below.

Treatment contrasts. In (1) and (2), δ^{AJ} represents a treatment contrast (a summary effect) between J and A. The δ^{AJ} ($J = $ B, C, $\ldots$) are regarded as fixed parameters and are the parameters of primary interest.

Heterogeneity. In (1) and (2), β_{di}^{AJ} represents heterogeneity in the J–A contrast between studies within designs. The heterogeneity terms $\boldsymbol{\beta}_{di}$ are taken as random effects

$$\boldsymbol{\beta}_{di} \sim N(\mathbf{0}, \boldsymbol{\Sigma}) \qquad (3)$$

as in the conventional random-effects model for meta-analysis. Model (3) without constraint on $\boldsymbol{\Sigma}$ (the "unstructured" model) allows each contrast to have a different heterogeneity variance. The positive definiteness of $\boldsymbol{\Sigma}$ ensures the "second-order consistency conditions" of Lu and Ades (2009). However, there are rarely enough studies to identify the unstructured model, and it is usual to assume that all treatment contrasts have the same heterogeneity variance τ^2. Hence, it is usual to assume that

$$\boldsymbol{\Sigma} = \tau^2 \mathbf{P} \qquad (4)$$

where $\mathbf{P}$ is a matrix with all diagonal entries equal to 1 and all off-diagonal entries equal to 0.5 (Higgins and Whitehead 1996).

Inconsistency. In (1) and (2), ω_d^{AJ} represents inconsistency in the J–A contrast between designs. The inconsistency terms ω_d^{AJ} are taken to be fixed parameters (White et al. 2012), although random inconsistency terms are also possible (Lumley 2002; Jackson et al. 2014). We can include a maximal set of inconsistency parameters, which White et al. (2012) term the "design-by-treatment interaction model", or a smaller set (Lu and Ades 2006). For the consistency model, we set $\omega_d^{AJ} = 0$ for all d, J.

Within-study error. In (1) and (2), ε_{di}^{AJ} is a within-study error term. We assume $\varepsilon_{di} \sim N(0, \mathbf{S}_{di})$, where $\mathbf{S}_{di}$ is assumed to be known.

Two missing-data problems can arise. First, design d may contain the reference treatment A but not some other treatments. Here we use the likelihood implied by (2) for the observed subvector of $\mathbf{y}_{di}$. A harder problem arises when design d excludes A. Here we still use (2), but either we add a reference treatment arm with a very small amount of data (the "augmented" approach) or we apply the model only to the contrasts that are actually estimated in each particular design (the "standard" approach), as shown in section 3.1.

3 The network commands

3.1 Data formats

I illustrate the data formats with the smoking data that were given by Hasselblad (1998) and used by Lu and Ades (2006) and White (2011). The raw data comprise the number of individuals randomized to a treatment (n) and the number of those individuals who are quitting smoking (d) for each arm of each study. The four treatments are here coded A, B, C, and D. For brevity, I show only the first four studies. The raw data can be stored in a long format, as follows:

```
. list if study<=4, noobs clean
    study    treat     d      n
        1        A     9    140
        1        C    23    140
        1        D    10    138
        2        B    11     78
        2        C    12     85
        2        D    29    170
        3        A    79    702
        3        B    77    694
        4        A    18    671
        4        B    21    535
```

Alternatively, the data can be stored in a wide format:

```
. list if study<=4, noobs clean
    study   dA    nA    dB    nB    dC    nC    dD    nD
        1    9   140     .     .    23   140    10   138
        2    .     .    11    78    12    85    29   170
        3   79   702    77   694     .     .     .     .
        4   18   671    21   535     .     .     .     .
```

Note that studies 1 and 2 are three-arm studies and the others are two-arm studies.

The **network** suite uses three data formats: augmented, standard, and pairs formats. They are illustrated here with the log odds-ratio as the effect measure.

In the augmented format, all treatments are compared with a reference treatment (here, treatment A), and studies without the reference treatment (for example, study 2) have a reference treatment arm created with a small amount of data (White 2011). In the listing below, arm A has been created in study 2 with 0.001 observations and mean 0.156. Usually, augmentation has a negligible effect on results (see section 4.1), but in some models, unidentified parameters can be estimated with large standard errors.

```
. list study _y* _S_B_B - _S_C_D if study<=4, noobs clean
   study    _y_B     _y_C    _y_D    _S_B_B    _S_B_C    _S_B_D    _S_C_C    _S_C_D
       1       .     1.05    0.13         .         .         .      0.17      0.12
       2   -0.12    -0.12    0.11   7589.01   7588.90   7588.90   7589.00   7588.90
       3   -0.02        .       .      0.03         .         .         .         .
       4    0.39        .       .      0.11         .         .         .         .
. * _S_D_D omitted to save space
```

In the standard format, each study has its own reference treatment to which the other treatments are compared. Unlike for the augmented format, the treatment contrast (for example, variable _y_1 in the output below) represents different contrasts in different studies, and variable _contrast_1 specifies the contrast. Three-arm studies also have values for _y_2 representing the contrast specified by _contrast_2.

```
. list study _contrast* _y* _S* if study<=4, noobs clean abbreviate(12)
   study   _contrast_1   _contrast_2    _y_1    _y_2    _S_1_1    _S_1_2    _S_2_2
       1       C - A        D - A      1.05    0.13      0.17      0.12      0.23
       2       C - B        D - B      0.00    0.23      0.20      0.11      0.15
       3       B - A                  -0.02       .      0.03         .         .
       4       B - A                   0.39       .      0.11         .         .
```

In the above formats, there is one record for each study. In the pairs format, used by Chaimani et al. (2013), there is one record for each possible contrast in each study, meaning that two-arm studies have a single record but three-arm studies have three records. Again, variable _contrast labels the contrasts:

```
. list study _contrast _y _stderr if study<=4, noobs clean abbreviate(12)
    study   _contrast      _y   _stderr
        1       C - A    1.05      0.41
        1       D - A    0.13      0.48
        1       D - C   -0.92      0.40
        2       C - B    0.00      0.45
        2       D - B    0.23      0.38
        2       D - C    0.22      0.37
        3       B - A   -0.02      0.17
        4       B - A    0.39      0.33
```

3.2   The network setup command

network setup imports data from a set of studies reporting count data (events, total number) or quantitative data (mean, standard deviation, total number) for two or more treatments. The data may be in long format (one record per treatment per study) or in wide format (one record per study) and may be imported into the augmented, standard, or pairs format.

After running network setup, the dataset contains various settings that are required by subsequent network commands; the settings are stored as characteristics and may be viewed using network query. In particular, each treatment has a code (typically A, B, C, but numerical codes are possible) and a name (typically a descriptive string). Subsequent analyses always use the treatment codes, while descriptive and graphical commands by default use the treatment names.

Syntax

For count data:

network setup *eventvar nvar* $\left[\,if\,\right]$ $\left[\,in\,\right]$, studyvar(*varname*) $\left[\,\text{or}\,|\,\text{rr}\,|\,\text{rd}\,|\,\text{hr}\right.$
   zeroadd(#) *common_options* $\left.\right]$

For quantitative data:

network setup *meanvar sdvar nvar* $\left[\,if\,\right]$ $\left[\,in\,\right]$, studyvar(*varname*) $\left[\,\text{md}\,|\,\text{smd}\right.$
   common_options $\left.\right]$

If the data are in wide format, then *eventvar nvar* or *meanvar sdvar nvar* are stubs for variable names and `trtvar(varname)` is not specified. For example, in the wide format in section 3.1, `network setup d n, studyvar(study)` would be appropriate.

If the data are in long format, then *eventvar nvar* or *meanvar sdvar nvar* are the variable names and `trtvar(varname)` is required. For example, in the long format in section 3.1, `network setup d n, studyvar(study) trtvar(treat)` would be appropriate.

common_options are the following:

<u>trtvar</u>(*varname*) <u>armv</u>ars(drop|keep[(*varlist*)]) trtlist(*string*) alpha

   <u>num</u>codes <u>noc</u>odes format(augmented|standard|pairs) <u>genp</u>refix(*string*)

   <u>gens</u>uffix(*string*) ref(*string*) augment(#) augmean(#) augsd(#)

   <u>augo</u>verall

Options describing the data

studyvar(*varname*) specifies the study variable. `studyvar()` is required.

trtvar(*varname*) specifies the treatment variable (implies the long format).

armvars(drop|keep[(*varlist*)]) is relevant only when the data are in long format and there are extra arm-level variables in the data. In this case, the easiest option is `armvars(drop)` to drop all extra arm-level variables.

Options specifying how treatments are coded

trtlist(*string*) specifies the list of treatment names to be used, which is useful if you want to omit some treatments, for example, for a sensitivity analysis. It is also useful to specify how the treatments will be coded (first treatment will be A, B, etc.). The default is to use all treatments found in alphabetical order; except that when `trtvar()` is numeric, the default is to use all treatments in numerical order.

alpha forces treatments to be coded in alphabetical order. This is the default except in long format when `trtvar()` is numeric with value labels.

numcodes codes treatments as numbers 1, 2, 3, ... or (if more than 9 treatments) 01, 02, 03, ... The default is to code treatments as letters A, B, C, ... or (if more than 26 treatments) AA, AB, AC, ...

nocodes uses the current treatment names as treatment codes. Treatment names are modified only if this is needed to make them valid Stata names. This option becomes increasingly awkward as treatment names become longer.

Options for count data

or specifies that the treatment effect be measured by the log odds-ratio (the default).

rr specifies that the treatment effect be measured by the log risk-ratio.

rd specifies that the treatment effect be measured by the risk difference.

hr specifies that the treatment effect be measured by the log hazard-ratio (also called the log rate-ratio). In this case, *nvar* must be the total person-time at risk, not the number of individuals.

zeroadd(#) specifies the number of successes and (except with the hr option) failures added to all arms of any study that contains a zero cell in any arm. The default is zeroadd(0.5).

Options for quantitative data

md specifies that the treatment effect be measured by the mean difference (the default).

smd specifies that the treatment effect be measured by the standardized mean difference, defined as the mean difference divided by the standard deviation, where the latter is computed pooled across all study arms. The formulas for Hedges's g in White and Thomas (2005) are used. These are unbiased estimators and involve corrections for small numbers of degrees of freedom. The covariance between g_1 and g_2 is taken as $J(\nu)^2 \left\{ \frac{\nu}{(\nu-2)N_0} + g_1 g_2 V(\nu) \right\}$, where ν is the degrees of freedom used to estimate the pooled standard deviation, N_0 is the sample size in the common reference group, and $V(\nu)$ and $J(\nu)$ are defined in White and Thomas (2005).

Use of the standardized mean difference has problems (Greenland, Schlesselman, and Criqui 1986). A new alternative is given by Lu, Brazier, and Ades (2013) and Lu, Kounali, and Ades (2014).

sdpool(on | off) specifies whether the standard deviation is pooled across arms in computing variances. The default, which follows metan, is sdpool(off) with md and sdpool(on) with smd. For multiarm studies, sdpool(on) pools across all arms.

Options for the format after setting up

format(augmented | standard | pairs) specifies the required format.

genprefix(*string*) specifies the prefix to be used before the default variable names (for example, y for treatment contrasts). The default is genprefix(_), where treatment contrasts are named _y*, etc.

gensuffix(*string*) specifies the suffix to be used after the default variable names. The default is no suffix.

Network meta-analysis

Options for augmented format

ref(*trtname*) specifies the name of the reference treatment.

augment(*#*) specifies the number of individuals to use when augmenting missing reference treatment arms. The default is augment(0.001).

augmean(*#*) specifies the mean outcome to use when augmenting missing reference treatment arms. The default is for each augmented study to use the weighted average of its arm-specific means.

augsd(*#*) applies only for quantitative data and indicates the standard deviation to use when augmenting missing reference treatment arms. The default is for each augmented study to use the weighted average of arm-specific standard deviations.

augoverall changes the default behavior for augmean() and augsd() to use the overall mean and standard deviation across all studies.

3.3   The network map command

network map draws a map of a network; that is, it shows which treatments are directly compared against which other treatments, and roughly how much information is available for each treatment and for each treatment comparison. network map works by calling networkplot (Chaimani et al. 2013) and has all the facilities of that command for displaying quantity and quality of evidence through weighting and coloring. network map's contribution is to offer more options for treatment placement, although better methods are available (Rücker and Schwarzer Forthcoming).

Syntax

network map [*if*] [*in*] [,

<u>circ</u>le[(*#*)] | <u>squa</u>re[(*#*)] | <u>tria</u>ngular[(*#*)] | <u>rand</u>om[(*#*)] <u>centre</u>

loc(*matname*) replace <u>impr</u>ove <u>list</u>loc <u>trtc</u>odes *graph_options*

networkplot_options]

Options

<u>circ</u>le[(*#*)] specifies that the treatments be placed around a circle. This is the most commonly used system and the default. The optional argument specifies the number of locations; the default is the number of treatments.

<u>squa</u>re[(*#*)] specifies that the treatments be placed in a square lattice. The optional argument specifies the number of rows and columns; the default is the square root of the number of treatments (rounded up).

328

triangular[(#)] specifies that the treatments be placed in a triangular lattice. The optional argument specifies the number of rows and the maximum row lengths; the default is approximately the square root of the number of treatments.

random[(#)] specifies that the treatments be randomly placed. The optional argument specifies the number of locations; the default is the number of treatments.

centre, only for use with circle(#), specifies to also place a treatment in the center.

loc(*matname*) specifies a treatment location matrix. This specifies where the treatments are placed on the map. The matrix should have at least as many rows as treatments and three columns containing the x coordinate, the y coordinate, and the clock position for each label. If the matrix does not exist, or if replace is specified, then a new matrix is created. If loc() is not specified, then a new matrix is created and stored in _network_map_location.

Note that the rows of *matname* are taken as the locations of the treatments in alphabetical order; row names are ignored. Thus, network map, loc(M) and network map if useit, loc(M) may place the treatments differently.

replace specifies that a new treatment location matrix be created.

improve requests an iterative procedure to improve the placement of the treatments. The algorithm swaps pairs of treatments if this reduces the number of line crossings. With this option, it is useful to increase the number of locations above the default; for example, in section 4.2 with eight treatments, we use triangular(5).

listloc specifies to print the treatment location matrix.

trtcodes specifies to use treatment codes rather than full treatment names.

graph_options are any of the options documented in [G-3] ***twoway_options***.

networkplot_options are any of the options documented in help networkplot, such as edgeweight and nodeweight (controlling which edges are thicker than others and which nodes are larger than others) and edgecolor (allowing edges to be colored according to evidence quality).

3.4   The network meta command

network meta defines and fits a consistency or inconsistency model. It can handle data in any of the three network formats. If data are in the augmented or standard formats, then the models are fit using mvmeta; if data are in the pairs format, then the models are fit using metareg. mvmeta or metareg must be installed. After fitting the model, the mvmeta or metareg command used can be recalled by pressing *F9*. network meta stores the results for use in network forest and network rank.

For inconsistency models, the design-by-treatment interaction model of Higgins et al. (2012) is used, unless the luades option is specified (see below). A Wald test for inconsistency is defined and performed; the command to test for inconsistency can be recalled by pressing *F8*.

Results using the three formats should be almost identical, except that results using the pairs format are wrong in the presence of multiarm studies; in this case, `network meta` issues a warning and stops but can be overruled with the `force` option.

Syntax

network <u>meta</u> <u>c</u>onsistency | <u>i</u>nconsistency $\begin{bmatrix} if \end{bmatrix}$ $\begin{bmatrix} in \end{bmatrix}$ $\begin{bmatrix} , \text{ \underline{reg}ress}(varlist) \end{bmatrix}$

<u>lua</u>des $\begin{bmatrix} (trtlist) \end{bmatrix}$ force <u>nowar</u>nings *mvmeta_options* $\big]$

Options

<u>reg</u>ress(*varlist*) specifies covariates for network meta-regression. Every treatment contrast is allowed to depend on the covariate(s) listed. This option is currently only allowed in the augmented format.

<u>lua</u>des$\begin{bmatrix} (trtlist) \end{bmatrix}$, for the inconsistency model, specifies the model of Lu and Ades (2006) as formalized by White et al. (2012). With only two-arm studies, this is the same as the design-by-treatment interaction model of White et al. (2012). With multi-arm studies, the Lu–Ades model is smaller than the design-by-treatment interaction model and depends on the treatment ordering. The optional argument specifies an ordering of the treatments. This option is only available in the augmented format.

force (not recommended) forces model fitting when `network meta` detects a difficulty. This could be a disconnected network; no degrees of freedom for inconsistency when an inconsistency model is specified; or no degrees of freedom for heterogeneity when a random-effects model is specified.

<u>nowar</u>nings (not recommended) suppresses warning messages.

mvmeta_options are any of the options documented in `help mvmeta`, such as `bscov()`. The default is to assume a common heterogeneity variance. This is implemented using `bscov(exch 0.5)`, a new shorthand for `bscov(prop P)`, where P is the matrix defined in (4).

3.5   The network rank command

`network rank` is used to rank treatments. It works only when the augmented format has been used to fit the model. Details are given in White (2011).

Syntax

network <u>rank</u> min | max $\begin{bmatrix} if \end{bmatrix}$ $\begin{bmatrix} in \end{bmatrix}$ $\begin{bmatrix} , \text{ \underline{trt}codes } mvmeta\_pbest\_options \end{bmatrix}$

Use `network rank min` if the best treatment is that with the lowest (most negative) treatment effect, and use `network rank max` if the best treatment is that with the highest (most positive) treatment effect.

Options

`trtcodes` specifies to use treatment codes rather than full treatment names.

mvmeta_pbest_options are any of the suboptions available for the `pbest()` option of `mvmeta`, but note that `network meta` makes sensible default choices for `if`, `in`, `zero`, and `id()`, which it would be unwise to change. The following options are likely to be useful:

   `all` reports probabilities for all ranks. The default is to report only the probabilities of being the best treatment.

   `reps(#)` sets the number of replicates; larger numbers reduce Monte Carlo error.

   `seed(#)` sets the random-number seed for reproducibility.

   `bar` draws a bar graph of ranks.

   `line` draws a line graph of ranks.

   `cumulative` makes the bar or line graph show cumulative ranks.

   `predict` ranks the true effects in a future study with the same covariates, thus allowing for heterogeneity as well as parameter uncertainty, as in the calculation of prediction intervals (Higgins, Thompson, and Spiegelhalter 2009). The default behavior is instead to rank linear predictors and does not allow for heterogeneity.

   `meanrank` tabulates the mean rank and the SUCRA (Salanti, Ades, and Ioannidis 2011). The SUCRA is the rescaled mean rank: it is 1 when a treatment is certain to be the best and 0 when a treatment is certain to be the worst.

   `saving(`*filename*`[ , replace ])` writes the draws from the posterior distribution (indexed by the identifier and the replication number) to *filename*. The `replace` option allows an existing *filename* to be overwritten.

   `clear` loads the rank data into memory and specifies the commands needed to reproduce the table and graph.

   `mcse` adds the Monte Carlo standard errors to the tables.

3.6   The network sidesplit command

`network sidesplit` fits the node-splitting model of Dias et al. (2010), for whom a "node" is a treatment contrast, for example, B versus A. I call this "side-splitting" because treatment contrasts are sides in the network map. To split the side B versus A, different parameters are used for the contrast of B versus A in studies containing both A and B (the direct parameter) and in other studies (the indirect parameter). The two parameters are estimated jointly and reported together with their difference and a test of whether the true difference is 0.

   `network sidesplit` currently only works with data in the augmented format.

Multi-arm studies

Multi-arm studies complicate this procedure. Suppose there is a study of A versus B versus C, and we split the B versus A contrast. The first part of table 1 shows the parameterization of the side-splitting model as proposed by Dias et al. (2010). With this model specification, the C versus A contrast is assumed to be the same in "direct" and "indirect" studies, while the C versus B contrast is allowed to differ. This model can be conceptualized by regarding B as a different treatment—say, B*—in the direct studies, where B* is ω units higher than B. But there is no reason why we should not reverse the roles of A and B; then the side-splitting model regards A, rather than B, as a different treatment A*, which is ω units lower than A. This gives the parameterization in the second block of table 1: here the C versus B contrast is assumed to be the same in direct and indirect designs, while the C versus A contrast is allowed to differ.

Table 1. Parameterizations of side-splitting models

Study	B versus A	C versus A	C versus B
Split B versus A as in Dias et al. (2010)			
ABC (direct)	$\delta^{AB} + \omega$	δ^{AC}	$\delta^{AC} - \delta^{AB} - \omega$
AC (indirect)		δ^{AC}	
BC (indirect)			$\delta^{AC} - \delta^{AB}$
Split A versus B as in Dias et al. (2010)			
ABC (direct)	$\delta^{AB} + \omega$	$\delta^{AC} + \omega$	$\delta^{AC} - \delta^{AB}$
AC (indirect)		δ^{AC}	
BC (indirect)			$\delta^{AC} - \delta^{AB}$
Symmetrical alternative			
ABC (direct)	$\delta^{AB} + \omega$	$\delta^{AC} + \omega/2$	$\delta^{AC} - \delta^{AB} - \omega/2$
AC (indirect)		δ^{AC}	
BC (indirect)			$\delta^{AC} - \delta^{AB}$

I propose a small change that treats A and B symmetrically (last block of table 1). Here, instead of allocating ω in the direct studies either fully to the C versus A contrast or fully to the C versus B contrast, it is shared between them. This model can be conceptualized by regarding both A and B as different treatments in the direct studies, where B* is $\omega/2$ units higher than B and A* is $\omega/2$ units lower than A. This method is intermediate between the two alternative ways (splitting B versus A and splitting A versus B) to implement the method of Dias et al. (2010). The symmetrical method is the default but can be changed using the `nosymmetric` option.

Syntax

network <u>sidesplit</u> *trtcode1 trtcode2*|all $\big[$ *if* $\big]$ $\big[$ *in* $\big]$ $\big[$, show <u>nosym</u>metric tau
    mvmeta_options $\big]$

network sidesplit *trtcode1 trtcode2* fits a single node-splitting model, while network
sidesplit all fits all appropriate node-splitting models.

Options

show shows the mvmeta calculation(s) and results.

nosymmetric uses the node-splitting model as originally specified by Dias et al. (2010),
    rather than the symmetrizing modification described above.

tau specifies to additionally display tau, the standard deviation of the between-studies
    heterogeneity.

mvmeta_options are any of the options documented in help mvmeta.

3.7   The network forest command

network forest draws a forest plot of network meta-analysis data, extending the idea
of Hawkins et al. (2009). For each contrast for which there is direct evidence (that is,
which is estimated within one or more studies), the forest plot displays the following
results:

1. "Studies": each study contributing direct evidence, grouped by design (that is,
   set of treatments in a study);

2. "Pooled within design": the pooled treatment effect in each design, estimated by
   the model most recently fit using network meta inconsistency; and

3. "Pooled overall": the overall treatment effect, estimated by the model most recently fit using network meta consistency.

Each of results 1, 2, and 3 is displayed as a point estimate and 95% confidence interval
(or other confidence level determined by set level or the level(#) option). The
marker representing each point estimate has size proportional to the inverse square
of the standard error. Because pooled estimates (results 2 and 3) allow for between-
studies heterogeneity, they may have wider confidence intervals and smaller markers
than study-specific estimates (result 1).

Syntax

network <u>fore</u>st $\left[\,if\,\right]$ $\left[\,in\,\right]$ $\left[\,,\,\right.$ <u>consi</u>stency(off) <u>incon</u>sistency(off) <u>l</u>ist
   clear <u>colo</u>rs(*string*) <u>contrasto</u>ptions(*string*) <u>trtc</u>odes contrastpos(*#*)
   <u>ncolu</u>mns(*#*) <u>colu</u>mns(*string*) force <u>diam</u>ond group(design | type) eform
   graph_options $\left.\right]$

Options controlling the summary treatment contrasts

consistency(off) omits the "Pooled overall" summaries from the forest plot.

inconsistency(off) omits the "Pooled within design" summaries from the forest plot.

Options controlling output

list lists the data for the forest plot.

clear clears the current data from memory so that the data for the forest plot can be
   loaded into memory. The forestplot command can then be recalled by pressing
   F9.

Options controlling graph appearance

colors(*string*) specifies up to three colors for the "Studies" results, "Pooled within design" results, and "Pooled overall" results, respectively. The default is colors(blue
   green red).

contrastoptions(*string*) are options for the text identifying the contrasts (for example,
   "C vs. B"). Any marker label options are possible, for example,
   contrastoptions(mlabsize(large) mlabcolor(red)).
   See [G-3] *marker_label_options*.

trtcodes specifies use of treatment codes rather than full treatment names.

contrastpos(*#*) specifies the value of the horizontal axis at which the text identifying
   the contrasts (for example, "C vs. B") is centered.

ncolumns(*#*) specifies the number of columns for the display. The default is automatically determined so that the number of rows per column is approximately 10 times
   the number of columns.

columns(*string*) specifies how to assign contrasts to columns. columns(xtile) assigns
   contrasts to columns in order using xtile and can lead to very unbalanced columns
   (that is, much more forest in one column than another). columns(smart) assigns
   contrasts to columns to optimize balance without keeping the logical order of the
   contrasts (so, for example, column 1 may contain "B vs. A" and "D vs. A" while
   column 2 contains "C vs. A"). The default is columns(smart).

force, only relevant when xlabel() is specified with a numlist, forces confidence intervals to not be truncated within the specified range. By default, confidence intervals are truncated within the range implied by xlabel(); truncated confidence intervals are indicated by arrows.

diamond specifies that summaries ("Pooled by design" and "Pooled overall") be displayed as diamonds. This is useful for monochrome printing.

group(design | type) specifies that, within comparisons, the forest plot may be ordered by design (showing the summary for each design after the studies for that design) or by type (showing all the studies and then all the summaries). The default is group(design) if inconsistency results are shown and group(type) otherwise.

eform labels the horizontal axis using the exponential of the values.

graph_options are any of the options documented in [G-3] ***twoway_options***, such as xlabel(), xline(0), or legend(pos(3) col(1)). The option most often needed is msize(*markersizestyle*) to change the marker sizes; the default is msize(*0.2), so try, for example, msize(*0.15) or msize(*0.3).

3.8 Utility commands

network import imports a dataset, either of pairwise comparisons using the syntax

> network <u>import</u> [*if*] [*in*], <u>study</u>var(*varname*) <u>treat</u>(*trtvar1 trtvar2*)
>   <u>eff</u>ect(*varname*) <u>stderr</u>(*varname*) [*options*]

where treat(*trtvar1 trtvar2*) specifies that the record compares *trtvar2* with *trtvar1*, effect(*varname*) specifies the point estimate for this comparison, and stderr(*varname*) specifies its standard error; or of comparisons with a reference treatment using the syntax

> network <u>import</u> [*if*] [*in*], <u>study</u>var(*varname*) <u>eff</u>ect(*effectstub*)
>   <u>var</u>iance(*varstub*) ref(*string*) [*options*]

where variables *effectstub_** contain the point estimates of the comparisons with reference, variables *varstub_*_** contain their variances and covariances, and ref() is the reference treatment used. See help network import for the complete list of options and their descriptions.

network convert converts between the three formats described. The syntax is

> network <u>con</u>vert augmented | standard | pairs [, large(*#*) ref(*trtcode*)]

The options are only relevant when converting to augmented format (see help convert for details). large(*#*) specifies the value used for the variance of contrasts with the reference treatment in studies without the reference treatment; the default is large(10000). ref(*trtcode*) specifies a new reference treatment.

network query displays the current network settings.

`network unset` removes the network settings.

`network table` tabulates network data. Data are reformatted and displayed using `tabdisp`. The syntax is

> network <u>tab</u>le $\begin{bmatrix} if \end{bmatrix}$ $\begin{bmatrix} in \end{bmatrix}$, $\begin{bmatrix} \underline{\text{trtc}}\text{odes} \enspace tabdisp\_options \end{bmatrix}$

where `trtcodes` specifies to use treatment codes rather than treatment names, and *tabdisp_options* are any of the options documented in `help tabdisp` except `cellvar()`. For example, `cellwidth(#)` may be useful to increase the column width to accommodate treatment names, and `stubwidth(#)` may be useful to increase the width of the study name column.

`network pattern` shows which treatments are used in which studies. This is done using the utility `misspattern`, which can display general patterns of missing data. The syntax is

> network <u>pattern</u> $\begin{bmatrix} if \end{bmatrix}$ $\begin{bmatrix} in \end{bmatrix}$ $\begin{bmatrix} , \enspace \underline{\text{trtc}}\text{odes} \enspace misspattern\_options \end{bmatrix}$

where `trtcodes` specifies to use treatment codes rather than treatment names, and *misspattern_options* are any of the options documented in `help misspattern`.

3.9 Requirements

Various parts of the `network` package require `mvmeta` version 3.1 or greater (White 2011), `metareg` (Harbord and Higgins 2008), and `networkplot` version 1.2 or greater (Chaimani et al. 2013).

4 Examples

4.1 Smoking network

I demonstrate the `network` package using the smoking data (Lu and Ades 2006), starting with the data in wide format, with `study` and `trt` as identifiers, `d` containing the number of events, and `n` containing the total number. In this version of the data, the treatments are coded 1–4 with labels "No contact", "Self help", "Individual counselling", and "Group counselling". `network setup` produces a dataset ready for `mvmeta`, coding the treatments A–D, using treatment A as reference and using the odds ratio as the measure of effect.

```
. use smoking
(Smoking data from Lu & Ades (2006))

. network setup d n, studyvar(study) trtvar(trt)
Treatments used
  A (reference):            No contact
  B:                        Self help
  C:                        Individual counselling
  D:                        Group counselling
```

```
Measure                               Log odds ratio
Studies
   ID variable:                       study
   Number used:                       24
   IDs with zero cells:               9 20
   - count added to all their cells:  .5
   IDs with augmented reference arm:  2 21 22 23 24
   - observations added:              0.001
   - mean in augmented observations:  study-specific mean
Network information
   Components:                        1 (connected)
   D.f. for inconsistency:            7
   D.f. for heterogeneity:            16
Current data
   Data format:                       augmented
   Design variable:                   _design
   Estimate variables:                _y*
   Variance variables:                _S*
   Command to list the data:          list study _y* _S*, noo sepby(_design)
```

Next, we tabulate the data, graph the patterns (shown in figure 1), and draw a network map (shown in figure 2):

```
. network table
```

			Treatment and Statistic					
study	− Group −		− Indivi −		− No con −		− Self h −	
	d	n	d	n	d	n	d	n
1	10	138	23	140	9	140		
2	29	170	12	85			11	78
3					79	702	77	694
4					18	671	21	535
5					8	116	19	146
6			363	714	75	731		
7			9	205	2	106		
8			237	1561	58	549		
9			9.5	49	.5	34		
10			31	98	3	100		
11			26	95	1	31		
12			17	77	6	39		
13			134	1031	95	1107		
14			35	504	15	187		
15			73	675	78	584		
16			54	888	69	1177		
17			107	761	64	642		
18			8	90	5	62		
19			34	237	20	234		
20	9.5	21			.5	21		
21			16	43			20	49
22	32	127					7	66
23	20	74	12	76				
24	3	26	9	55				

```
. network pattern
```

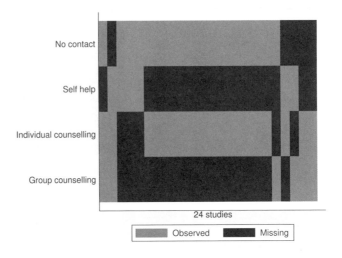

Figure 1. Network pattern for smoking data

```
. network map
```

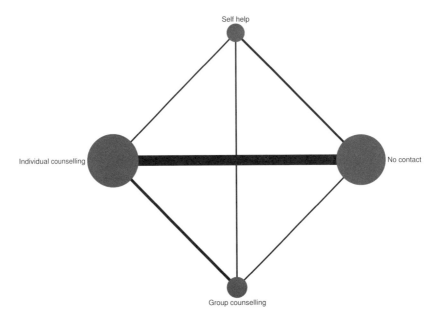

Figure 2. Network map for smoking data

Note that the treatment names are used when possible. They are abbreviated in the network table; this can be improved by using the `cellwidth()` option. The output shows that each pair of treatments is directly compared but that much of the data compares treatments A and C.

Next, we fit the consistency model:

```
. network meta consistency
Command is: mvmeta _y _S, bscovariance(exch 0.5) longparm suppress(uv mm)
> vars(_y_B _y_C _y_D)
Note: using method reml
Note: using variables _y_B _y_C _y_D
Note: 24 observations on 3 variables
Note: variance-covariance matrix is proportional to .5*I(3)+.5*J(3,3,1)

initial:      log likelihood = -60.906947
rescale:      log likelihood = -60.906947
rescale eq:   log likelihood = -60.694018
Iteration 0:  log likelihood = -60.694018
Iteration 1:  log likelihood = -59.279414
Iteration 2:  log likelihood = -59.252153
Iteration 3:  log likelihood = -59.252035
Iteration 4:  log likelihood = -59.252035

Multivariate meta-analysis
Variance-covariance matrix = proportional .5*I(3)+.5*J(3,3,1)
Method = reml                          Number of dimensions    =      3
Restricted log likelihood = -59.252035 Number of observations  =     24
```

	Coef.	Std. Err.	z	P>|z|	[95% Conf. Interval]	
_y_B						
_cons	.3984173	.3310951	1.20	0.229	-.2505171	1.047352
_y_C						
_cons	.7023359	.1991037	3.53	0.000	.3120999	1.092572
_y_D						
_cons	.8658433	.3762801	2.30	0.021	.1283477	1.603339

```
Estimated between-studies SDs and correlation matrix:
             SD        _y_B       _y_C       _y_D
_y_B   .67445374        1          .          .
_y_C   .67445374       .5          1          .
_y_D   .67445374       .5         .5          1
mvmeta command stored as F9
```

Note that treatment codes, not names, are used. The `mvmeta` model used is displayed (and stored in *F9*) so that the user can modify it if desired. The estimated log odds-ratio for intervention B compared with A is 0.398, etc. We use these results to find the probabilities that each treatment is the best (that is, has the highest odds) under the consistency model and to plot the rankogram (Salanti, Ades, and Ioannidis 2011), shown in figure 3.

```
. network rank max, line cumulative xlabel(1/4) seed(37195)
> tabdispoptions(cellwidth(15))
Command is: mvmeta, noest pbest(max  in 1, zero id(study) line cumulative
> xlabel(1/4) seed(37195) tabdispoptions(cellwidth(15)) stripprefix(_y_)
> zeroname(A) rename(A = No contact, B = Self help, C = Individual counselling,
> D = Group counselling))
Option line specified -> option all assumed

Estimated probabilities (%) of each treatment being the best (and other ranks)
- assuming the maximum parameter is the best
- using 1000 draws
- allowing for parameter uncertainty
```

study and Rank	Treatment			
	No contact	Self help	Individual coun	Group counselli
1				
Best	0.0	6.0	31.4	62.6
2nd	0.4	18.7	54.9	26.0
3rd	13.6	62.2	13.7	10.5
Worst	86.0	13.1	0.0	0.9

```
mvmeta command is stored in F9
```

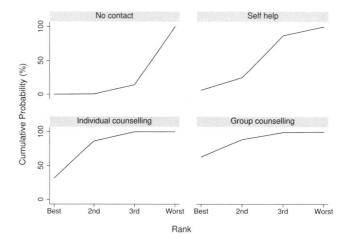

Figure 3. Rankogram for smoking data

The table and graph show, for example, that group counselling has a 62.6% probability of being the best treatment and about a 90% probability of being one of the two best treatments.

We now fit the inconsistency model:

```
. network meta inconsistency
Command is: mvmeta _y _S, bscovariance(exch 0.5) longparm suppress(uv mm)
> eq(_y_C: des_ACD des_BC des_BCD, _y_D: des_AD des_BCD des_BD des_CD)
> vars(_y_B _y_C _y_D)
Note: using method reml
Note: regressing _y_B on (nothing)
Note: regressing _y_C on des_ACD des_BC des_BCD
Note: regressing _y_D on des_AD des_BCD des_BD des_CD
Note: 24 observations on 3 variables
Note: variance-covariance matrix is proportional to .5*I(3)+.5*J(3,3,1)

initial:       log likelihood = -50.816796
rescale:       log likelihood = -50.816796
rescale eq:    log likelihood = -50.816796
Iteration 0:   log likelihood = -50.816796
Iteration 1:   log likelihood = -50.089407
Iteration 2:   log likelihood = -50.088702
Iteration 3:   log likelihood = -50.088702

Multivariate meta-analysis
Variance-covariance matrix = proportional .5*I(3)+.5*J(3,3,1)
Method = reml                                Number of dimensions   =      3
Restricted log likelihood = -50.088702       Number of observations =     24
```

	Coef.	Std. Err.	z	P>|z|	[95% Conf. Interval]	
_y_B						
_cons	.3299969	.4675425	0.71	0.480	-.5863696	1.246363
_y_C						
des_ACD	.3468324	.8821287	0.39	0.694	-1.382108	2.075773
des_BC	-.5261706	1.004485	-0.52	0.600	-2.494926	1.442584
des_BCD	-.3732418	1.013837	-0.37	0.713	-2.360325	1.613841
_cons	.7044606	.2347922	3.00	0.003	.2442763	1.164645
_y_D						
des_AD	3.393989	1.889991	1.80	0.073	-.3103244	7.098303
des_BCD	.4267799	1.3027	0.33	0.743	-2.126466	2.980026
des_BD	1.244879	1.323322	0.94	0.347	-1.348784	3.838543
des_CD	.8178211	1.123139	0.73	0.467	-1.383491	3.019133
_cons	.1285276	.8825026	0.15	0.884	-1.601146	1.858201

```
Estimated between-studies SDs and correlation matrix:
            SD        _y_B       _y_C       _y_D
_y_B  .74313772          1          .          .
_y_C  .74313772         .5          1          .
_y_D  .74313772         .5         .5          1
Testing for inconsistency:
 ( 1)  [_y_C]des_ACD = 0
 ( 2)  [_y_D]des_AD = 0
 ( 3)  [_y_C]des_BC = 0
 ( 4)  [_y_C]des_BCD = 0
 ( 5)  [_y_D]des_BCD = 0
 ( 6)  [_y_D]des_BD = 0
 ( 7)  [_y_D]des_CD = 0

        chi2( 7) =     5.11
      Prob > chi2 =    0.6464
mvmeta command stored as F9; test command stored as F8
```

The global test for inconsistency gives a p-value of 0.65, giving no evidence of inconsistency.

To see that the results are the same using the standard format, we convert the format and repeat the consistency analysis:

```
. network convert standard
Converting augmented to standard ...

. network meta consistency
Command is: mvmeta _y _S, bscovariance(exch 0.5) commonparm noconstant
> suppress(uv mm) eq(_y_1: _trtdiff1_B _trtdiff1_C _trtdiff1_D, _y_2:
> _trtdiff2_B _trtdiff2_C _trtdiff2_D) vars(_y_1 _y_2)
Note: using method reml
Note: regressing _y_1 on _trtdiff1_B _trtdiff1_C _trtdiff1_D
Note: regressing _y_2 on _trtdiff2_B _trtdiff2_C _trtdiff2_D
Note: 24 observations on 2 variables
Note: variance-covariance matrix is proportional to .5*I(2)+.5*J(2,2,1)

initial:       log likelihood = -34.471727
rescale:       log likelihood = -34.471727
rescale eq:    log likelihood = -34.259312
Iteration 0:   log likelihood = -34.259312
Iteration 1:   log likelihood = -32.844434
Iteration 2:   log likelihood = -32.817174
Iteration 3:   log likelihood = -32.817057
Iteration 4:   log likelihood = -32.817057

Multivariate meta-analysis
Variance-covariance matrix = proportional .5*I(2)+.5*J(2,2,1)
Method = reml                          Number of dimensions   =       2
Restricted log likelihood = -32.817057 Number of observations =      24
```

	Coef.	Std. Err.	z	P>|z|	[95% Conf. Interval]	
_y_1						
_trtdiff1_B	.3984487	.3311044	1.20	0.229	-.250504	1.047401
_trtdiff1_C	.7023556	.1991095	3.53	0.000	.3121081	1.092603
_trtdiff1_D	.8658902	.3762928	2.30	0.021	.1283699	1.603411

```
The above coefficients also apply to the following equations:
    _y_2: _trtdiff2_B _trtdiff2_C _trtdiff2_D

Estimated between-studies SDs and correlation matrix:
            SD         _y_1        _y_2
_y_1 .67446638          1           .
_y_2 .67446638         .5           1
mvmeta command stored as F9
```

Although the parameterization is different, the numerical results are almost identical. Differences arise in the fifth or sixth decimal place because of the tiny approximation introduced by augmentation.

Having fit both consistency and inconsistency models, we produce a forest plot (shown in figure 4):

```
. network forest, msize(*0.15) diamond name(smoke_forest, replace) eform
> xlabel(0.1 1 10 100)
group(design) assumed
```

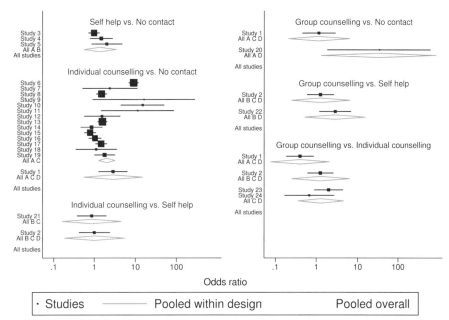

Test of consistency: chi2(7)=5.11, P=0.646

Figure 4. Forest plot for smoking network

The forest plot shows the individual study results, grouped by treatment contrast and design. It is clear that there is substantial heterogeneity between studies of C ("Individual counselling") versus A ("No contact"). Pooled results within design (from the inconsistency model) are shown as diamonds. Where there is only one study of a given design—for example, study 2 is the only "B C D" study at the bottom left—the point estimate pooled within design is the same as the single study's result, but the confidence interval is wider because the heterogeneity is assumed to be the same as in the other studies (here, estimated primarily from the "A C" studies). Overall pooled results are also shown as diamonds. The similarity of the "Pooled within design" and "Pooled overall" results again supports the consistency model.

We next explore inconsistency by side-splitting. We first split the side A–C, using the original method of Dias et al. (2010) for which the results depend on whether we split A–C or C–A (because A and C are contained in a three-arm study),

```
. network sidesplit A C, nosymmetric
```

	Coef.	Std. Err.	z	P>|z|	[95% Conf.	Interval]
direct	.7482566	.2088504	3.58	0.000	.3389174	1.157596
indirect	.317684	.5560699	0.57	0.568	-.7721931	1.407561
difference	.4305726	.5769152	0.75	0.455	-.7001604	1.561306

```
. network sidesplit C A, nosymmetric
```

	Coef.	Std. Err.	z	P>|z|	[95% Conf.	Interval]
direct	-.6840217	.21289	-3.21	0.001	-1.101278	-.2667649
indirect	-.937374	.636093	-1.47	0.141	-2.184093	.3093454
difference	.2533524	.6619598	0.38	0.702	-1.044065	1.55077

and using the new method for which the results are the same (apart from a change of sign) for splitting A–C or C–A:

```
. network sidesplit A C
```

	Coef.	Std. Err.	z	P>|z|	[95% Conf.	Interval]
direct	.7215198	.2132355	3.38	0.001	.303586	1.139454
indirect	.5736366	.6371481	0.90	0.368	-.6751508	1.822424
difference	.1478833	.6675922	0.22	0.825	-1.160573	1.45634

```
. network sidesplit C A
```

	Coef.	Std. Err.	z	P>|z|	[95% Conf.	Interval]
direct	-.7215198	.2132355	-3.38	0.001	-1.139454	-.3035859
indirect	-.5736366	.6371482	-0.90	0.368	-1.822424	.6751509
difference	-.1478833	.6675923	-0.22	0.825	-1.45634	1.160574

Finally, we split each node in turn—this can be slow because each split involves fitting a model:

```
. network sidesplit all
```

Side	Direct		Indirect		Difference		
	Coef.	Std. Err.	Coef.	Std. Err.	Coef.	Std. Err.	P>|z|
A B	.3266968	.4431874	.5101193	.5274799	-.1834225	.6881715	0.790
A C	.7215198	.2132355	.5736366	.6371481	.1478833	.6675922	0.825
A D	.8201395	.7576194	.8893868	.4420005	-.0692474	.8743754	0.937
B C	-.0756829	.5752542	.5163241	.4300301	-.592007	.7180019	0.410
B D	.6231378	.5693373	.2974687	.604145	.3256691	.8305243	0.695
C D	-.075668	.411104	.8722707	.70663	-.9479387	.8161917	0.245

These results again support consistency.

4.2 Thrombolytics network

The thrombolytics network (Lu and Ades 2006) includes eight treatments. It is more typical than the smoking network in that many possible pairs of treatments are not directly compared. We use these data, with just treatment codes, to demonstrate network map. We first set up the data:

```
. use thromb, clear
(Thrombolytics network meta-analysis from Lu & Ades (2006), corrected)
. network setup r n, studyvar(study) trtvar(treat)
Treatments used
    A (reference):                   A
    B:                               B
    C:                               C
    D:                               D
    E:                               E
    F:                               F
    G:                               G
    H:                               H
Measure                              Log odds ratio
Studies
    ID variable:                     study
    Number used:                     28
    IDs with zero cells:             [none]
    IDs with augmented reference arm: 17 18 19 20 21 22 23 24 25 26 27 28
    - observations added:            0.001
    - mean in augmented observations: study-specific mean
Network information
    Components:                      1 (connected)
    D.f. for inconsistency:          8
    D.f. for heterogeneity:          15
Current data
    Data format:                     augmented
    Design variable:                 _design
    Estimate variables:              _y*
    Variance variables:              _S*
    Command to list the data:        list study _y* _S*, noo sepby(_design)
```

The network map (figure 5) follows a standard format, but there are many line crossings that obscure the structure.

```
. network map
Graph command stored in F9
```

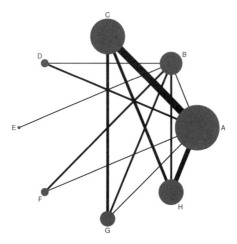

Figure 5. Network map for thrombolytics data: Default

We try improving the graph, using the `improve` option to repeatedly switch treatment locations to minimize the number of line crossings. We first base the map on the circle (see figure 6):

```
. network map, circle improve
Improving locations ...
loop 1 score 24= 21 19 17 16=.. 11== 7==....=... 4....=.=..... 3=......=.=
> ........==......==
loop 2 score 3=........=..=.....=........=.=...=..=......=.=........==......==
Stopping because loop 2 gave no improvement
Evaluating optimal locations ...
   1-3 and 2-8: score 1
   1-6 and 2-4: score 1
   1-7 and 2-8: score 1
Graph command stored in F9
```

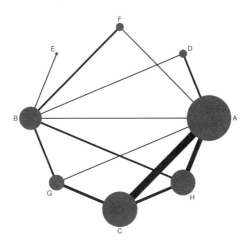

Figure 6. Network map for thrombolytics data: Circle with `improve` option

The numbers in the output count the line crossings (with co-incident lines scored as 10 crossings); "=" indicates a switch of treatment locations that doesn't improve the score (and is done); and "." indicates a switch of treatment locations that would make the score worse (and so is not done). The final output shows, for example, that lines 1–3 (A–C) and 2–8 (B–H) cross.

To get a map with no crossings, we next use the `improve` option starting with a triangular grid of side 5, which has 23 locations and hence 15 gaps (see figure 7):

```
. network map, triangular(5) improve
Improving locations ...
loop 1 score 117==. 106 103 58 55== 53.= 42....== 41... 38=............=....=...
> 36.= 35.... 28.. 17............. 15=.=... 5.========.===....=....= 4.==..=====
> ....===.....=.==.====.........==.........=..=.........==.........=..=....=....
> .================..=.....===============. 3..=...===============....=...=======
> ========.=...=.===============...==..===============....=== 1===============
> ....==.===============....==..===============....==..===============......=.=
> ===============......=.===============....=...===============....=...==========
> =====....=..===============
loop 2 score 1=........=.=....=.......=.................=....=.....=.....==.=.
> .==...=== 0
Stopping after achieving score of 0
Evaluating optimal locations ...
Graph command stored in F9
```

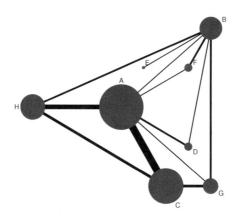

Figure 7. Network map for thrombolytics data: 5×5 triangular grid with `improve` option

That's good—there are no line crossings. Finally, we tidy the map by retrieving the treatment locations from matrix `_network_map_location`, moving treatments C, D, and E, and moving the label for treatment F (see figure 8):

```
. matrix loc2 = _network_map_location[1..8,.]
. * move C left to level of A
. matrix loc2[3,1]=loc2[1,1]

. * move D up to level of A
. matrix loc2[4,2]=loc2[1,2]

. * move E up to level of B
. matrix loc2[5,2]=loc2[2,2]

. * move F label to 10 o´clock
. matrix loc2[6,3]=10

. network map, loc(loc2)
Graph command stored in F9
```

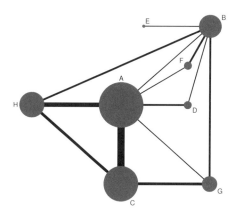

Figure 8. Network map for thrombolytics data: Figure 7 after moving treatments C, D, and E and the label for treatment F

Finally, we fit the consistency and inconsistency models (results not shown) and draw a forest plot (shown in figure 9). The most striking feature of this plot is that the two H versus B studies disagree strongly with the results of the consistency model, even though the overall test of inconsistency has a p-value of 0.38. A similar result is found using `network sidesplit all`.

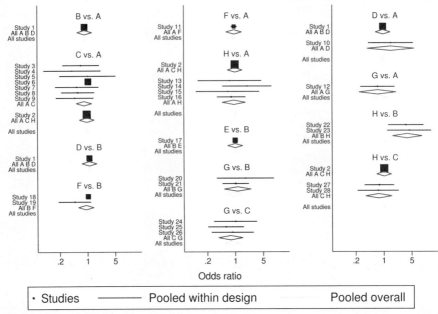

Figure 9. Forest plot for thrombolytics data

4.3 Quantitative data

I finally show how quantitative data may be used, using fictitious data on three treatments in long format:

```
. list, noobs clean
    study    trt    sbpmean    sbpsd    count
        1      P        150       15      100
        1      A        160       14      100
        1      B        162       16      100
        2      P        170       22       73
        2      A        175       20       77
        3      A        160       19       25
        3      B        154       19       25
```

Because the treatments are already encoded (P meaning placebo), we set them up using the `nocodes` option, and we set placebo as the reference:

```
. network setup sbpmean sbpsd count, studyvar(study) trtvar(trt) nocodes ref(P)
Treatments used
   A:                                   A
   B:                                   B
   P (reference):                       P

Measure                                 Mean difference
   Standard deviation pooling:          off

Studies
   ID variable:                         study
   Number used:                         3
   IDs with augmented reference arm:    3
   - observations added:                0.001
   - mean in augmented observations:    study-specific mean
   - SD in augmented observations:      study-specific within-arms SD

Network information
   Components:                          1 (connected)
   D.f. for inconsistency:              2
   D.f. for heterogeneity:              0

Current data
   Data format:                         augmented
   Design variable:                     _design
   Estimate variables:                  _y*
   Variance variables:                  _S*
   Command to list the data:            list study _y* _S*, noo sepby(_design)
```

The data can now be tabulated and listed:

```
. network table
```

	Treatment and Statistic								
	A			B			P		
study	sbpmean	sbpsd	count	sbpmean	sbpsd	count	sbpmean	sbpsd	count
1	160	14	100	162	16	100	150	15	100
2	175	20	77				170	22	73
3	160	19	25	154	19	25			

```
. list study _*
```

	study	_design	_y_A	_S_A_A	_y_B	_S_B_B	_S_A_B
1.	1	A B P	10	4.21	12	4.81	2.25
2.	2	A P	5	11.824942	.	.	.
3.	3	A B	3	361014.42	-3	361014.42	360999.98

Further analyses would proceed as above, except that there are no degrees of freedom for heterogeneity, so a random-effects inconsistency model cannot be fit.

5   Discussion

The main limitation of the analysis methods presented here is that they are two-stage methods that rely on a normal approximation to the distribution of the estimated study-specific treatment effects. This approximation can be problematic with count data, especially with small counts. The main alternative is a Bayesian analysis that avoids this approximation. Comparisons of frequentist and Bayesian results suggest that frequentist results can be somewhat biased toward the null.

A second limitation of the analysis methods presented here is that the models are restricted to those that can be fit with `mvmeta`, whereas Bayesian models can allow for further levels of hierarchical modeling.

The key advantage of the methods presented here is their relative simplicity and speed, and hence the opportunity for the user to use a variety of analyses; for example, it would be easy to repeat the analyses omitting one or more studies. The data formats provided are sufficiently flexible to interface with other user-written software, and in particular, with the routines of Chaimani et al. (2013), which are available from http://www.mtm.uoi.gr.

The `network` suite is work in progress, and I would be delighted to hear suggestions for improvements and new features, or even for others to write new subcommands. Possible future features include the ability to create WinBUGS code and fit the model in a Bayesian way, a feature already available in R (van Valkenhoef et al. 2012b,a); the random inconsistency model (Jackson et al. 2014); and methods to explore inconsistency (White et al. 2012).

6   Acknowledgments

This work was supported by the Medical Research Council (Unit Programme number U105260558). I thank Anna Chaimani for tailoring `networkplot` to work with `network`, and I thank everyone who has made comments on this software for helping me to improve it.

7 References

Chaimani, A., J. P. T. Higgins, D. Mavridis, P. Spyridonos, and G. Salanti. 2013. Graphical tools for network meta-analysis in Stata. *PLOS ONE* 8: e76654.

Dias, S., N. J. Welton, D. M. Caldwell, and A. E. Ades. 2010. Checking consistency in mixed treatment comparison meta-analysis. *Statistics in Medicine* 29: 932–944.

Greenland, S., J. J. Schlesselman, and M. H. Criqui. 1986. The fallacy of employing standardized regression coefficients and correlations as measures of effect. *American Journal of Epidemiology* 123: 203–208.

Harbord, R. M., and J. P. T. Higgins. 2008. Meta-regression in Stata. *Stata Journal* 8: 493–519. (Reprinted in this collection on pp. 85–111.)

Hasselblad, V. 1998. Meta-analysis of multitreatment studies. *Medical Decision Making* 18: 37–43.

Hawkins, N., D. A. Scott, B. S. Woods, and N. Thatcher. 2009. No study left behind: A network meta-analysis in non–small-cell lung cancer demonstrating the importance of considering all relevant data. *Value in Health* 12: 996–1003.

Higgins, J. P. T., D. Jackson, J. K. Barrett, G. Lu, A. E. Ades, and I. R. White. 2012. Consistency and inconsistency in network meta-analysis: Concepts and models for multi-arm studies. *Research Synthesis Methods* 3: 98–110.

Higgins, J. P. T., S. G. Thompson, and D. J. Spiegelhalter. 2009. A re-evaluation of random-effects meta-analysis. *Journal of the Royal Statistical Society, Series A* 172: 137–159.

Higgins, J. P. T., and A. Whitehead. 1996. Borrowing strength from external trials in a meta-analysis. *Statistics in Medicine* 15: 2733–2749.

Jackson, D., J. K. Barrett, S. Rice, I. R. White, and J. P. T. Higgins. 2014. A design-by-treatment interaction model for network meta-analysis with random inconsistency effects. *Statistics in Medicine* 33: 3639–3654.

Lu, G., and A. E. Ades. 2004. Combination of direct and indirect evidence in mixed treatment comparisons. *Statistics in Medicine* 23: 3105–3124.

———. 2006. Assessing evidence inconsistency in mixed treatment comparisons. *Journal of the American Statistical Association* 101: 447–459.

———. 2009. Modeling between-trial variance structure in mixed treatment comparisons. *Biostatistics* 10: 792–805.

Lu, G., J. E. Brazier, and A. E. Ades. 2013. Mapping from disease-specific to generic health-related quality-of-life scales: A common factor model. *Value in Health* 16: 177–184.

Lu, G., D. Kounali, and A. E. Ades. 2014. Simultaneous multioutcome synthesis and mapping of treatment effects to a common scale. *Value in Health* 17: 280–287.

Lumley, T. 2002. Network meta-analysis for indirect treatment comparisons. *Statistics in Medicine* 21: 2313–2324.

Miladinovic, B., A. Chaimani, I. Hozo, and B. Djulbegovic. 2014. Indirect treatment comparison. *Stata Journal* 14: 76–86. (Reprinted in this collection on pp. 311–320.)

Mills, E. J., K. Thorlund, and J. P. A. Ioannidis. 2013. Demystifying trial networks and network meta-analysis. *British Medical Journal* 346: f2914.

Rücker, G., and G. Schwarzer. Forthcoming. Automated drawing of network plots in network meta-analysis. *Research Synthesis Methods*.

Salanti, G., A. E. Ades, and J. P. A. Ioannidis. 2011. Graphical methods and numerical summaries for presenting results from multiple-treatment meta-analysis: An overview and tutorial. *Journal of Clinical Epidemiology* 64: 163–171.

Salanti, G., J. P. T. Higgins, A. E. Ades, and J. P. A. Ioannidis. 2008. Evaluation of networks of randomized trials. *Statistical Methods in Medical Research* 17: 279–301.

van Valkenhoef, G., G. Lu, B. de Brock, H. Hillege, A. E. Ades, and N. J. Welton. 2012a. Automating network meta-analysis. *Research Synthesis Methods* 3: 285–299.

van Valkenhoef, G., T. Tervonen, B. de Brock, and H. Hillege. 2012b. Algorithmic parameterization of mixed treatment comparisons. *Statistics and Computing* 22: 1099–1111.

White, I. R. 2009. Multivariate random-effects meta-analysis. *Stata Journal* 9: 40–56. (Reprinted in this collection on pp. 232–248.)

———. 2011. Multivariate random-effects meta-regression: Updates to mvmeta. *Stata Journal* 11: 255–270. (Reprinted in this collection on pp. 249–264.)

White, I. R., J. K. Barrett, D. Jackson, and J. P. T. Higgins. 2012. Consistency and inconsistency in network meta-analysis: Model estimation using multivariate meta-regression. *Research Synthesis Methods* 3: 111–125.

White, I. R., and J. Thomas. 2005. Standardized mean differences in individually randomized and cluster-randomized trials, with applications to meta-analysis. *Clinical Trials* 2: 141–151.

The Stata Journal (2015)
15, Number 4, st0411, forthcoming

Visualizing assumptions and results in network meta-analysis: The network graphs package

Anna Chaimani
Department of Hygiene and Epidemiology
University of Ioannina School of Medicine
Ioannina, Greece
achaiman@cc.uoi.gr

Georgia Salanti
Department of Hygiene and Epidemiology
University of Ioannina School of Medicine
Ioannina, Greece
gsalanti@cc.uoi.gr

Abstract. Network meta-analysis has been established in recent years as a particularly useful evidence synthesis tool. However, it is still challenging to develop understandable and concise ways to present data, assumptions, and results from network meta-analysis to inform decision making and evaluate the credibility of the results. In this article, we provide a suite of commands with graphical tools to facilitate the understanding of data, the evaluation of assumptions, and the interpretation of findings from network meta-analysis.

Keywords: st0411, clusterank, ifplot, intervalplot, mdsrank, netfunnel, netleague, netweight, networkplot, sucra, network graphs, network meta-analysis, mixed-treatment comparison, multiple treatments, ranking, inconsistency, graphical tools

1 Introduction

Network meta-analysis integrates direct and indirect evidence in a collection of trials providing information on the relative effects of three or more competing interventions for the same outcome. Despite the increasing popularity of network meta-analysis (Lee 2014; Nikolakopoulou et al. 2014), it is still the subject of skepticism, particularly among researchers with a less technical background. This phenomenon can be partly explained by the lack of flexible tools implemented in a user-friendly software environment providing comprehensive ways to understand and present the different steps of a network meta-analysis.

Graphical illustrations have been extensively used in statistical analyses because they offer a concise way to describe complex data structures, assumptions, and outputs (Anscombe 1973). Many different plots are available for the presentation of standard pairwise meta-analysis (Anzures-Cabrera and Higgins 2010; Bax et al. 2009; Chaimani, Mavridis, and Salanti 2014), but their adaptation into network meta-analysis, when feasible, needs to account for the presence of multiple treatment comparisons. For instance, forest plots in large networks cannot easily accommodate study-level data and multiple summary estimates because of space limitations. Funnel plots are meaningless without an appropriate "adjustment" because they include studies that evaluate different pairs of interventions (Chaimani et al. 2013a; Chaimani and Salanti 2012).

Summary statistics from network meta-analysis are estimated under the assumption of consistency, which implies that all different sources of evidence are in statistical agreement (Caldwell, Ades, and Higgins 2005; Salanti 2012). Several sophisticated methods have been developed for the assessment of inconsistency in a network of interventions (Dias et al. 2010; Higgins et al. 2012; Lu and Ades 2006). However, less advanced methods, such as the "loop-specific approach" (Bucher et al. 1997), can be more easily conceived and interpreted by clinicians, especially through graphical depictions, and could be considered in conjunction with other approaches.

Transparent and concise presentation of findings might be challenging for networks that include a large number of treatments (for example, more than 10). Reporting of results can be supplemented by graphs that show the relative ranking of treatments (Salanti, Ades, and Ioannidis 2011) and allow for information on two outcomes (Chaimani et al. 2013a), which can be useful for decision makers.

In Stata, simple indirect comparisons (using information from a single loop) can be performed using the `indirect` command (Miladinovic et al. 2014), and full network meta-analysis can be performed using the `mvmeta` (White 2011; White et al. 2012) and `network`[1] (White 2013) packages. We have previously presented commands that can be used in conjunction with `mvmeta` to check assumptions and produce results (Chaimani et al. 2013a). In this article, we extend and elaborate on our previous work, and we present a suite of nine commands that produce graphs that describe the evidence base, evaluate the assumptions, and present the results obtained from a network meta-analysis.

2   Example datasets

We illustrate the use of the nine commands using three published networks. The first network includes 27 studies comparing the effectiveness (change in glycated hemoglobin A1c [HbA1c]) and tolerability (weight gain) of six noninsulin antidiabetic drugs and placebo (Phung et al. 2010). The second network comprises 23 trials that evaluate four antiplatelet regimens and placebo for the prevention of serious vascular events after transient ischemic attack or stroke (Thijs, Lemmens, and Fieuws 2008). The last network consists of 22 trials comparing the safety of five antihypertensive treatments and placebo with respect to the incidence of diabetes (Elliott and Meyer 2007).

---

1. `network` calls `mvmeta` and is equivalent.

3   The network graphs package

3.1   Notation and setting up the data

Input variables

There are two groups of commands in the `network graphs` package. The first group (`networkplot`, `netweight`, `ifplot`, and `netfunnel`) requires as input variables with study-level data that are stored in a dataset. The second group of commands (`intervalplot`, `netleague`, `sucra`, `mdsrank`, and `clusterank`) takes as input summary data produced by the `mvmeta` or `network` package that run network meta-analysis (see White [2013; 2011] and White et al. [2012] for a description of input data for `mvmeta` and `network`).

The input variables for the first group of commands require a format that assumes one treatment comparison per row. For a two-arm trial, this amounts to inputting the treatments being compared, the effect sizes, and other characteristics for the single treatment comparison observed in that study. For multiarm trials, all comparisons enter the data as different rows, with the study code (*id*) denoting which observations belong to the same study.

More specifically, the input variables for the `networkplot`, `netweight`, `ifplot`, and `netfunnel` commands include the following:

id: The variable specifying the ID numbers of the studies, with repetitions for multiarm studies. Each k-arm trial contributes $\{k \times (k-1)\}/2$ rows to the dataset.

t1 and *t2*: The variables containing the codes of two treatments being compared in every observation of the dataset. These variables can be numeric or string.

ES: The variable with the effect sizes for every treatment comparison defined by *t1* and *t2*. For ratio measures (odds ratio, risk ratio, hazard ratio, etc.), the effect sizes in *ES* should be in logarithmic scale (for example, log odds-ratio or log risk-ratio). In all example datasets, we have estimated *ES* as *t1* versus *t2*.

seES: The variable containing the standard errors of the effect sizes in *ES*.

The second group of commands—`intervalplot`, `sucra`, `netleague`, `mdsrank`, and `clusterank`—are commands used to present and interpret the output of network meta-analysis; hence, they need as input estimates that can be produced by a network meta-analysis routine (such as `network` or WinBUGS [Lunn et al. 2000] codes). For example:

LCI, *UCI*, *LPI*, and *UPI* (inputs for `intervalplot`) are variables that contain the lower and upper limits of the confidence and predictive intervals for the effect sizes of all possible pairwise comparisons estimated in network meta-analysis.

outcome1 and *outcome2* (inputs for `clusterank`) are variables containing the estimated values of a ranking measure for two different outcomes.

Options

In this section, we describe options that are available in multiple commands in the network graphs package. Other options that are unique to each command are described in the respective sections below and in the help files.

mvmetaresults specifies to derive the input directly from mvmeta or network (the default). This option is available in the commands intervalplot, netleague, and sucra, which also allow for different inputs (not derived from mvmeta).

nomvmeta specifies not to derive the input from mvmeta or network. This option is available in the commands intervalplot, netleague, and sucra.

labels(*string*) specifies names for competing treatments in the network, which are displayed in the output results and the corresponding plots. The treatments should be given in numerical or alphabetical order separated by a space. For example, specifying the option labels(Placebo Aspirin "Pla+Asp" ...) assigns the name Placebo to the treatment coded as 1 or A, the name Aspirin to the treatment coded 2 or B, and so on. For the commands intervalplot, netleague, and sucra when the option nomvmeta has not been specified, the first treatment in labels() should be the treatment assumed to be the reference when running mvmeta or network, and the following treatments should be given in numerical or alphabetical order. The labels() option is available for the commands networkplot, ifplot, intervalplot, netleague, sucra, and mdsrank.

eform specifies to display the results on the exponential scale. This option is available for the commands ifplot, intervalplot, and netleague.

keep specifies to store the results as additional variables at the end of the dataset. This option is available for the commands ifplot and intervalplot.

title(*string*) specifies a title for the produced graph. This option is available for the commands networkplot, netweight, intervalplot, and sucra.

aspect(*#*) specifies the aspect ratio for the region of the graph. This option is available for the commands networkplot and netweight.

notable suppresses display of the output. This option is available for the commands netweight, ifplot, intervalplot, and sucra.

noplot suppresses display of the produced graph. This option is available for the commands netweight, ifplot, intervalplot, sucra, and mdsrank.

scatteroptions(*string*) specifies standard options allowed for scatterplots. This option is available for the commands netfunnel, mdsrank, and clusterank.

cilevel(*integer*) specifies the level of statistical significance for the estimated confidence intervals. This option is available for the ifplot and intervalplot commands.

xtitle(*string*) and xlabel(*string*) specify a title and the values, respectively, that are displayed for the horizontal axis. This option is available for the commands netfunnel and intervalplot.

Dialog boxes

The dialog boxes are an alternative way to run a command within Stata; they might be preferable for users who are not familiar with Stata language. Dialog boxes have been developed for all commands of the network graphs package and can be accessed by typing db *commandname* in the Stata Command window.

3.2 The networkplot command

Description of networkplot

The networkplot command plots a network of interventions using nodes and edges. Nodes represent the competing treatments, and edges represent the available direct comparisons between pairs of treatments. networkplot allows for weighting and coloring options for both nodes and edges according to prespecified characteristics. The use of weighting and coloring schemes can reveal important differences in the characteristics of treatments or comparisons. Sometimes, differences in comparisons can be an indication of potential violation of the assumption underlying network meta-analysis (Jansen and Naci 2013; Salanti 2012).

Syntax for networkplot

networkplot *varlist* [*if*] [*in*] [, <u>labels</u>(*string*) <u>title</u>(*string*) <u>aspect</u>(*#*)

　　<u>now</u>eight <u>nodeweight</u>(*weightvar* sum | mean) <u>edgeweight</u>(*weightvar* sum | mean)

　　<u>nodecolor</u>(*string*) <u>edgecolor</u>(*edge_color* | by *groupvar* *pool_method*)

　　<u>bylevels</u>(*#*) <u>bycolors</u>(*string*) <u>edgepattern</u>(*string*) <u>plotregion</u>(*string*)

　　<u>edgescale</u>(*#*) <u>nodescale</u>(*#*)]

In *varlist*, the variables *t1* and *t2* (see section 3.1) should be specified.

Options for networkplot

labels(*string*), title(*string*), aspect(*#*); see section 3.1.

noweight specifies that all nodes and edges are of equal size and thickness (when nodeweight() and edgeweight() have not been specified). By default, both nodes and edges are weighted according to the number of studies involved in each treatment and comparison, respectively.

nodeweight(*weightvar* sum|mean) specifies a variable according to which nodes are weighted. sum specifies to use the sum of the values in *weightvar* to weight the nodes, while mean specifies to use the mean of the values.

edgeweight(*weightvar* sum|mean) specifies a variable according to which edges are weighted. sum specifies to use the sum of the values in *weightvar* to weight the edges, while mean specifies to use the mean of the values.

nodecolor(*string*) specifies the color for all nodes. The default is nodecolor(blue).

edgecolor(*edge_color*|by *groupvar pool_method*) specifies the color for edges.

> *edge_color* specifies the color for all edges. The default color is black.

> by specifies that edges be colored according to the bias level for each comparison.

> *groupvar* is a variable that contains bias scores for each observation (that is, study data such as effect size). It can be a numeric variable coded as 1 = low risk, 2 = unclear risk, and 3 = high risk, or it can be a string variable with values l[ow], u[nclear], and h[igh]. By default, a three-level *groupvar* is assumed, allowing for green, yellow, and red lines representing comparisons at low, unclear, and high risk of bias, respectively. More or fewer than three levels with user-specified colors are also allowed if the options bylevels() and bycolors() have been specified.

> *pool_method* specifies the method by which to estimate the risk of bias level for the summary estimate of each comparison. It can be one of the following:

>> mode specifies to use the most prevalent bias level within each comparison as the comparison-specific bias level (the default).

>> mean specifies to use the average bias level of each comparison as the comparison-specific bias level.

>> wmean *weightvar* specifies to use the weighted (according to the *weightvar*) average bias level of each comparison as the comparison-specific bias level.

>> max specifies to use the maximum bias level observed within each comparison as the comparison-specific bias level.

>> min specifies to use the minimum bias level observed within each comparison as the comparison-specific bias level.

bylevels(#) specifies the number of levels for the *groupvar* in edgecolor() when there are more than three bias levels. In this case, the *groupvar* variable can only be numeric. bycolors() is required with bylevels().

bycolors(*string*) specifies the colors for each bias level of the *groupvar* in the respective order. bylevels() is required with bycolors().

edgepattern(*string*) specifies the pattern for all edges.

plotregion(*string*) specifies options for the region of the network plot.

`edgescale(#)` specifies a real number that is used to scale all edges.

`nodescale(#)` specifies a real number that is used to scale all nodes.

Example of networkplot using the antidiabetics network

The dataset of the antidiabetics network contains information from 24 study comparisons. For each observation, negative values of the standardized mean difference (*ES*) favor the first treatment (*t1*) in the respective comparison over the second treatment (*t2*). The treatments have been coded as numbers [1 = placebo, 2 = sulfonylurea, 3 = dipeptidyl peptidase-4 (DPP-4) inhibitor, 4 = thiazolidinedione, 5 = glucagon-like peptide-1 (GLP-1) analog, 6 = alpha-glucosidase inhibitors (AGI), 7 = glinine]. For each study, the variable `blinding` contains the level of risk of bias with respect to double-blinding. Note that the fifth study is a three-arm study and so is represented by three rows.

```
. use antidiabetics_efficacy_wide
. list in 1/7, clean noobs
    id   t1   t2          ES        seES   blinding
     1    3    1   -.8701839    .1102315          1
     2    3    2    .1362578    .0427965          1
     3    3    1   -.6923013    .1341008          2
     4    3    1   -.3768538    .1208566          2
     5    2    1   -.7710395    .1149981          1
     5    5    1   -.7710395    .1149981          1
     5    5    2           0    .0909091          1
```

Using the `networkplot` command, we produce the plot of the network (see figure 1). This gives information on the network structure, the number of studies evaluating each intervention, the precision of the direct estimate for each pairwise comparison, and the average bias level for every comparison with respect to blinding.

```
. generate invvarES = 1/(seES^2)
. networkplot t1 t2, edgeweight(invvarES) edgecolor(by blinding mean)
> edgescale(1.2) aspect(0.8) labels(Placebo Sulfonylurea "DPP-4 inhibitor"
> Thiazolidinedione "GLP-1 analog" AGI Glinine)
```

The resulting graph, shown in figure 1, implies that the comparison DPP-4 inhibitor versus sulfonylurea provides the most precise direct estimate in the network (variance 0.001) and suggests that there is only one comparison (DPP-4 inhibitor versus thiazolidinedione) at high risk of bias.

Figure 2 shows how the above command can be executed via the `networkplot` dialog box, accessed by typing `db networkplot` in the Command window.

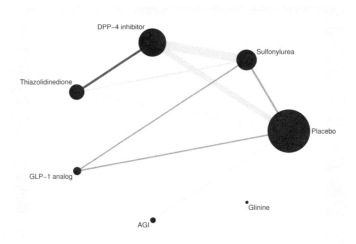

Figure 1. Plot of the antidiabetics network. The size of the nodes is proportional to the number of studies evaluating each intervention, and the thickness of the edges is proportional to the precision (the inverse of the variance) of each direct comparison. The color of the edges represents the average bias level of the corresponding comparisons with respect to blinding.

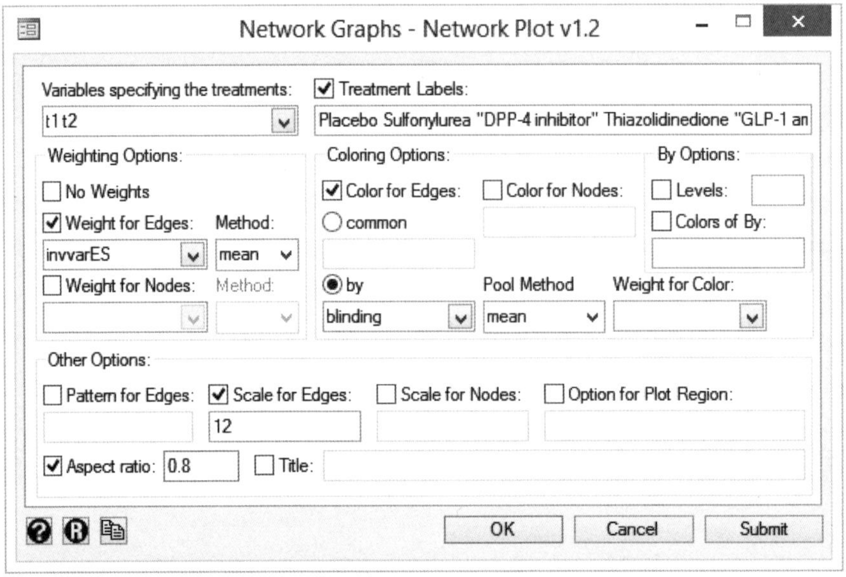

Figure 2. Example of using the `networkplot` dialog box for the antidiabetics network

3.3   The netweight command

Contributions of direct comparisons to the network

When performing a network meta-analysis, each direct comparison contributes to the estimation of each network meta-analytic summary effect by a different weight (Krahn, Binder, and König 2013; Lu et al. 2011). Identifying comparisons with large or small contributions is of great interest and enhances the understanding of the evidence flow. The contributions of the different pieces of evidence within a network have also been used in the evaluation of the quality of evidence from network meta-analysis (Salanti et al. 2014).

These contributions are obtained as complicated functions of a) the structure of the network and b) the variances of each pairwise direct summary effect. We obtain the contributions by fitting the network meta-analysis model in two stages:

1. Estimation of the direct relative effects using standard pairwise meta-analysis.

2. Synthesis of the direct relative effects assuming consistency to estimate the network relative effects.

This approach results in the estimation of vector $\widehat{\boldsymbol{\mu}}$ that contains the network estimates for the basic parameters (that is, a subset of the available comparisons sufficient to estimate all the comparisons via consistency) (Krahn, Binder, and König 2013; Lu et al. 2011). Defining with $\mathbf{X}$ the design matrix expressing the linear relationships between the available direct comparisons and the basic parameters, the vector $\widehat{\boldsymbol{\mu}}$ is

$$\widehat{\boldsymbol{\mu}} = \left\{ \mathbf{X}' \left(\widehat{\mathbf{v}}^{\mathbf{D}} \right)^{-1} \mathbf{X} \right\} \mathbf{X}' (\widehat{\mathbf{v}}^{\mathbf{D}})^{-1} \widehat{\boldsymbol{\mu}}^{\mathbf{D}}$$

where $\widehat{\boldsymbol{\mu}}^{\mathbf{D}}$ and $\widehat{\mathbf{v}}^{\mathbf{D}}$ are the vectors including the available direct estimates and their variances, respectively. The matrix $\mathbf{H} = \{\mathbf{X}'(\widehat{\mathbf{v}}^{\mathbf{D}})^{-1}\mathbf{X}\}\mathbf{X}'(\widehat{\mathbf{v}}^{\mathbf{D}})^{-1}$ maps the direct estimates to the network estimates (often called "the hat matrix") and contains the weight of each direct comparison in the estimation of every basic parameter. The matrix $\mathbf{H}$ can be extended to incorporate the weights of the direct comparisons to every possible network summary effect via the consistency equations. Then, its elements h_{ij} ($i = 1, \ldots, N$ and $j = 1, \ldots, N'$ with N being the number of all possible pairwise comparisons and N' being the number of all available direct comparisons) can be expressed as the percentage contribution of the column-defining direct comparison to the row-defining network estimate [which is $|h_{ij}|/(|h_{i1}| + \cdots + |h_{iN'}|)$]. The percentage contribution of each direct comparison j to the entire network is $(\sum_{i=1}^{N} |h_{ij}|)/(\sum_{j=1}^{N'} \sum_{i=1}^{N} |h_{ij}|)$.

Description of netweight

The netweight command calculates all direct pairwise summary effect sizes with their variances, creates the design matrix, and estimates the percentage contribution of each direct comparison to the network summary estimates and in the entire network. Then, it produces the contribution plot that uses weighted squares to represent the respective contributions. The command can also combine the estimated contributions with a particular trial-level characteristic (for example, the risk of bias of the studies) and produce a bar graph showing the percentage of information in each network estimate that corresponds to the different levels of the characteristic.

Currently, `netweight` does not account for the correlation in direct effect sizes from multiarm trials.

Syntax for netweight

`netweight` *varlist* [*if*] [*in*] [, <u>title</u>(*string*) <u>aspect</u>(*#*) <u>notable</u> <u>noplot</u>
    random fixed <u>tau2</u>(*#*) <u>bargraph</u>(by *groupvar pool_method*) order
    <u>bylevels</u>(*#*) <u>bycolors</u>(*string*) <u>scale</u>(*string*) <u>novalues</u> <u>nostudies</u>
    <u>col</u>or(*string*) <u>symbol</u>(*string*) <u>noym</u>atrix <u>novm</u>atrix <u>noxm</u>atrix <u>nohm</u>atrix]

In *varlist*, the variables *ES*, *seES*, *t1*, and *t2* (see section 3.1) should be specified.

Options for netweight

`title`(*string*), `aspect`(*#*), `notable`, `noplot`; see section 3.1.

`random` specifies that the random-effects model be used for the estimation of all direct pairwise summary effects (the default).

`fixed` specifies that the fixed-effect model be used for the estimation of all direct pairwise summary effects.

`tau2`(*#*) specifies a real nonnegative number to use as the heterogeneity variance common for all comparisons.

`bargraph`(by *groupvar pool_method*) specifies that a bar graph be drawn instead of the contribution plot. The bars are colored according to the bias level (see the `edgecolor`() option of the `networkplot` command), and their length is proportional to the percentage contribution of each direct comparison to the network estimates (Salanti et al. 2014).

For a description of *groupvar* and *pool_method*, see the `edgecolor`() option for the `networkplot` command, in section 3.2 above.

`order` specifies to order the direct comparisons (on the horizontal axis) presented in the bar graph according to their bias level. This option is ignored unless `bargraph`() is specified.

`bylevels`(*#*) specifies the number of levels for the *groupvar* in `bargraph`() when there are more than three bias levels. In this case, the *groupvar* variable can only be numeric. `bycolors`() is required with `bylevels`().

bycolors(*string*) specifies the colors for each bias level of the *groupvar* in the respective order. bylevels() is required with bycolors().

scale(*string*) specifies a real number that is used to scale the weighted squares in the contribution plot.

novalues suppresses the display of the percentage contributions in the contribution plot.

nostudies suppresses the display of the number of included studies in each comparison in the contribution plot.

color(*string*) specifies the color of the weighted squares in the contribution plot.

symbol(*string*) specifies an alternative symbol for the weighted squares in the contribution plot.

noymatrix suppresses the display of the vector containing the direct relative effects.

novmatrix suppresses the display of the matrix containing the variances of the direct relative effects.

noxmatrix suppresses the display of the design matrix.

nohmatrix suppresses the display of the hat matrix that maps the direct estimates into the network estimates.

Example for netweight using the antiplatelet regimens network

The dataset of the antiplatelet regimens network has 29 observations. For each pairwise comparison, a log odds-ratio (*ES*) smaller than 0 suggests that the first treatment in that particular comparison (*t1*) is more effective than the second treatment (*t2*).

For this network, we have no information on the risk of bias of the included studies. To illustrate the use of netweight, we considered that the risk of bias might be associated with the year of study publication (represented by the variable year), and we classified the trials into three categories: studies published before 1990 (considered at high risk of bias), studies between 1990 and 2000 (at unclear risk), and studies after 2000 (at low risk).

The data are in the same format as described above for the antidiabetics network. The treatments have been coded as numbers (1 = placebo, 2 = thienopyridines+aspirin, 3 = aspirin, 4 = dipyridamole + aspirin, 5 = thienopyridines).

```
. use antiplatelet.dta, clear
. list in 1/5, clean noobs
    id   t1   t2          ES        seES   year        rob
     1    4    3   -.0622905    .2781967   1983       high
     1    3    1    -.450053    .2575157   1983       high
     1    4    1   -.5123435    .2592596   1983       high
     2    5    3    .2051014    .1378609   2003        low
     3    5    3   -.1010332    .0706103   1996    unclear
```

We use the `netweight` command to produce the contribution plot of the network shown in figure 3:

```
. netweight ES seES t1 t2, aspect(0.9) notable
```

Note: Effect sizes of 29 observations were reversed (stored in matrix e(R))

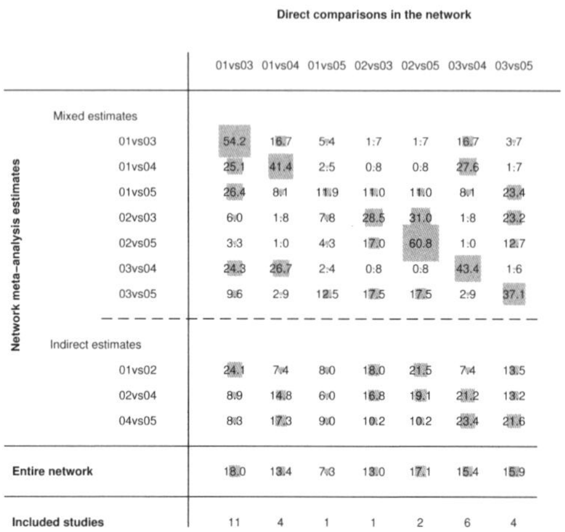

Direct comparisons in the network

Network meta-analysis estimates	01vs03	01vs04	01vs05	02vs03	02vs05	03vs04	03vs05
Mixed estimates							
01vs03	54.2	16.7	5.4	1.7	1.7	16.7	3.7
01vs04	25.1	41.4	2.5	0.8	0.8	27.6	1.7
01vs05	26.4	8.1	11.9	11.0	11.0	8.1	23.4
02vs03	6.0	1.8	7.8	28.5	31.0	1.8	23.2
02vs05	3.3	1.0	4.3	17.0	60.8	1.0	12.7
03vs04	24.3	26.7	2.4	0.8	0.8	43.4	1.6
03vs05	9.6	2.9	12.5	17.5	17.5	2.9	37.1
Indirect estimates							
01vs02	24.1	7.4	8.0	18.0	21.5	7.4	13.5
02vs04	8.9	14.8	6.0	16.8	19.1	21.2	13.2
04vs05	8.3	17.3	9.0	10.2	10.2	23.4	21.6
Entire network	18.0	13.4	7.3	13.0	17.1	15.4	15.9
Included studies	11	4	1	1	2	6	4

Figure 3. Contribution plot for the antiplatelet regimens network. The size of the squares is proportional to the percentage contribution of the column-defining direct comparison to the row-defining network estimate.

The contribution plot of the network suggests that the comparison of placebo (treatment 1) versus aspirin (treatment 3) has the largest contribution in the entire network (18%).

Figure 4 shows how the above command can be executed via the `netweight` dialog box, accessed by typing `db netweight` in the Command window.

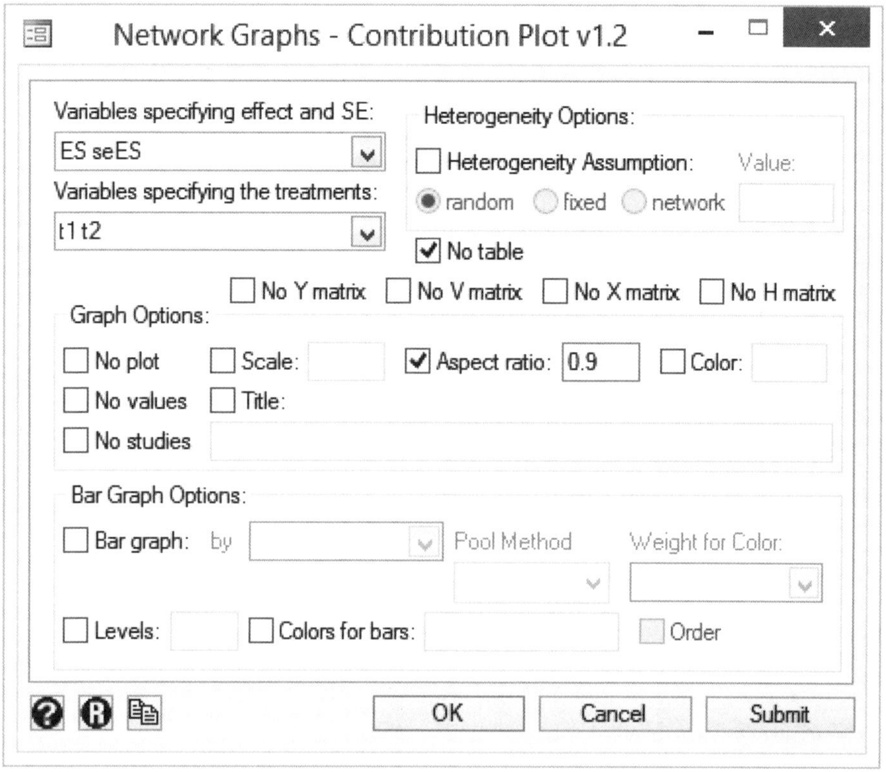

Figure 4. Example of using the `netweight` dialog box for the antiplatelet regimens network

Then, we produce a bar graph, shown in figure 5, showing how much information comes from high, unclear, and low risk of bias studies for each network estimate:

```
. netweight ES seES t1 t2, bargraph(by rob mean) notable
```

```
Note: Effect sizes of 29 observations were reversed (stored in matrix e(R))
```

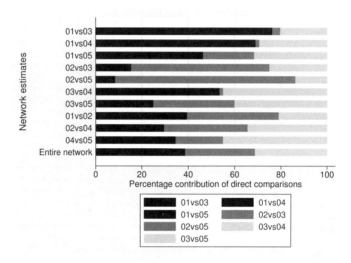

Figure 5. Bar graph showing the risk of bias for each network estimate and the entire network with respect to trial publication year for the antiplatelet regimens network. For each direct estimate, the average bias level of the included studies has been assumed as the comparison-specific level of bias.

The graph suggests that the estimated summary effects for the comparisons placebo versus aspirin, placebo versus dipyridamole + aspirin, and aspirin versus dipyridamole + aspirin derive most of their information from older studies.

Figure 6 shows how the above command can be executed via the `netweight` dialog box, accessed by typing `db netweight` in the Command window.

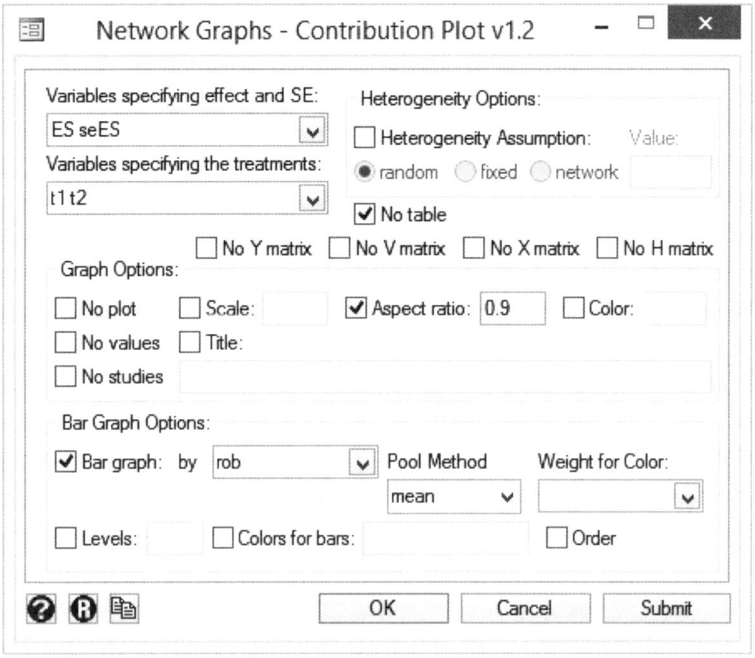

Figure 6. Another example of using the `netweight` dialog box for the antiplatelet regimens network

3.4 The ifplot command

Loop-specific approach for inconsistency

The "loop-specific approach" evaluates inconsistency separately in every closed loop of a network of interventions (Bucher et al. 1997). More specifically, in a network with L total number of loops, the inconsistency factor within each loop l $(l = 1, \ldots, L)$ is estimated as

$$\widehat{w}_l^{XY} = |\widehat{\mu}_{XY}^D - \widehat{\mu}_{XY}^{I_l}|$$

where $\widehat{\mu}_{XY}^D$ and $\widehat{\mu}_{XY}^{I_l}$ are the estimated direct and indirect (using information only from studies in the l loop) summary effects for the XY comparison. The variance of $\widehat{w}_l^{XY}$ is

$$\operatorname{var}\left(\widehat{w}_l^{XY}\right) = \widehat{v}_{XY}^D + \widehat{v}_{XY}^{I_l}$$

where $\widehat{v}_{XY}^D$ and $\widehat{v}_{XY}^{I_l}$ are the variances of the direct and indirect estimates. Note that there is only one direct estimate for each XY comparison, whereas different loops may give different indirect estimates (that is, $\widehat{\mu}_{XY}^{I_1} \neq \cdots \neq \widehat{\mu}_{XY}^{I_L}$). The choice of the comparison according to which inconsistency is estimated within each loop does not affect the results; that is, in a loop XYZ, it is $\widehat{w}_l^{XY} = \widehat{w}_l^{XZ} = \widehat{w}_l^{YZ} = \widehat{w}_l$.

Loops in which the lower confidence interval limit of the inconsistency factor does not reach the zero line are considered to present statistically significant inconsistency. However, the absence of statistically significant inconsistency is not evidence against the presence of inconsistency because of the multiple and correlated tests that are undertaken and the low power of the method (Song et al. 2012; Veroniki et al. 2014; Veroniki et al. 2013).

Description of ifplot

The `ifplot` command identifies all triangular and quadratic loops in a network of interventions and estimates the respective inconsistency factors and their uncertainties. Then, it produces the inconsistency plot that presents for each loop the estimated inconsistency factor and its confidence interval (truncated to 0). The command allows for different assumptions for the between-studies variance (that is, loop-specific, comparison-specific, or network-specific) and different estimators (for example, method of moments or restricted maximum likelihood).

Multiarm trials

The loop-specific approach does not account for the correlation in the effect sizes induced by multiarm trials. This is expected to emphasize the consistency in the loops because `ifplot` treats as independent the comparisons from the same trial, which are consistent by definition. The `ifplot` command slightly mitigates this by dropping one of the direct comparisons from the multiarm trials when it appears in a particular loop. Among the $\{k \times (k-1)\}/2$ comparisons belonging to the same trial that can be excluded from a loop, `ifplot` chooses the comparison with the largest number of studies within the loop. Inconsistency is not identifiable in loops formed only by multiarm trials, and hence such loops are excluded.

Syntax for ifplot

`ifplot` *varlist* [*if*] [*in*] [, labels(*string*) eform keep notable noplot
    cilevel(*integer*) tau2(loop | comparison | #) mm reml eb random fixed
    summary details plotoptions(*string*) xlabel(#,#,...,#) separate]

In *varlist*, the variables *ES*, *seES*, *t1*, *t2*, and *id* (see section 3.1) should be specified.

Options for ifplot

`labels(`*string*`)`, `eform`, `keep`, `notable`, `noplot`, `cilevel(`*integer*`)`; see section 3.1.

`tau2(loop | comparison | #)` specifies the assumption for the heterogeneity variance.

> `loop` specifies a common heterogeneity for all comparisons within each loop but different heterogeneities across loops (the default).

> `comparison` specifies different comparison-specific heterogeneities for each loop.

> `#` specifies a real nonnegative number for the network-specific heterogeneity variance common for all loops and comparisons.

`mm` specifies that the method of moments (DerSimonian and Laird 1986) (the default) be used for the estimation of the loop-specific heterogeneity (when `tau2(loop)` has been specified). When `tau2(comparison)` has been specified, this is the only possible method for estimating heterogeneity.

`reml` specifies that the restricted maximum likelihood method (Viechtbauer 2005) be used for the estimation of the loop-specific heterogeneity (when `tau2(loop)` has been specified).

`eb` specifies that the empirical Bayes method (Morris 1983) be used for the estimation of the loop-specific heterogeneity (when `tau2(loop)` has been specified).

`random` specifies that the random-effects model (the default) be used for all comparisons (when `tau2(comparison)` has been specified). When `tau2(loop)` or `tau2(#)` has been specified, this is the only possible model.

`fixed` specifies that the fixed-effect model be used for all comparisons (for use when `tau2(comparison)` has been specified). `fixed` is equal to `tau2(0)`.

`summary` specifies to display all direct and indirect summary estimates for every loop.

`details` specifies to display all comparisons that are dropped when there are multiarm studies in a loop.

`plotoptions(`*string*`)` specifies standard options of `metan` that handle the appearance of the inconsistency plot.

`xlabel(#,#,...,#)` specifies the values to display on the horizontal axis. The input values must be comma-delimited.

`separate` specifies to display inconsistency factors separately for loops with and without evidence of statistical inconsistency in the inconsistency plot and the output results.

Example for ifplot using the antiplatelet regimens network

The data for this network are described in section 3.3.

We use the `ifplot` command to assess the presence of inconsistency in every closed loop of the network. The resulting inconsistency plot is shown in figure 7.

```
. ifplot ES seES t1 t2 id, eform plotopt(texts(180)) xlabel(1,1.3,1.8) notab
* 3 triangular loops found
```

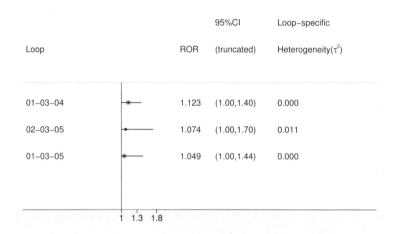

Loop	ROR	95%CI (truncated)	Loop–specific Heterogeneity(τ^2)
01–03–04	1.123	(1.00,1.40)	0.000
02–03–05	1.074	(1.00,1.70)	0.011
01–03–05	1.049	(1.00,1.44)	0.000

Figure 7. Inconsistency plot for the antiplatelet regimens network showing for each loop the ratio of odds ratios between direct and indirect estimates. The within-loop heterogeneities have been estimated using the method of moments estimator.

Using the loop-specific approach, none of the three loops in the network was found to present statistically significant inconsistency, although large values are included in the confidence intervals of the inconsistency factors.

Figure 8 shows how the above command can be executed via the `ifplot` dialog box, accessed by typing `db ifplot` in the Command window.

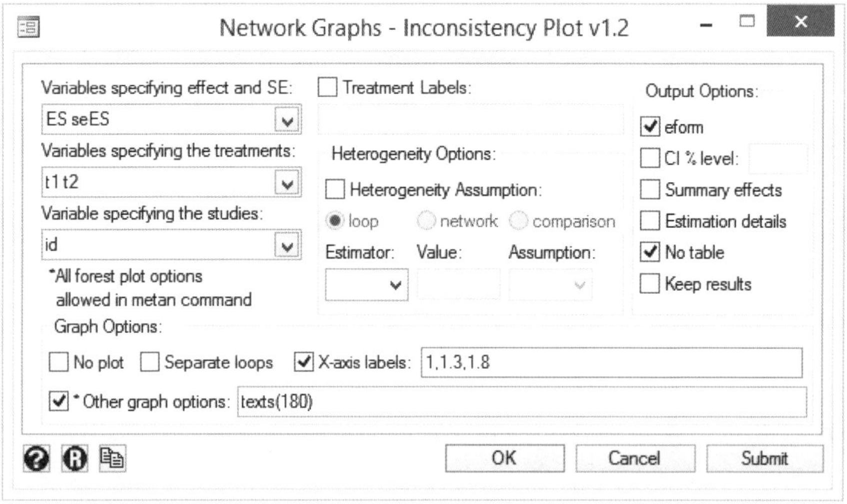

Figure 8. Example of using the `ifplot` dialog box for the antiplatelet regimens network

3.5   The netfunnel command

Comparison-adjusted funnel plot

Differences in the relative effects between small and large trials in a network of interventions often challenge the interpretation of the pairwise summary effects and need exploration (Chaimani et al. 2013b; Moreno et al. 2011). Extending the funnel plot from pairwise to network meta-analysis must account for different treatment comparisons being included: each comparison has its own summary effect. Hence, all the studies in the network do not have a common reference line of symmetry. In the comparison-adjusted funnel plot, the horizontal axis shows the difference of each i-study's estimate y_{iXY} from the direct summary effect for the respective comparison $(y_{iXY} - \widehat{\mu}_{XY}^{D})$, while the vertical axis presents a measure of dispersion of y_{iXY}. In the absence of small-study effects, all studies are expected to lie symmetrically around the 0 line of the comparison-adjusted funnel plot (Chaimani et al. 2013a; Chaimani and Salanti 2012).

Obtaining meaningful conclusions from this graph requires the definition of all comparisons across studies in a consistent direction, such as active intervention versus inactive, newer treatment versus older (Salanti et al. 2010), or sponsored versus nonsponsored.

Description of netfunnel

The `netfunnel` command plots a comparison-adjusted funnel plot for assessing small-study effects within a network of interventions. The command can plot observations using a different color for each comparison. The default direction for all comparisons in the network is in alphabetical or numerical order based on the codes of the treatments that appear in the dataset (for example, A versus B, B versus C, A versus C, or 1 versus 2, 2 versus 3, 1 versus 3). As explained above, investigators should order the treatments in a meaningful way rather than using the default ordering; for example, one might order the codes of the treatments from oldest to newest.

Syntax for netfunnel

netfunnel *varlist* [*if*] [*in*] [, scatteroptions(*string*) xtitle(*string*)

   xlabel(*string*) fixed random bycomparison addplot(*string*) ytitle(*string*)

   ylabel(*string*) noci noalphabetical]

In *varlist*, the variables *ES*, *seES*, *t1*, and *t2* (see section 3.1) should be specified. The order of *t1* and *t2* specifies the direction of the comparisons in the plot; specifying *t1* first means that the effect sizes have been estimated as *t1* versus *t2*, whereas specifying *t1* second means that the direction of the effect sizes is *t2* versus *t1*. This affects the appearance of the comparison-adjusted funnel plot only when the option `noalphabetical` has not been specified; however, the interpretation of the graph always depends on the direction of the relative effects.

Options for netfunnel

scatteroptions(*string*), xtitle(*string*), xlabel(*string*); see section 3.1.

fixed specifies that the fixed-effect model be used to estimate the direct summary effects (the default).

random specifies that the random-effects model be used to estimate the direct summary effects.

bycomparison specifies that different colors be used for different pairwise comparisons.

addplot(*string*) requests the addition of other twoway graphs specified in *string*.

ytitle(*string*) and ylabel(*string*) specify a title and the values, respectively, that are displayed for the vertical axis.

noci specifies to not include the pseudo 95% confidence interval lines for the difference $(y_{iXY} - \widehat{\mu}_{XY}^{D})$ in the plot.

noalphabetical specifies to include the comparisons in the plot as specified by the user (that is, *t1* versus *t2* for all studies) and not in alphabetical order (that is, A versus B, B versus C, and so on; the default).

Example for netfunnel using the antidiabetics network

The data for this network are described in section 3.2.

We use the `netfunnel` command to assess whether small and large trials tend to give different efficacy results. We focus on the comparisons of all active treatments against placebo, which might be more prone to small-study effects. As pointed out in section 3.1, all effect sizes (*ES*) in the dataset have been estimated as *t1* versus *t2* (placebo [1] appears only in *t2*); hence, all comparisons including placebo have been estimated as active treatment versus placebo. The resulting plot in figure 9 appears symmetric, implying the absence of small-study effects in the network.

```
. use antidiabetics_efficacy_wide.dta, clear
. netfunnel ES seES t1 t2 if t2==1, noalphabetical ylabel(0 0.1 0.2 0.3)

Comparisons in the plot:
    1.      7 vs 1
    2.      6 vs 1
    3.      5 vs 1
    4.      4 vs 1
    5.      3 vs 1
    6.      2 vs 1
```

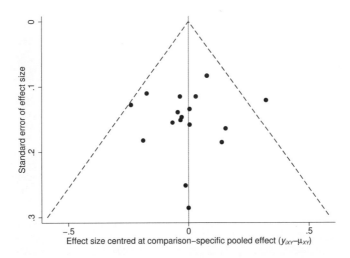

Figure 9. Comparison-adjusted funnel plot of the placebo-controlled antidiabetic trials

The comparison-adjusted funnel plot of figure 9 appears symmetric, implying the absence of small-study effects in the network.

Figure 10 shows how the above command can be executed via the `netfunnel` dialog box, accessed by typing `db netfunnel` in the Command window.

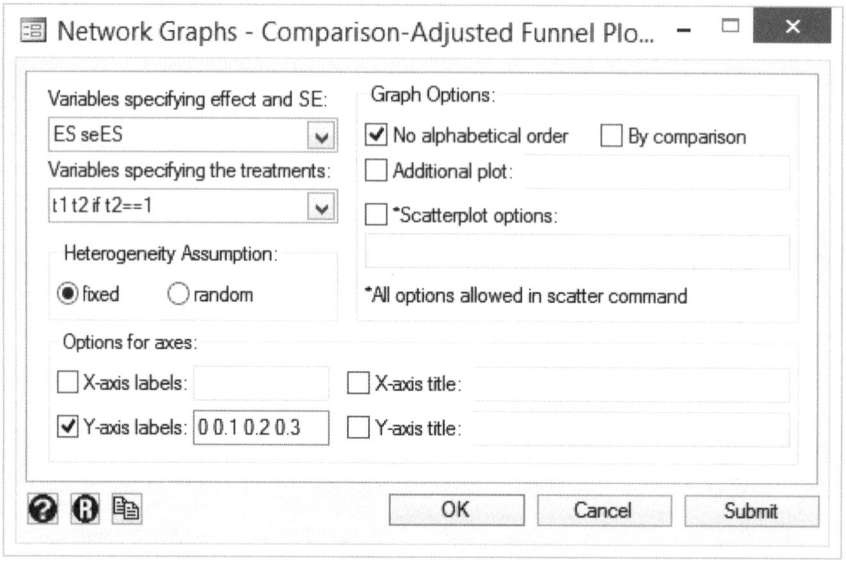

Figure 10. Example of using the `netfunnel` dialog box for the antidiabetics network

3.6   The intervalplot command

Predictive intervals in meta-analysis

The extent of uncertainty in the estimated treatment effects in meta-analysis is reflected not only by the confidence intervals but also by the predictive intervals that incorporate the extent of heterogeneity (Higgins, Thompson, and Spiegelhalter 2009; Riley, Higgins, and Deeks 2011). The predictive interval is the interval within which the relative treatment effect of a future study is expected to lie. For the relative effect of treatments X versus Y, we estimate the predictive interval as

$$\widehat{\mu}_{XY} \pm t^{\alpha}_{\mathrm{d.f.}} \sqrt{\widehat{\tau}^2 + \widehat{v}_{XY}}$$

where $\widehat{\tau}^2$ is the estimated heterogeneity variance. A common choice for the degrees of freedom (d.f.) of the t distribution in a standard pairwise meta-analysis is d.f. $= S - 2$ (with S being the total number of studies). For the case of network meta-analysis, this can be modified into d.f. $= S - N^{'} - 1$ (with $N^{'}$ being the number of available direct comparisons) (Cooper, Hedges, and Valentine 2009).

Description of intervalplot

The `intervalplot` command plots the estimated effect sizes and their uncertainties for all pairwise comparisons in a network meta-analysis. More specifically, `intervalplot` produces a forest plot where the horizontal lines representing the confidence intervals are extended to simultaneously show the predictive intervals. The treatment effects and their uncertainties can be estimated within Stata using the `mvmeta` (or `network`) package or by using other software.

Syntax for intervalplot

`intervalplot` [*varlist*] [*if*] [*in*] [, mvmetaresults nomvmeta
    labels(*treatment_labels*|comparison) eform keep title(*string*) notable
    noplot xtitle(*string*) xlabel(*string*) predictions separate
    reference(*reference_treatment*) novalues null(#) nulloptions(*line_options*)
    fcicolor(*string*) scicolor(*string*) fcipattern(*string*) scipattern(*string*)
    symbol(*string*) range(*string*) labtitle(*string*) valuestitle(*string*)
    symbolsize(*string*) textsize(*string*) lwidth(*string*) margin(# # # #)]

The network meta-analysis summary effects and their uncertainties are required as input for this command. These can be provided in two different ways:

1. By running `intervalplot` directly after performing network meta-analysis with the `mvmeta` or `network` command. In this case, the option `mvmetaresults` (the default) may be specified, and the *varlist* should be omitted.

2. By including the network meta-analysis estimates for all comparisons we want to plot and their lower/upper confidence limits as variables in a dataset (for example, variables *ES*, *LCI*, and *UCI*; see section 3.1). Optionally, we can include the variables *LPI* and *UPI*, the predictive lower and upper limits. In this case, the option `nomvmeta` is required.

Options for intervalplot

`mvmetaresults`, `nomvmeta`, `labels(`*treatment_labels*`|comparison)`, `eform`, `keep`, `title(`*string*`)`, `notable`, `noplot`, `xtitle(`*string*`)`, `xlabel(`*string*`)`; see section 3.1.

`predictions` specifies that the predictive intervals be added in the plot (when the option `nomvmeta` has not been specified). In the default plot, only the confidence intervals are displayed.

`separate` specifies that results in the plot be classified according to the comparator treatment (possible when `labels()` has been specified).

reference(*reference_treatment*) specifies that only the relative effects of each treatment versus the *reference_treatment* be displayed (when labels() has been specified). Note that the reference treatment specified in intervalplot does not necessarily need to be the reference treatment of the analysis.

novalues suppresses the display of the numerical estimates and their uncertainties in the plot.

null(*#*) specifies the value for the line of no effect.

nulloptions(*line_options*) specifies options for the line of no effect.

fcicolor(*string*), scicolor(*string*), fcipattern(*string*), scipattern(*string*), and symbol(*string*) specify the color, the pattern, and the symbol for the confidence and predictive intervals, respectively (when specified). The defaults are fcicolor(black) and scicolor(cranberry), fcipattern(solid) and scipattern(dash), and fcisymbol(diamond) and symbol(circle).

range(*string*) specifies the range for the horizontal axis.

labtitle(*string*) and valuestitle(*string*) specify titles for the treatment comparisons and the values of the confidence and predictive intervals, respectively.

symbolsize(*string*) and textsize(*string*) specify the sizes for the symbols of the effect sizes and the text, respectively. The default is symbolsize(small).

lwidth(*string*) specifies the width for the horizontal lines representing the confidence and predictive intervals.

margin(*# # # #*) specifies the margins for the region of the plot.

Example for intervalplot using the antihypertensives network

The dataset for this network consists of 48 observations equal to the total number of study arms. The treatments (*t*) have been coded as numbers (1 = placebo, 2 = beta-blockers, 3 = diuretics, 4 = calcium-channel blockers, 5 = angiotensin-converting-enzyme inhibitors, 6 = angiotensin-receptor blockers), and the data are in the appropriate format for network (see White [2013; 2011] and White et al. [2012] for a detailed description), where r is the number of events and n is the total participants in each arm.

```
. use antihypertensives.dta, clear
. list in 1/5, clean noobs
    id      r       n    t
     1     45     410    5
     1     70     405    2
     1     32     202    4
     2    119    4096    5
     2    154    3954    4
```

Using the `network setup` command (White 2013), we prepared the data in a format suitable to run network meta-analysis in Stata.[2]

Then, we performed network meta-analysis via the command `network meta` to obtain the network estimates of the relative treatment effects. We then run the command `intervalplot` to estimate the predictive intervals and see the result in figure 11.

```
. network setup r n, studyvar(id) trtvar(t) numcodes ref(1)
  (output omitted)
. network meta c
  (output omitted)
. intervalplot, eform pred null(1) labels(Placebo BB Diuretics CCB ACE ARB)
> separate margin(10 40 5 5) notable
  The intervalplot command assumes that the saved results from mvmeta or network
> meta commands have been derived from the current dataset
```

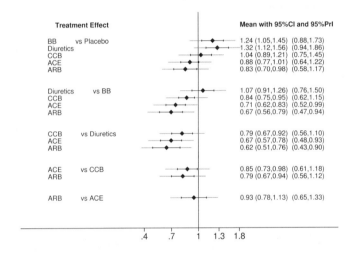

Figure 11. Predictive intervals plot for the antihypertensives network. The graph presents the network estimates for all pairwise comparisons. Black horizontal lines represent the confidence intervals, and gray lines represent the predictive intervals.

Conclusions about some comparisons seem to be substantially affected by the estimated (common) heterogeneity in the network; although their confidence intervals suggest an association, the respective predictive interval crosses the line of no effect and suggests that future studies might favor either treatment (for example, diuretics versus placebo or calcium-channel blockers versus diuretics).

2. Note that when only study-level data (that is, effect size and standard error for each study) are available, the `network import` and `network convert` augment commands should be used instead of `network setup`.

Figure 12 shows how the above command can be executed via the `intervalplot` dialog box, accessed by typing `db intervalplot` in the Command window.

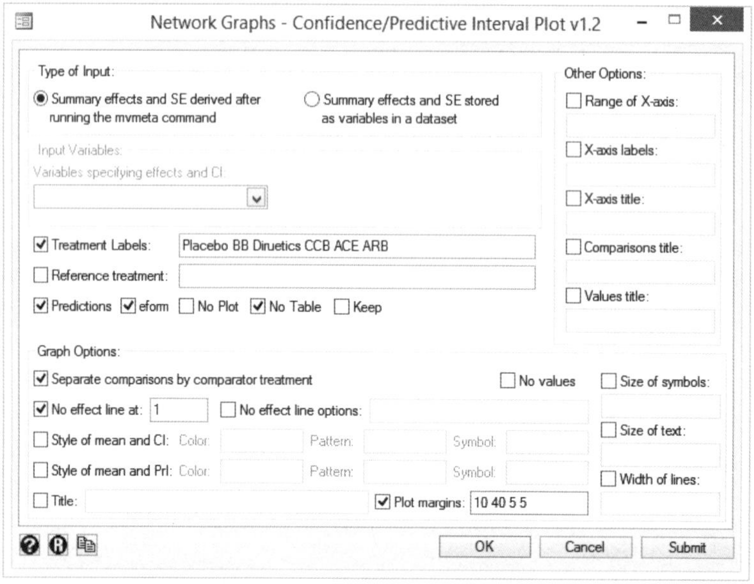

Figure 12. Example of using the `intervalplot` dialog box for the antihypertensives network

3.7   The netleague command

Description of netleague

The `netleague` command creates a "league table" showing in the off-diagonal cells the relative treatment effects for all possible pairwise comparisons estimated in a network meta-analysis (Cipriani et al. 2009). The diagonal cells include the names of the competing treatments in the network, which can be sorted according to a prespecified order.

Syntax for netleague

```
netleague [varlist] [, mvmetaresults nomvmeta labels(string) eform
   sort(string) export(string) nokeep]
```

The network meta-analysis summary effects and their uncertainties can be provided as input in two different ways:

1. By running `netleague` directly after performing network meta-analysis with the `mvmeta` or `network` command. In this case, the option `mvmetaresults` (the default) may be specified, and the *varlist* should be omitted.

2. By including these summary estimates as variables in the dataset; the variables *ES*, *seES*, *t1*, and *t2* (see section 3.1) should be specified in *varlist*. In this case, the option `nomvmeta` is required.

Options for netleague

`mvmetaresults`, `nomvmeta`, `labels(`*string*`)`, `eform`; see section 3.1.

`sort(`*string*`)` specifies the order for treatments from top to bottom in the league table. When the option `nomvmeta` has not been specified, `sort()` requires the option `labels()`. When the option `nomvmeta` has been specified, the names of treatments should be given as displayed in the dataset in variables *t1* and *t2*. By default, the treatments are ordered alphabetically from bottom to top.

`export(`*string*`)` specifies the path of an Excel file where the league table is exported.

`nokeep` specifies not to store the league table at the end of the dataset.

Example for netleague using the antihypertensives network

The data for this network are described in section 3.6.

Using the `network setup` and `network meta` commands, we perform network meta-analysis to obtain the network estimates of the relative treatment effects (see section 3.6). We then directly run the `netleague` command to produce the league table of the network. We order the treatments according to their relative rankings based on the surface under the cumulative ranking curves (SUCRA) percentages.

```
. netleague, labels(Placebo BB Diuretics CCB ACE ARB)
> sort(ARB ACE Placebo CCB BB Diuretics) eform

 Warning: The existing dataset is stored as a temporary file
 Warning: To save any changes applied at this temporary file in a specific
> directory you need to use the ´Save as´ menu

 The league table has been stored at the end of the dataset
```

The league table for the first four treatments of the network is

```
. list _ARB _ACE _Placebo _CCB in 1/4, clean noobs noheader
              ARB     1.07 (0.89,1.29)    1.21 (1.02,1.44)    1.26 (1.07,1.49)
  0.93 (0.78,1.13)                 ACE    1.13 (0.99,1.29)    1.18 (1.02,1.36)
  0.83 (0.70,0.98)    0.88 (0.77,1.01)             Placebo    1.04 (0.89,1.21)
  0.79 (0.67,0.94)    0.85 (0.73,0.98)    0.96 (0.82,1.12)                 CCB
```

Figure 13 shows how the above command can be executed via the `netleague` dialog box, accessed by typing `db netleague` in the Command window.

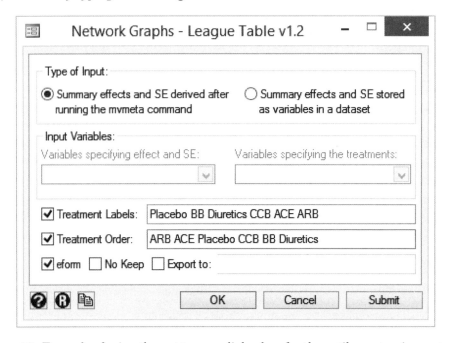

Figure 13. Example of using the `netleague` dialog box for the antihypertensives network

Note that the user can choose between the upper and lower triangle of the table, which provide the network estimates in the opposite direction (that is, row- versus column-defining treatment or column- versus row-defining treatment).

3.8 The sucra command

Ranking probabilities for competing treatments

When performing a network meta-analysis, it is common to estimate the ranking probabilities p_{tr} for each treatment t being at order r. Then, the competing treatments can be classified using the cumulative probabilities p_{tr}^{cum} that treatment t is ranked among the first r places. Two relative ranking measures that account for the uncertainty in treatment order are as follows:

1. The SUCRA that expresses the percentage of effectiveness or safety each treatment has compared with an "ideal" treatment always ranked first without uncertainty (Salanti, Ades, and Ioannidis 2011) [given by $(\sum_{r=1}^{T-1} p_{tr}^{\text{cum}})/(T-1)$, with T being the total number of treatments].

2. The mean rank, which is the mean of the distribution of the ranking probabilities [equal to $\sum_{r=1}^{T}(p_{tr} \times r)$].

Description of sucra

The sucra command gives the SUCRA percentages and mean ranks, and produces rankograms (line plots of the probabilities versus ranks) and cumulative ranking plots (line plots of the cumulative probabilities versus ranks) for all treatments in a network of interventions.

Syntax for sucra

sucra [*varlist*] [, mvmetaresults nomvmeta labels(*string*) title(*string*)

   notable noplot compare(*varlist*) stats(*string*) rprobabilities(*string*)

   reverse rankograms names(*string*) lcolol(*string*) lpattern(*string*)]

The input in *varlist* is the ranking probabilities for all treatments and ranks from a network meta-analysis. This can be provided in three different ways:

1. By running sucra after performing network meta-analysis with the mvmeta command (where the option pbest(min|max, zero all reps() gen()) has been added; see the help file for mvmeta) or with network rank min|max, zero all reps() gen() (see the help file for network). These commands add in the data $t \times r$ new variables, each one corresponding to the probability of the rth rank for each treatment. These new variables are typically named prob*r*_*t* (where prob is the prefix we have specified in gen() of pbest(), r is the rank, and t is the treatment) and automatically appear at the end of the dataset. If sucra is run after mvmeta or network, the option mvmetaresults may be specified (the default), and all the variables containing the probability for a treatment being at a particular rank should be specified in *varlist*.

2. By providing the columns of the treatment-by-ranking probabilities matrix as variables in the dataset. In this case, all variables containing the ranking probabilities for each treatment should be specified in *varlist*, and the option nomvmeta is required.

3. By specifying the path of a .txt file where the ranking probabilities are stored (for example, after running network meta-analysis in WinBUGS). See table 1 for an example of this .txt file. The path of the file should be specified in option stats(), and the variable representing the ranking probabilities (for example, prob) should

be specified in option `rprobabilities()`; the *varlist* should be omitted. The command assumes that each node prob$[t, r]$ $(t, r = 1, \ldots, 6)$ represents the probability of treatment t being at order r, and the opposite (that is, prob$[r, t]$) when the option `reverse` has been specified. The option `nomvmeta` is also required.

Table 1. Example of a `.txt` file including the ranking probabilities for the antihypertensives network as estimated from WinBUGS

node	mean	sd	MC error	2.5%	median	97.5%	start	sample
prob[1,1]	0.00659	0.08091	1.825E−4	0.0	0.0	0.0	1001	300000
prob[1,2]	0.071	0.2568	7.242E−4	0.0	0.0	1.0	1001	300000
prob[1,3]	0.6391	0.4803	0.001344	0.0	1.0	1.0	1001	300000
prob[1,4]	0.2769	0.4475	0.001472	0.0	0.0	1.0	1001	300000
prob[1,5]	0.00617	0.07831	1.631E−4	0.0	0.0	0.0	1001	300000
prob[1,6]	2.333E−4	0.01527	2.808E−5	0.0	0.0	0.0	1001	300000
prob[2,1]	0.0	0.0	1.054E−13	0.0	0.0	0.0	1001	300000
prob[2,2]	7.0E−5	0.008366	1.512E−5	0.0	0.0	0.0	1001	300000
prob[2,3]	8.867E−4	0.02976	5.734E−5	0.0	0.0	0.0	1001	300000
prob[2,4]	0.01184	0.1082	2.45E−4	0.0	0.0	0.0	1001	300000
prob[2,5]	0.7448	0.436	0.001502	0.0	1.0	1.0	1001	300000
prob[2,6]	0.2424	0.4285	0.001486	0.0	0.0	1.0	1001	300000
prob[3,1]	3.333E−6	0.001826	3.339E−6	0.0	0.0	0.0	1001	300000
prob[3,2]	3.333E−5	0.005773	1.05E−5	0.0	0.0	0.0	1001	300000
prob[3,3]	3.333E−4	0.01825	3.458E−5	0.0	0.0	0.0	1001	300000
prob[3,4]	0.00566	0.07502	1.505E−4	0.0	0.0	0.0	1001	300000
prob[3,5]	0.2372	0.4254	0.001465	0.0	0.0	1.0	1001	300000
prob[3,6]	0.7567	0.4291	0.001489	0.0	1.0	1.0	1001	300000
prob[4,1]	0.001077	0.03279	6.232E−5	0.0	0.0	0.0	1001	300000
prob[4,2]	0.01836	0.1343	3.052E−4	0.0	0.0	0.0	1001	300000
prob[4,3]	0.2728	0.4454	0.001423	0.0	0.0	1.0	1001	300000
prob[4,4]	0.6957	0.4601	0.001464	0.0	1.0	1.0	1001	300000
prob[4,5]	0.01135	0.1059	2.342E−4	0.0	0.0	0.0	1001	300000
prob[4,6]	6.167E−4	0.02483	4.823E−5	0.0	0.0	0.0	1001	300000
prob[5,1]	0.2761	0.4471	0.001272	0.0	0.0	1.0	1001	300000
prob[5,2]	0.6638	0.4724	0.001367	0.0	1.0	1.0	1001	300000
prob[5,3]	0.05498	0.2279	5.915E−4	0.0	0.0	1.0	1001	300000
prob[5,4]	0.005007	0.07058	1.447E−4	0.0	0.0	0.0	1001	300000
prob[5,5]	6.333E−5	0.007958	1.44E−5	0.0	0.0	0.0	1001	300000
prob[5,6]	3.333E−6	0.001826	3.339E−6	0.0	0.0	0.0	1001	300000
prob[6,1]	0.7162	0.4508	0.001309	0.0	1.0	1.0	1001	300000
prob[6,2]	0.2467	0.4311	0.001153	0.0	0.0	1.0	1001	300000
prob[6,3]	0.03186	0.1756	4.231E−4	0.0	0.0	1.0	1001	300000
prob[6,4]	0.004833	0.06935	1.409E−4	0.0	0.0	0.0	1001	300000
prob[6,5]	3.6E−4	0.01897	3.588E−5	0.0	0.0	0.0	1001	300000
prob[6,6]	3.333E−5	0.005773	1.15E−5	0.0	0.0	0.0	1001	300000

Options for sucra

mvmetaresults, nomvmeta, labels(*string*), title(*string*), notable, noplot; see section 3.1.

compare(*varlist*) specifies a second set of variables containing ranking probabilities. These can be, for example, the ranking probabilities for the same treatments but for different outcomes. compare() will add a second ranking plot to the existing ranking plot for each treatment.

stats(*string*) specifies the path of the file that has ranking probabilities.

rprobabilities(*string*) specifies the variable representing the ranking probabilities.

reverse specifies that the node probabilities are prob(r, t) instead of prob(t, r) with t being treatment and r being order (when option stats() has also been specified).

rankograms specifies that rankograms be drawn instead of cumulative ranking probability plots.

names(*string*) specifies a label name for the first (specified in *varlist*) and second (specified in compare(*varlist*)) set of ranking probabilities. An example for relative ranking results from different outcomes would be names("Effectiveness" "Acceptability"). These label names are displayed in the output results and ranking plots.

lcolol(*string*) and lpattern(*string*) specify the color and the pattern, respectively, of the lines in the ranking plots for the first and second (separated with a space) set of ranking probabilities. The default colors are lcolol(black cranberry), and the default patterns are lpattern(solid dash).

Example for sucra using the antihypertensives network

The data for this network are described in section 3.6.

Using the **network rank** command after **network setup** and **network meta** (see section 3.5), we performed network meta-analysis to obtain the ranking probabilities for all competing treatments in the network. These have been stored at the end of the dataset in the following format:

```
. network rank min, zero all reps(10000) gen(prob)
(output omitted)
. list probmin_zero-probmin__y_3 prob2_zero-prob2__y_3 in 1, clean table noobs
> abbreviate(14)

   probmin_zero  probmin__y_2  probmin__y_3  prob2_zero  prob2__y_2  prob2__y_3
            .17             0             0    4.540009           0           0
```

Each `probr_t` represents the probability of treatment t (with $t = $ `zero`, `_y_2`, `_y_3`) being at order r (with $r = $ `min`, $2, 3$). Note that `min` represents first rank (because the outcome is harmful), `zero` represents the reference treatment, and `_y_2` and `_y_3` are treatments coded as 2 and 3. For example, according to the above results, the reference treatment has a 4.54% probability of being second, that is, the variable `prob2_zero`.

Then, we run the `sucra` command to produce the rankograms:

```
. sucra prob*, labels(Placebo BB Diuretics CCB ACE ARB) rankog
Treatment Relative Ranking of Model 1
```

Treatment	SUCRA	PrBest	MeanRank
Placebo	54.9	0.2	3.3
BB	16.2	0.0	5.2
Diuretics	4.0	0.0	5.8
CCB	46.4	0.1	3.7
ACE	83.8	24.0	1.8
ARB	94.8	75.8	1.3

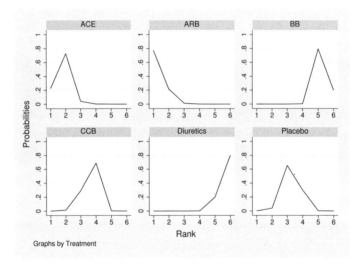

Graphs by Treatment

Figure 14. Rankograms for the antihypertensives network showing the probability for every treatment being at a particular order

Figure 15 shows how the above command can be executed via the `sucra` dialog box, accessed by typing `db sucra` in the Command window.

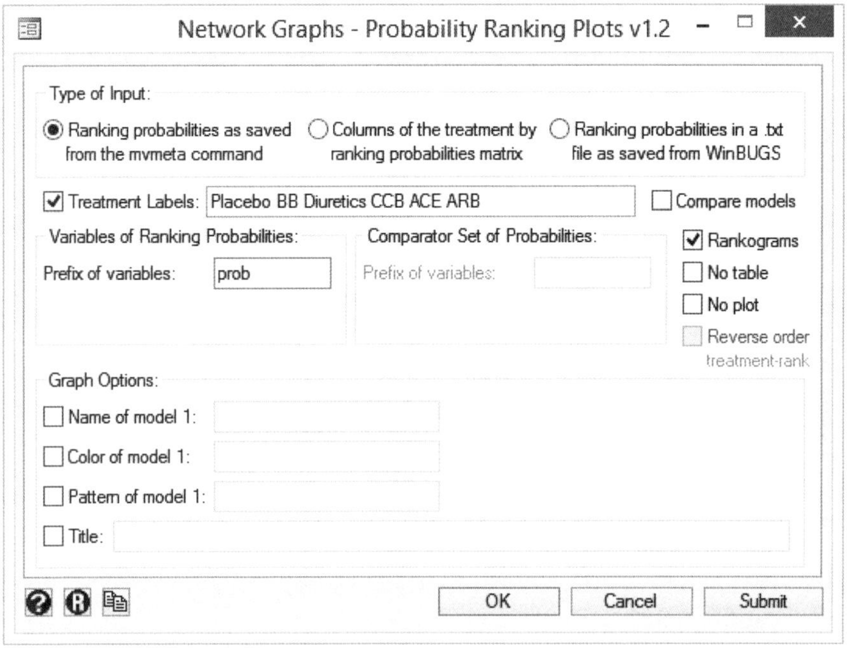

Figure 15. Example of using the `sucra` dialog box for the antihypertensives network

We again run `network rank` to estimate the predictive ranking probabilities (adding the option `predict` and choosing names `pred_prob` to denote the predictive probabilities stored in the dataset). Using the `sucra` command, we compare the cumulative ranking plots based on the estimated and predictive ranking probabilities.

The resulting curves (shown in figure 16) imply that the incorporation of the heterogeneity in the predictive ranking probabilities does not materially affect the results for the relative ranking of treatments.

```
. network rank min, zero all reps(10000) gen(pred_prob) predict
  (output omitted)
. sucra prob*, labels(Placebo BB Diuretics CCB ACE ARB) compare(pred_prob*)
> names("Estimated probabilities" "Predictive probabilities")
Treatment Relative Ranking of Estimated probabilities
```

Treatment	SUCRA	PrBest	MeanRank
Placebo	54.9	0.2	3.3
BB	16.2	0.0	5.2
Diuretics	4.0	0.0	5.8
CCB	46.4	0.1	3.7
ACE	83.8	24.0	1.8
ARB	94.8	75.8	1.3

```
Treatment Relative Ranking of Predictive probabilities
```

Treatment	SUCRA	PrBest	MeanRank
Placebo	56.0	3.7	3.2
BB	17.2	0.1	5.1
Diuretics	7.7	0.0	5.6
CCB	48.3	1.7	3.6
ACE	80.1	29.8	2.0
ARB	90.6	64.7	1.5

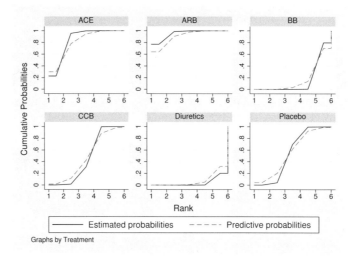

Figure 16. Cumulative probability curves for the antihypertensives network showing the estimated and predictive probabilities for each treatment being up to a specific rank

Figure 17 shows how the above command can be executed via the `sucra` dialog box, accessed by typing `db sucra` in the Command window.

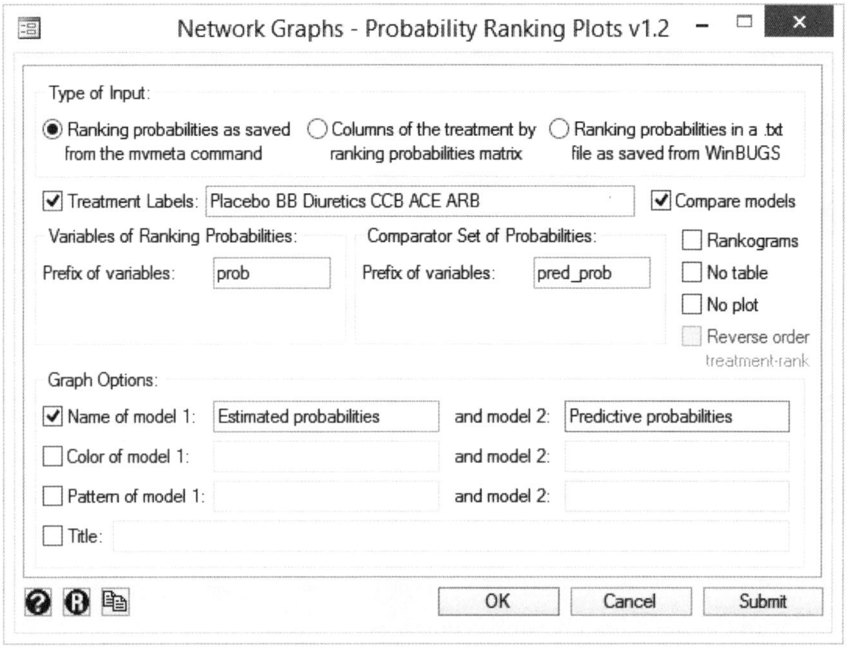

Figure 17. Another example of using the `sucra` dialog box for the antihypertensives network

3.9   The mdsrank command

Multidimensional scaling approach for ranking

A different approach to estimate the relative ranking is to use multidimensional scaling (MDS) techniques. To apply this method, the network estimates for all possible comparisons are treated as proximity data aiming to reveal their latent structure. In this way, the absolute value $|\widehat{\mu}_{XY}|$ defines the dissimilarity between the two treatments (X, Y) with $|\widehat{\mu}_{XX}| = 0$. Weighting the absolute effects sizes by their inverse standard errors or inverse variances ensures that the assumption of a common distribution between the elements of the matrix is plausible. Assuming that the rank of the treatments is the only dimension underlying the outcome, the purpose of the technique would be to reduce the $T \times T$ matrix into a $T \times 1$ vector. This vector involves the set of distances being as close as possible to the observed dissimilarities (that is, relative effects) and would represent the relative ranking of the treatments.

Description of mdsrank

The `mdsrank` command creates the squared matrix containing the pairwise relative effect sizes and plots the resulting values of the unique dimension for each treatment.

Syntax for mdsrank

mdsrank *varlist* [*if*] [*in*] [, <u>label</u>s(*string*) <u>noplot</u> <u>scat</u>teroptions(*string*)
    best(min|max)]

In *varlist*, the summary effect sizes from network meta-analysis and their uncertainties (variables *ES* and *seES*; see section 3.1), as well as the treatment comparisons they refer to (variables *t1* and *t2*; see section 3.1), should be specified.

Options for mdsrank

labels(*string*), noplot, scatteroptions(*string*); see section 3.1.

best(min|max) specifies whether larger-dimension scores correspond to a more favorable outcome with the treatment. The default is best(min).

Example for mdsrank using the antihypertensives network

The data for this network are described in section 3.6.

Using the `intervalplot` command (after `network meta`; see section 3.5), we obtain the estimated relative effects via the option `keep`.

```
. intervalplot, labels(Placebo BB Diuretics CCB ACE ARB) noplot notab keep
  The intervalplot command assumes that the saved results from mvmeta or network
> meta commands have been derived from the current dataset
```

These results have been stored at the end of the dataset. For example, the first five comparisons in the data give the results for all active treatments versus placebo.

```
. split _Comparison, par(" vs ") gen(t)
variables created as string:
t1  t2
. list _Comparison _Effect_Size _Standard_Error t1 t2 in 1/5, noobs clean

         _Comparison    _Effect~e    _Stand~r          t1        t2
         BB vs Placebo    .2121609    .0810291          BB   Placebo
   Diuretics vs Placebo    .2806841    .0842885   Diuretics   Placebo
        CCB vs Placebo    .0404351    .0786069         CCB   Placebo
        ACE vs Placebo   -.1233237    .0679924         ACE   Placebo
        ARB vs Placebo    -.191084    .0874827         ARB   Placebo
```

Then, we use the **mdsrank** command to estimate the relative ranking of treatments (shown in figure 18) using the MDS method:

```
. mdsrank _Effect_Size _Standard_Error t1 t2 if _Effect_Size!=., best(max)

Warning: The existing dataset is stored as a temporary file
Warning: To save any changes applied at this temporary file in a specific
> directory you need to use the 'Save as' menu
```

Treatment	Dim1	Rank
ACE	2.31	1
ARB	2.05	2
Placebo	0.41	3
CCB	0.19	4
BB	-2.34	5
Diuretics	-2.62	6

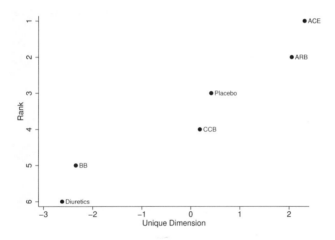

Figure 18. Relative ranking of treatments for the diabetes network based on the MDS approach. Larger values of the dimension correspond to higher ranks.

The relative ranking obtained from this approach is similar to that derived from the ranking probabilities, but they disagree in the order of the two best treatments.

Figure 19 shows how the above command can be executed via the `mdsrank` dialog box, accessed by typing `db mdsrank` in the Command window.

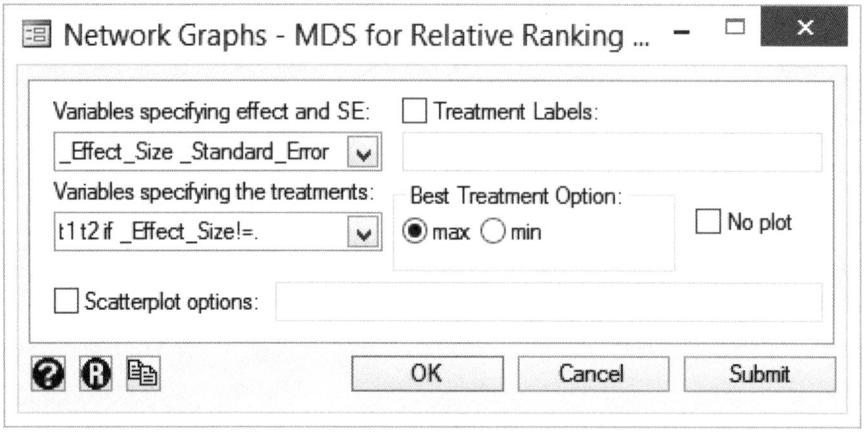

Figure 19. Example of using the `mdsrank` dialog box for the antihypertensives network

3.10   The clusterank command

Cluster analysis for more than one outcome

When performing a network meta-analysis, the competing treatments can be ranked according to their performance on one or more outcomes (for example, effectiveness and safety). However, the relative ranking for each outcome might be different, and this makes the choice of the "best" treatment challenging. A possible way to make inferences based on results for two outcomes is by using a two-dimensional plot and constructing groups of treatments with similar performance on both outcomes (Chaimani et al. 2013a). To form meaningful groups of treatments, hierarchical clustering methods have been used (Kaufman and Rousseeuw 2005).

Description of clusterank

The `clusterank` command performs hierarchical cluster analysis to group the competing treatments into meaningful groups. It requires the values of a ranking measure (for example, SUCRA percentages or MDS dimension) for two outcomes.

Optimal cluster analysis

The command chooses the appropriate metric (Euclidean, squared Euclidean, absolute-value distance, etc.) and linkage method (single, average, weighted, complete, ward, centroid, median) based on the cophenetic correlation coefficient, which measures how faithfully the output dendrogram represents the dissimilarities between observations (Handl, Knowles, and Kell 2005). The optimal level of dendrogram and the optimal number of clusters are chosen using an internal cluster validation measure, called clustering gain (Jung et al. 2003). This measure has been designed to have a maximum value when intracluster similarity is maximized and intercluster similarity is minimized.

Syntax for clusterank

clusterank *varlist* [, <u>scatter</u>options(*string*) best(min|max)

    <u>method</u>(*linkage_method distance_metric*) <u>cluster</u>s(*integer*) <u>dendr</u>ogram]

In *varlist*, the variables *outcome1* and *outcome2* (a ranking measure—for example, mean rank—for each outcome; see section 3.1) and *t* (optional, the names of the treatments; see section 3.1) should be specified.

Options for clusterank

scatteroptions(*string*); see section 3.1.

best(min|max) specifies whether larger or smaller values of the ranking measure correspond to better outcome with the treatment (applies to both outcomes). The default is best(max).

method(*linkage_method distance_metric*) specifies which linkage method and distance metric to use. By default, the method of hierarchical clustering is decided according to the cophenetic correlation coefficient (Handl, Knowles, and Kell 2005). For other options, see the help file for cluster (type help cluster in Stata).

clusters(*integer*) specifies the number of clusters used to group the treatments. By default, the optimal number of clusters is decided according to the "clustering gain" (Jung et al. 2003).

dendrogram specifies that the dendrogram of the hierarchical analysis be displayed instead of the clustered ranking plot.

Example for clusterank using the antidiabetics network

The data for this network are described in section 3.2.

Both datasets for efficacy and tolerability are in the appropriate format for network (where y is the mean change score in each study arm, sd is the respective standard deviation, and n is the number of total participants). After running network meta-analysis using the network meta command, we obtain the SUCRA (via the sucra command; see section 3.8) percentages for efficacy (*outcome1*) and tolerability (*outcome2*) to present the relative ranking of the seven interventions separately for each outcome.

Relative ranking for efficacy:

```
. use antidiabetics_efficacy_long.dta, clear
. network setup y sd n, studyvar(id) trtvar(t) ref(1) numcodes
  (output omitted)
. network meta c
  (output omitted)
. network rank min, zero all reps(10000) gen(eff_prob)
  (output omitted)
. sucra eff_prob*, labels(Placebo Sulfonylurea "DPP-4 inhibitor"
> Thiazolidinedione "GLP-1 analog" AGI Glinine) noplot
Treatment Relative Ranking of Model 1
```

Treatment	SUCRA	PrBest	MeanRank
Placebo	1.0	0.0	6.9
Sulfonylurea	56.3	2.8	3.6
DPP-4 inhibitor	46.4	1.3	4.2
Thiazolidinedione	64.3	7.7	3.1
GLP-1 analog	66.4	21.1	3.0
AGI	41.9	10.2	4.5
Glinine	73.8	57.0	2.6

Relative ranking for tolerability:

```
. use antidiabetics_tolerability.dta, clear
. network setup y sd n, studyvar(id) trtvar(t) ref(1) numcodes
  (output omitted)
. network meta c
  (output omitted)
. network rank min, zero all reps(10000) gen(tol_prob)
  (output omitted)
. sucra tol_prob*, labels(Placebo Sulfonylurea "DPP-4 inhibitor"
> Thiazolidinedione "GLP-1 analog" AGI Glinine) noplot
Treatment Relative Ranking of Model 1
```

Treatment	SUCRA	PrBest	MeanRank
Placebo	59.9	0.0	3.4
Sulfonylurea	24.0	0.0	5.6
DPP-4 inhibitor	57.4	0.0	3.6
Thiazolidinedione	2.8	0.0	6.8
GLP-1 analog	91.8	51.0	1.5
AGI	91.8	49.0	1.6
Glinine	23.3	0.0	5.6

We store the SUCRA percentages in a dataset and then run the `clusterank` command, with results shown in figure 20.

```
. use antidiabeticssucras.dta, clear

. list, clean noobs
                          t    outcome1    outcome2
                    Placebo           1        59.9
                Sulfonylurea        56.3          24
             DPP-4 inhibitor        46.4        57.4
           Thiazolidinedione        64.3         2.8
                GLP-1 analog        66.4        91.8
                         AGI        41.9        90.7
                     Glinine        73.8        23.3

. clusterank outcome1 outcome2 t

Best linkage method: averagelinkage
Best distance metric: Canberra

Cophenetic Correlation Coefficient c = 0.91

** Maximum value of clustering gain = 2722.18
** Optimal number of clusters =  4
```

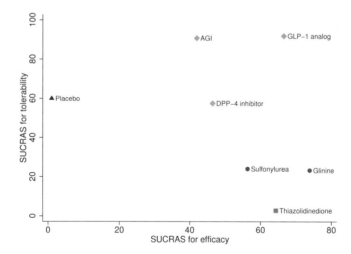

Figure 20. Clustered ranking plot for the antidiabetics network presenting jointly the relative ranking of treatments (based on the SUCRA percentages) for efficacy and tolerability. Treatments lying in the upper right corner are considered to perform well for both outcomes. Different plotting symbols represent different clusters of treatments.

Figure 20 shows three treatments performing similarly well on both outcomes that form the best group of treatments.

Figure 21 shows how the above command can be executed via the `clusterank` dialog box, accessed by typing `db clusterank` in the Command window.

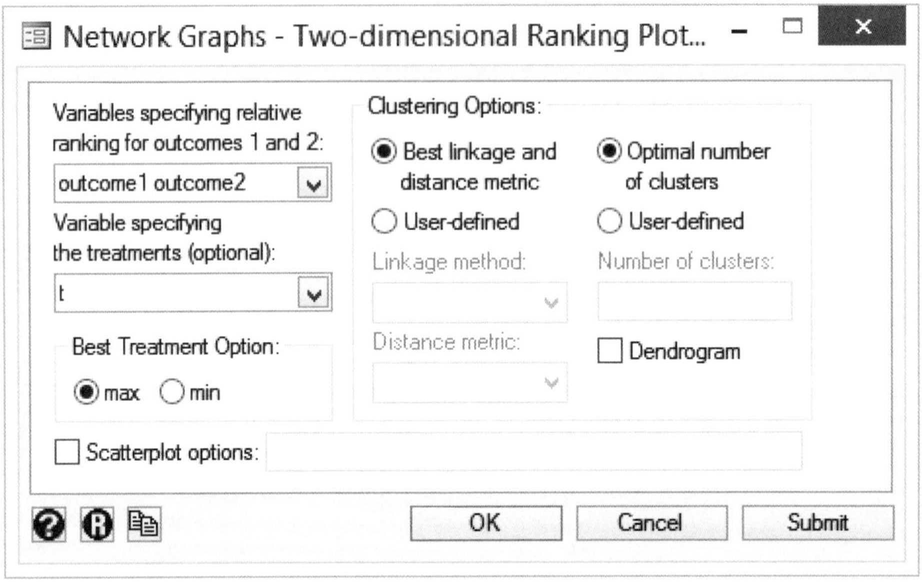

Figure 21. Example of using of the dialog box for `clusterank` for the antidiabetics network

4   Discussion

In this article, we introduced a package of nine commands that can be used to understand and present graphically the different steps in a network meta-analysis. Using these commands in conjunction with `mvmeta` (White 2011) or `network` (White 2013) can simplify the technicalities of the network meta-analysis procedure for dichotomous or continuous data; the commands might be less convenient for other types of data (for example, time-to-event data).

The suggested graphs aim to summarize in a comprehensive way the most important findings from this complex statistical tool and facilitate their interpretation. The commands' usefulness might be limited under specific conditions; for example, inconsistency plots cannot be applied in networks without closed loops, and comparison-adjusted funnel plots are not informative for networks with only one or two studies within each pairwise comparison.

5 Acknowledgments

This work was supported by the European Research Council (IMMA 260559 project). We thank Drs. Julian Higgins, Dimitris Mavridis, Panagiota Spyridonos, and Ian White for their helpful comments and suggestions on developing the commands.

6 References

Anscombe, F. J. 1973. Graphs in statistical analysis. *American Statistician* 27: 17–21.

Anzures-Cabrera, J., and J. P. T. Higgins. 2010. Graphical displays for meta-analysis: An overview with suggestions for practice. *Research Synthesis Methods* 1: 66–80.

Bax, L., N. Ikeda, N. Fukui, Y. Yaju, H. Tsuruta, and K. G. M. Moons. 2009. More than numbers: the power of graphs in meta-analysis. *American Journal of Epidemiology* 169: 249–255.

Bucher, H. C., G. H. Guyatt, L. E. Griffith, and S. D. Walter. 1997. The results of direct and indirect treatment comparisons in meta-analysis of randomized controlled trials. *Journal of Clinical Epidemiology* 50: 683–691.

Caldwell, D. M., A. E. Ades, and J. P. T. Higgins. 2005. Simultaneous comparison of multiple treatments: Combining direct and indirect evidence. *British Medical Journal* 331: 897–900.

Chaimani, A., J. P. T. Higgins, D. Mavridis, P. Spyridonos, and G. Salanti. 2013a. Graphical tools for network meta-analysis in Stata. *PLOS ONE* 8: e76654.

Chaimani, A., D. Mavridis, and G. Salanti. 2014. A hands-on practical tutorial on performing meta-analysis with Stata. *Evidence-Based Mental Health* 17: 111–116.

Chaimani, A., and G. Salanti. 2012. Using network meta-analysis to evaluate the existence of small-study effects in a network of interventions. *Research Synthesis Methods* 3: 161–176.

Chaimani, A., H. S. Vasiliadis, N. Pandis, C. H. Schmid, N. J. Welton, and G. Salanti. 2013b. Effects of study precision and risk of bias in networks of interventions: A network meta-epidemiological study. *International Journal of Epidemiology* 42: 1120–1131.

Cipriani, A., T. A. Furukawa, G. Salanti, J. R. Geddes, J. P. T. Higgins, R. Churchill, N. Watanabe, A. Nakagawa, I. M. Omori, H. McGuire, M. Tansella, and C. Barbui. 2009. Comparative efficacy and acceptability of 12 new-generation antidepressants: A multiple-treatments meta-analysis. *Lancet* 373: 746–758.

Cooper, H., L. V. Hedges, and J. C. Valentine. 2009. *The Handbook of Research Synthesis and Meta-Analysis*. 2nd ed. New York: Russell Sage Foundation.

DerSimonian, R., and N. Laird. 1986. Meta-analysis in clinical trials. *Controlled Clinical Trials* 7: 177–188.

Dias, S., N. J. Welton, D. M. Caldwell, and A. E. Ades. 2010. Checking consistency in mixed treatment comparison meta-analysis. *Statistics in Medicine* 29: 932–944.

Elliott, W. J., and P. M. Meyer. 2007. Incident diabetes in clinical trials of antihypertensive drugs: A network meta-analysis. *Lancet* 369: 201–207.

Handl, J., J. Knowles, and D. B. Kell. 2005. Computational cluster validation in postgenomic data analysis. *Bioinformatics* 21: 3201–3212.

Higgins, J. P. T., D. Jackson, J. K. Barrett, G. Lu, A. E. Ades, and I. R. White. 2012. Consistency and inconsistency in network meta-analysis: Concepts and models for multi-arm studies. *Research Synthesis Methods* 3: 98–110.

Higgins, J. P. T., S. G. Thompson, and D. J. Spiegelhalter. 2009. A re-evaluation of random-effects meta-analysis. *Journal of the Royal Statistical Society, Series A* 172: 137–159.

Jansen, J. P., and H. Naci. 2013. Is network meta-analysis as valid as standard pairwise meta-analysis? It all depends on the distribution of effect modifiers. *BMC Medicine* 11: 159.

Jung, Y., H. Park, D.-Z. Du, and B. L. Drake. 2003. A decision criterion for the optimal number of clusters in hierarchical clustering. *Journal of Global Optimization* 25: 91–111.

Kaufman, L., and P. J. Rousseeuw. 2005. *Finding Groups in Data: An Introduction to Cluster Analysis.* Hoboken, NJ: Wiley.

Krahn, U., H. Binder, and J. König. 2013. A graphical tool for locating inconsistency in network meta-analyses. *BMC Medical Research Methodology* 13: 35.

Lee, A. W. 2014. Review of mixed treatment comparisons in published systematic reviews shows marked increase since 2009. *Journal of Clinical Epidemiology* 67: 138–143.

Lu, G., and A. E. Ades. 2006. Assessing evidence inconsistency in mixed treatment comparisons. *Journal of the American Statistical Association* 101: 447–459.

Lu, G., N. J. Welton, J. P. T. Higgins, I. R. White, and A. E. Ades. 2011. Linear inference for mixed treatment comparison meta-analysis: A two-stage approach. *Research Synthesis Methods* 2: 43–60.

Lunn, D. J., A. Thomas, N. Best, and D. J. Spiegelhalter. 2000. WinBUGS – A Bayesian modelling framework: Concepts, structure and extensibility. *Statistics and Computing* 10: 325–337.

Miladinovic, B., A. Chaimani, I. Hozo, and B. Djulbegovic. 2014. Indirect treatment comparison. *Stata Journal* 14: 76–86. (Reprinted in this collection on pp. 311–320.)

Moreno, S. G., A. J. Sutton, A. E. Ades, N. J. Cooper, and K. R. Abrams. 2011. Adjusting for publication biases across similar interventions performed well when compared with gold standard data. *Journal of Clinical Epidemiology* 64: 1230–1241.

Morris, C. N. 1983. Parametric empirical Bayes inference: Theory and applications. *Journal of the American Statistical Association* 78: 47–55.

Nikolakopoulou, A., A. Chaimani, A. A. Veroniki, H. S. Vasiliadis, C. H. Schmid, and G. Salanti. 2014. Characteristics of networks of interventions: A description of a database of 186 published networks. *PLOS ONE* 9: e86754.

Phung, O. J., J. M. Scholle, M. Talwar, and C. I. Coleman. 2010. Effect of noninsulin antidiabetic drugs added to metformin therapy on glycemic control, weight gain, and hypoglycemia in type 2 diabetes. *Journal of the American Medical Association* 303: 1410–1418.

Riley, R. D., J. P. T. Higgins, and J. J. Deeks. 2011. Interpretation of random effects meta-analyses. *British Medical Journal* 342: d549.

Salanti, G. 2012. Indirect and mixed-treatment comparison, network, or multiple-treatments meta-analysis: Many names, many benefits, many concerns for the next generation evidence synthesis tool. *Research Synthesis Methods* 3: 80–97.

Salanti, G., A. E. Ades, and J. P. A. Ioannidis. 2011. Graphical methods and numerical summaries for presenting results from multiple-treatment meta-analysis: An overview and tutorial. *Journal of Clinical Epidemiology* 64: 163–171.

Salanti, G., S. Dias, N. J. Welton, A. E. Ades, V. Golfinopoulos, M. Kyrgiou, D. Mauri, and J. P. A. Ioannidis. 2010. Evaluating novel agent effects in multiple-treatments meta-regression. *Statistics in Medicine* 29: 2369–2383.

Salanti, G., C. D. Giovane, A. Chaimani, D. M. Caldwell, and J. P. T. Higgins. 2014. Evaluating the quality of evidence from a network meta-analysis. *PLOS ONE* 9: e99682.

Song, F., A. Clark, M. O. Bachmann, and J. Maas. 2012. Simulation evaluation of statistical properties of methods for indirect and mixed treatment comparisons. *BMC Medical Research Methodology* 12: 138.

Thijs, V., R. Lemmens, and S. Fieuws. 2008. Network meta-analysis: Simultaneous meta-analysis of common antiplatelet regimens after transient ischaemic attack or stroke. *European Heart Journal* 29: 1086–1092.

Veroniki, A. A., D. Mavridis, J. P. T. Higgins, and G. Salanti. 2014. Characteristics of a loop of evidence that affect detection and estimation of inconsistency: A simulation study. *BMC Medical Research Methodology* 14: 106.

Veroniki, A. A., H. S. Vasiliadis, J. P. T. Higgins, and G. Salanti. 2013. Evaluation of inconsistency in networks of interventions. *International Journal of Epidemiology* 42: 332–345.

Viechtbauer, W. 2005. Bias and efficiency of meta-analytic variance estimators in the random-effects model. *Journal of Educational and Behavioral Statistics* 30: 261–293.

White, I. R. 2011. Multivariate random-effects meta-regression: Updates to mvmeta. *Stata Journal* 11: 255–270. (Reprinted in this collection on pp. 249–264.)

———. 2013. A suite of Stata programs for network meta-analysis. UK Stata Users Group meeting proceedings. http://repec.org/usug2013/white.uk13.pptx.

White, I. R., J. K. Barrett, D. Jackson, and J. P. T. Higgins. 2012. Consistency and inconsistency in network meta-analysis: Model estimation using multivariate meta-regression. *Research Synthesis Methods* 3: 111–125.

Part 7

Advanced methods: glst, metamiss, sem, gsem, metacumbounds, metasim, metapow, and metapowplot

The most commonly used methods for fixed- and random-effects meta-analysis are based on weighted averages of the intervention effect estimates from different studies. More advanced methods require computational routines that are greatly facilitated by the sophisticated estimation procedures available to users writing Stata commands. The last set of articles in this collection includes advanced meta-analysis commands that use these procedures.

The `glst` command (Orsini, Bellocco, and Greenland 2006) is extremely useful to those wanting to conduct meta-analyses of observational data reported as dose–response associations (for example, associations between alcohol consumption and cardiovascular mortality, or between consumption of beta-carotene and lung cancer). Individual papers often report such associations as risk ratios or odds ratios comparing two or more exposure levels with a baseline category. The `glst` command uses the method of Greenland and Longnecker (1992) to convert such data to an estimate of the dose–response relationship—the risk ratio (or odds ratio) per unit increase in exposure. `glst` can also estimate a summary linear trend across multiple studies. Alternatively, a dose–response meta-analysis can be derived by using the log of the dose–response estimate, with its standard error, as input to the `metan` command.

The `metamiss` command performs meta-analysis with binary outcomes in which results from some studies are missing (White and Higgins 2009). Results adjusted for bias are obtained via assumptions about the informative missingness odds-ratio.

Palmer and Sterne (Forthcoming) describe how to fit fixed- and random-effects meta-analysis models using the `sem` and `gsem` commands, introduced in Stata 12 and 13 respectively, for structural equation modeling. These commands can fit meta-analysis models because they allow the variance of the residuals to be constrained to obtain correct standard errors for the summary estimates. The authors also show how to fit fixed- and random-effects multivariate meta-analysis and meta-regression models using these commands.

Miladinovic, Hozo, and Djulbegovic (2013) present the `metacumbounds` command for estimating trial-sequential monitoring boundaries of Gordon Lan and DeMets (1983) in cumulative meta-analysis. It is based on the `ldbounds` R package (Casper and Perez 2014).

Crowther et al. (2013) describe the `metasim`, `metapow`, and `metapowplot` commands to estimate the probability that the conclusions of a meta-analysis will change given the inclusion of a hypothetical new study. This is based on the methodology of Sutton et al. (2007).

1   References

Casper, C., and O. A. Perez. 2014. ldbounds: Lan–DeMets method for group sequential boundaries. R package version 1.1-1. http://CRAN.R-project.org/package=ldbounds.

Crowther, M. J., S. R. Hinchliffe, A. Donald, and A. J. Sutton. 2013. Simulation-based sample-size calculation for designing new clinical trials and diagnostic test accuracy studies to update an existing meta-analysis. *Stata Journal* 13: 451–473. (Reprinted in this collection on pp. 476–498.)

Gordon Lan, K. K., and D. L. DeMets. 1983. Discrete sequential boundaries for clinical trials. *Biometrika* 70: 659–663.

Greenland, S., and M. P. Longnecker. 1992. Methods for trend estimation from summarized dose–reponse data, with applications to meta-analysis. *American Journal of Epidemiology* 135: 1301–1309.

Miladinovic, B., I. Hozo, and B. Djulbegovic. 2013. Trial sequential boundaries for cumulative meta-analyses. *Stata Journal* 13: 77–91. (Reprinted in this collection on pp. 462–475.)

Orsini, N., R. Bellocco, and S. Greenland. 2006. Generalized least squares for trend estimation of summarized dose–response data. *Stata Journal* 6: 40–57. (Reprinted in this collection on pp. 404–421.)

Palmer, T. M., and J. A. C. Sterne. Forthcoming. Fitting fixed- and random-effects meta-analysis models using structural equation modeling with the sem and gsem commands. *Stata Journal* (Reprinted in this collection on pp. 435–461.)

Sutton, A. J., N. J. Cooper, D. R. Jones, P. C. Lambert, J. R. Thompson, and K. R. Abrams. 2007. Evidence-based sample size calculations based upon updated meta-analysis. *Statistics in Medicine* 26: 2479–2500.

White, I. R., and J. P. T. Higgins. 2009. Meta-analysis with missing data. *Stata Journal* 9: 57–69. (Reprinted in this collection on pp. 422–434.)

The Stata Journal (2006)

6, Number 1, st0096, pp. 40–57

Generalized least squares for trend estimation of summarized dose–response data

Nicola Orsini
Karolinska Institutet
Stockholm, Sweden
nicola.orsini@ki.se

Rino Bellocco
Karolinska Institutet
Stockholm, Sweden
and Department of Statistics
University of Milano-Bicocca
rino.bellocco@ki.se

Sander Greenland
Department of Epidemiology and Department of Statistics
University of California
Los Angeles, CA
lesdomes@ucla.edu

Abstract. This paper presents a command, `glst`, for trend estimation across different exposure levels for either single or multiple summarized case–control, incidence-rate, and cumulative incidence data. This approach is based on constructing an approximate covariance estimate for the log relative risks and estimating a corrected linear trend using generalized least squares. For trend analysis of multiple studies, `glst` can estimate fixed- and random-effects meta-regression models.

Keywords: st0096, glst, dose–response data, generalized least squares, trend, meta-analysis, meta-regression

1 Introduction

Epidemiological studies often assess whether the observed relationship between increasing (or decreasing) levels of exposure and the risk (or odds) of diseases follows a linear dose–response pattern. Methods for trend estimation of single and multiple summarized dose–response studies (Berlin, Longnecker, and Greenland 1993) are particularly useful when the full original data are not available.

To demonstrate these methods, our paper uses different types of dose–response data arising from published case–control, incidence-rate, and cumulative incidence data (also see [ST] **epitab**). Summarized data are typically reported as a series of dose-specific relative risks, with one category serving as the common referent group. The term *relative risk* (RR) will be used as a generic term for the risk ratio (cumulative incidence data), rate ratio (incidence-rate data), and odds ratio (case–control data).

Table 1 shows a summary of case–control data investigating the association between the consumption of alcohol and the risk of breast cancer, first presented by

Rohan and McMichael (1988), in which it appears that risk of breast cancer increases with increasing levels of alcohol intake.

Table 1. Case–control data on alcohol and breast cancer risk (Rohan and McMichael 1988)

Alcohol (g/d)	Assigned dose (g/d)	No. of cases	No. of controls	Total subjects	Adjusted RR (95% CI)
0	0	165	172	337	1.0 (Referent)
<2.5	2	74	93	167	0.80 (0.51–1.27)
2.5–9.3	6	90	96	186	1.16 (0.73–1.85)
>9.3	11	122	90	212	1.57 (0.99–2.51)

Table 2 shows a summary of incidence-rate data investigating the association between the long-term intake of dietary fiber and risk of coronary heart disease among women, first presented by Wolk et al. (1999), which supports the hypothesis that higher fiber intake reduces the risk of coronary heart disease.

Table 2. Incidence-rate data on fiber intake and coronary heart disease risk (Wolk et al. 1999)

Quintile of fiber intake	Assigned dose (g/d)	No. of cases	Person-years	Adjusted RR (95% CI)
1	11.5	148	134,707	1.0 (Referent)
2	14.3	127	133,824	0.98 (0.77–1.24)
3	16.4	114	130,654	0.92 (0.71–1.18)
4	18.8	107	124,522	0.87 (0.66–1.15)
5	22.9	95	117,808	0.77 (0.57–1.04)

Table 3 shows a summary of cumulative incidence data investigating the association between high-fat dairy food intake and risk of colorectal cancer, first presented by Larsson, Bergkvist, and Wolk (2005), which suggests that more servings per day of high-fat dairy food reduces the risk of colorectal cancer.

Table 3. Cumulative incidence data on high-fat dairy food and colorectal cancer risk (Larsson, Bergkvist, and Wolk 2005)

High-fat dairy (servings/d)	Assigned dose (servings/d)	No. of cases	Total subjects	Adjusted RR (95% CI)
<1.0	0.5	110	8,103	1.0 (Referent)
1.0– <2.0	1.5	212	17,538	0.75 (0.60–0.96)
2.0– <3.0	2.5	211	15,304	0.74 (0.58–0.95)
3.0– <4.0	3.5	132	9,078	0.68 (0.52–0.90)
≥4.0	6.5	133	10,685	0.59 (0.44–0.79)

For each of these summarized tables, we have adjusted relative risks and confidence limits for each nonreference exposure level. The usual approach to trend estimation, namely, the expected change of the log relative risks for a unit change of the exposure level, is to fit a linear regression through the origin, where the response variable is the log relative risks, the assigned dose is the covariate, and the log relative risks are weighted by the inverse of their variances. This method is known as weighted least-squares (WLS) regression (see [R] **vwls**), and it assumes that the log relative risks are independent—an assumption that is never satisfied in practice. The log relative risks are correlated given that they are estimated using a common referent group, and this standard approach underestimates the variance of the slope (Greenland and Longnecker 1992). This problem can be particularly relevant in a meta-analysis of summarized dose–response data where each study slope (trend) is weighted by the inverse of the variance (Shi and Copas 2004).

An efficient estimation method for the slope of a single study is therefore proposed and implemented in the command `glst`, as described by Greenland and Longnecker (1992). This method is then incorporated in the estimation of fixed and random-effects meta-regression models for the analysis of multiple studies.

The rest of the article is organized as follows: section 2 introduces the dose–response model and the estimation method; section 3 describes the syntax of the command `glst`; section 4 presents some practical examples based on published data; section 5 compares the corrected and uncorrected methods for trend estimation; and section 6 contains final comments.

2   Method

2.1   Log-linear dose–response model for a single study

It is possible to analyze the shape of the dose–response relationship between reported log relative risks and the exposure levels by estimating a log-linear dose–response regression model (Greenland and Longnecker 1992; Berlin, Longnecker, and Greenland 1993;

Shi and Copas 2004). Assuming that the exposure variable takes value 0 in the reference category, the estimated log relative risk in the reference category is set to zero (log 1); therefore, no intercept models are used. The matrix notation is

$$\mathbf{y} = \mathbf{X}\beta + \mathbf{e} \tag{1}$$

$$\mathbf{y} = \begin{bmatrix} y_1 \\ \vdots \\ y_i \\ \vdots \\ y_n \end{bmatrix} \quad \mathbf{X} = \begin{bmatrix} x_{11} & x_{12} & \cdots & x_{1p} \\ \vdots & \vdots & & \vdots \\ x_{i1} & x_{i2} & & x_{ip} \\ \vdots & \vdots & & \vdots \\ x_{n1} & x_{n2} & \cdots & x_{np} \end{bmatrix} \quad \beta = \begin{bmatrix} \beta_1 \\ \vdots \\ \beta_p \end{bmatrix} \quad \mathbf{e} = \begin{bmatrix} \varepsilon_1 \\ \vdots \\ \varepsilon_i \\ \vdots \\ \varepsilon_n \end{bmatrix}$$

where $\mathbf{y}$ is an $n \times 1$ vector of (reported) estimated log relative risks; $i = 1, 2, \ldots, n$ identifies nonreference exposure levels; $\mathbf{X}$ is an $n \times p$ matrix of nonstochastic covariates, where the first column, denoted by x_{i1}, identifies the exposure variable, and the remaining $p-1$ columns, for instance, may represent transformations of x_{i1}; β is a $p \times 1$ vector of unknown regression coefficients; and $\mathbf{e}$ is an $n \times 1$ vector of random errors, with expected value $E(\mathbf{e}) = 0$ and variance–covariance matrix $\text{Cov}(\mathbf{e}) = E(\mathbf{ee}')$ equal to the following symmetric matrix given by

$$\text{Cov}(\mathbf{e}) = \mathbf{\Sigma} = \begin{bmatrix} \sigma_{11} & & & & \\ \vdots & \ddots & & & \\ \sigma_{i1} & & \sigma_{ij} & & \\ \vdots & & & \ddots & \\ \sigma_{n1} & \cdots & \sigma_{nj} & \cdots & \sigma_{nn} \end{bmatrix}$$

Thus the response variable $\mathbf{y}$ has expected value $E(\mathbf{y}) = \mathbf{X}\beta$ and covariance matrix $\text{Cov}(\mathbf{y}) = \mathbf{\Sigma}$.

2.2   Generalized least squares

We use generalized least squares (GLS) to efficiently estimate the β vector of regression coefficients in (1). Assuming that the variance–covariance matrix of $\mathbf{e}$ is $\text{Cov}(\mathbf{e}) = \mathbf{\Sigma}$, this method involves minimizing $(\mathbf{y} - \mathbf{X}\beta)'\mathbf{\Sigma}^{-1}(\mathbf{y} - \mathbf{X}\beta)$ with respect to β. Suppose initially that the variance–covariance matrix $\mathbf{\Sigma}$ is known. In matrix notation, the resulting estimator $\mathbf{b}$ of the regression coefficients β is

$$\mathbf{b} = (\mathbf{X}'\mathbf{\Sigma}^{-1}\mathbf{X})^{-1}\mathbf{X}'\mathbf{\Sigma}^{-1}\mathbf{y} \tag{2}$$

and the estimated covariance matrix $\mathbf{v}$ of $\mathbf{b}$ is

$$\mathbf{v} = \widehat{\text{Cov}}(\mathbf{b}) = (\mathbf{X}'\mathbf{\Sigma}^{-1}\mathbf{X})^{-1} \tag{3}$$

A remarkable property of the GLS estimator is that for any choice of $\mathbf{\Sigma}$, the GLS estimate of β is unbiased; that is, $E(\mathbf{b}) = \beta$.

GLS estimation imposes no distributional assumption for the random errors, $\mathbf{e}$, whereas maximum likelihood (ML) estimation assumes a distribution, and the log likelihood of the sample observed is then maximized. Under the assumption that random errors are normally distributed with zero mean and variance–covariance matrix $\mathbf{\Sigma}$, i.e., $\mathbf{e} \sim N(0, \mathbf{\Sigma})$, the log-likelihood function can be written as the following:

$$l = -\frac{n}{2}\log(2\pi) - \frac{1}{2}\log|\mathbf{\Sigma}| - \frac{1}{2}\left\{ (\mathbf{y} - \mathbf{X}\beta)'\mathbf{\Sigma}^{-1}(\mathbf{y} - \mathbf{X}\beta) \right\} \tag{4}$$

Maximizing (4) with respect to β is equivalent to solving $\partial l/\partial \beta = 0$. The solution is the ML estimator of β, which under the normality assumption turns out to be the same as the GLS estimator given by (2).

2.3   Statistical inference

To construct confidence intervals and tests of hypotheses about β, we can make direct use of the GLS estimate, $\mathbf{b}$, and its estimated covariance matrix, $\mathbf{v}$. When the normality assumption of the random error $\mathbf{e}$ is introduced, the distributional properties of $\mathbf{y}$ and functions of $\mathbf{y}$ follow at once.

Because $\mathbf{y} \sim N(\mathbf{X}\beta, \mathbf{\Sigma})$, the vector $\mathbf{b}$, which is a linear function of $\mathbf{y}$, is therefore approximately normally distributed $\mathbf{b} \sim N(\beta, \mathbf{v})$.

A test of the null hypothesis, H_0: $\mathbf{b}_j = 0$ versus H_A: $\mathbf{b}_j \neq 0$, can be based on the following Wald statistic,

$$Z = \frac{\mathbf{b}_j}{\sqrt{\mathbf{v}_j}}$$

where $\mathbf{b}_j$ denotes the jth element of the vector $\mathbf{b}$ and $\mathbf{v}_j$ denotes the jth diagonal element of $\mathbf{v}$, with $j = 1, 2, \ldots, p$. The Z statistic can be compared with a standard normal distribution.

Wald test–type confidence intervals of β are computed using the large-sample approximation, the z distribution rather than the t distribution, because the estimates, $\mathbf{b}$, are based on a collection of n presumably large groups of subjects rather than n subjects (Grizzle, Starmer, and Koch 1969; Greenland 1987).

2.4   Covariances

In summarized dose–response data, the log relative risks, $\mathbf{y}$, are estimated using a common reference group. Therefore, the elements of $\mathbf{y}$ are not independent and the off-diagonal elements of $\mathbf{\Sigma}$ are not zero (Greenland and Longnecker 1992). This section describes the method and formulas needed to estimate all the elements of $\mathbf{\Sigma}$.

The diagonal element σ_{ii} of $\boldsymbol{\Sigma}$, the variance of the log relative risk y_i, is estimated from the normal theory–based confidence limits

$$\sigma_{ii} = \left[\left\{ \log(u_b) - \log(l_b) \right\} / (2 \times z_{\alpha/2}) \right]^2 \tag{5}$$

where u_b and l_b are, respectively, the upper and lower bounds of the reported relative risks, $\exp(y_i)$, and $z_{\alpha/2}$ denotes the $(1 - \alpha/2)$-level standard normal deviate (e.g., use 1.96 for 95% confidence interval).

Following the method proposed by Greenland and Longnecker (1992), one way to estimate the off-diagonal elements σ_{ij} of $\boldsymbol{\Sigma}$, with $i \neq j$, is to assume that the correlations between the unadjusted log relative risks are approximately equal to those of the adjusted log relative risks. Here, besides the log relative risks, their variances, and exposure levels, we also need to know for each exposure level the number of cases and the number of controls for case–control data (table 4), or the number of cases for incidence-rate data (table 5), or the number of cases and noncases for cumulative incidence data (table 6)—information usually available from the publication.

Table 4. Summary of case–control data

	Exposure levels						Total
	x_{01}	x_{11}	...	x_{i1}	...	x_{n1}	
Cases	A_0	A_1	...	A_i	...	A_n	$M_1 = \sum_{i=0}^{n} A_i$
Controls	B_0	B_1	...	B_i	...	B_n	$M_0 = \sum_{i=0}^{n} B_i$
Total	N_0	N_1	...	N_i	...	N_n	$M_1 + M_0$

The off-diagonal elements of $\boldsymbol{\Sigma}$ can be estimated using the following three-step procedure, where formulas used for steps 1 and 2 change according to the study type: case–control, incidence-rate, or cumulative incidence data.

For case–control data, where we model log odds-ratios, the off-diagonal elements σ_{ij} of $\boldsymbol{\Sigma}$ are computed as follows:

1. Fit cell counts A_i and B_i as modeled in table 4 (which has margin M_1 and N_i), such that

$$(A_i \times B_0)/(A_0 \times B_i) = \exp(y_i) \tag{6}$$

   where A_i is the fitted number of cases and B_i is the fitted number of controls at each exposure level (see iterative algorithm described in Greenland and Longnecker 1992, appendix 2).

2. For $i \neq j$, estimate the asymptotic correlation, r_{ij}, of y_i and y_j by

$$r_{ij} = s_0/(s_i s_j)^{1/2} \tag{7}$$

   where $s_0 = (1/A_0 + 1/B_0)$ and $s_i = (1/A_i + 1/B_i + 1/A_0 + 1/B_0)$.

3. Estimate the off-diagonal elements, σ_{ij}, of the asymptotic covariance matrix $\boldsymbol{\Sigma}$ by

$$\sigma_{ij} = r_{ij} \times (\sigma_i \sigma_j)^{1/2}$$

   where σ_i and σ_j are the variances of y_i and y_j, estimated using (5).

The above method can be easily extended to the analysis of incidence-rate and cumulative incidence data, upon redefinition of terms in (6) and (7).

Table 5. Summary of incidence-rate data

	\multicolumn{5}{c}{Exposure levels}					Total
	x_{01}	x_{11}	...	x_{i1}	...	x_{n1}	Total
Cases	A_0	A_1	...	A_i	...	A_n	$M_1 = \sum_{i=0}^{n} A_i$
Person-time	N_0	N_1	...	N_i	...	N_n	$M_0 = \sum_{i=0}^{n} N_i$

For instance, for incidence-rate data, where we model log incidence-rate ratios, fit cell counts A_i as modeled in table 5 such that $(A_i \times N_0)/(A_0 \times N_i) = \exp(y_i)$. In (7), we redefine $s_0 = (1/A_0)$ and $s_i = (1/A_i + 1/A_0)$.

Table 6. Summary of cumulative incidence data

	x_{01}	x_{11}	...	x_{i1}	...	x_{n1}	Total
Cases	A_0	A_1	...	A_i	...	A_n	$M_1 = \sum_{i=0}^{n} A_i$
Noncases	B_0	B_1	...	B_i	...	B_n	$M_0 = \sum_{i=0}^{n} B_i$
Total	N_0	N_1	...	N_i	...	N_n	$M_1 + M_0$

Then, for cumulative incidence data, where we model log risk-ratios, fit cell counts A_i as modeled in table 6 such that $(A_i \times N_0)/(A_0 \times N_i) = \exp(y_i)$. In (7), again s_0 and s_1 need to be computed differently: $s_0 = (1/A_0 - 1/N_0)$ and $s_i = (1/A_i - 1/N_i + 1/A_0 - 1/N_0)$.

2.5   Heterogeneity

The analysis of the estimated residual vector $\widehat{\mathbf{e}} = \mathbf{y} - \mathbf{Xb}$ is useful to evaluate how close reported and fitted log relative risks are at each exposure level. A statistic for the goodness of fit of the model is

$$Q = (\mathbf{y} - \mathbf{Xb})'\mathbf{\Sigma}^{-1}(\mathbf{y} - \mathbf{Xb}) \tag{8}$$

where Q has approximately, under the null hypothesis that the fitted model is correct, a χ^2 distribution with $n - p$ degrees of freedom. If the p-value derived from this statistic is small, we may infer that there is some problem with the model; e.g., perhaps heterogeneity is present or there is some unaccounted-for bias. If, however, the p-value is large, we can conclude only that the test did not detect a problem with the model, not that there is no problem. The Q statistic (like most fit statistics) has low power; i.e., its sensitivity to model problems is limited.

2.6   Log-linear dose–response model for multiple studies

The method discussed in the previous section can be applied to estimate the underlying trend from multiple summarized data. When dealing with multiple studies and multiple exposure levels, a more flexible method of trend estimation requires pooling the study data before estimating the dose–response model (Greenland and Longnecker 1992).

In a meta-analysis of dose–response studies, heterogeneity means that the shape or slope of the dose–response relationship varies among studies (Berlin, Longnecker, and Greenland 1993). The pool-first method increases the number of the log relative risks and dose values available for the analysis and it allows either to get a better fit of the dose–response relationship, by including fractional polynomials and splines in $\mathbf{X}$, or to identify sources of heterogeneity across studies, by including effect modifiers in $\mathbf{X}$.

Fixed-effects dose–response meta-regression model

Let $\mathbf{y}_k$ be the $n_k \times 1$ response vector and let $\mathbf{X}_k$ be the $n_k \times p$ covariates matrix for the kth study, with $k = 1, 2, \ldots, S$. The number of nonreference exposure levels, n_k, for the kth study might vary among the S studies. We pool the data by concatenating the matrices $\mathbf{y}_k$ and $\mathbf{X}_k$

$$\mathbf{y} = \begin{bmatrix} \mathbf{y}_1 \\ \vdots \\ \mathbf{y}_k \\ \vdots \\ \mathbf{y}_S \end{bmatrix} \quad \mathbf{X} = \begin{bmatrix} \mathbf{X}_1 \\ \vdots \\ \mathbf{X}_k \\ \vdots \\ \mathbf{X}_S \end{bmatrix}$$

so the outcome $\mathbf{y}$ will be an $n \times 1$ vector, where $n = \sum_{k=1}^{S} n_k$, and the linear predictor $\mathbf{X}$ will be an $n \times p$ matrix.

Using the pool-first method, the log-linear model

$$\mathbf{y} = \mathbf{X}\beta + \mathbf{e} \tag{9}$$

becomes a fixed-effects dose–response meta-regression model, where now the vector of random errors, $\mathbf{e}$, has expected value $E(\mathbf{e}) = 0$ and covariance $\text{Cov}(\mathbf{e}) = E(\mathbf{ee'})$ equal to the following symmetric $n \times n$ block-diagonal matrix,

$$\boldsymbol{\Sigma} = \begin{bmatrix} \boldsymbol{\Sigma}_1 & & & & \\ \vdots & \ddots & & & \\ \mathbf{0} & & \boldsymbol{\Sigma}_k & & \\ \vdots & & & \ddots & \\ \mathbf{0} & \dots & \mathbf{0} & \dots & \boldsymbol{\Sigma}_S \end{bmatrix} \tag{10}$$

where $\boldsymbol{\Sigma}_k$ is the $n_k \times n_k$ estimated covariance matrix for the kth study. We assume that the log relative risks are correlated within each study but uncorrelated across different studies.

The GLS estimators are given by (2) and (3), where the variance–covariance matrix is now given by (10). The summary slope (trend) across studies is a weighted average of each study slope with weighting matrix given by the inverse of $\boldsymbol{\Sigma}$.

A test for heterogeneity is again given by (8), where the variance–covariance matrix is given by (10). The Q statistic has approximately, under the null hypothesis, a χ^2 distribution with $n - p$ degrees of freedom.

The assumption implicit in a fixed-effects meta-regression model is that each study is estimating the same underlying trend. If heterogeneity is detected then it means that we could fit a better dose–response model, namely, one closer to the observed log relative risks, by either including in the linear predictor transformations of the dose variable and/or interaction terms between exposure dose levels and additional covariates, such as the study design. If important residual heterogeneity is still present after accounting for all known effect modifiers, a random-effects meta-regression dose–response model will be necessary to estimate a summary trend across studies (Berlin, Longnecker, and Greenland 1993).

Random-effects dose–response meta-regression model

We extend the fixed-effects dose–response model (9) to incorporate residual heterogeneity by including an additive random effect

$$\mathbf{y} = \mathbf{X}\beta + \mathbf{Z}\eta + \mathbf{e}$$

where $\mathbf{Z}$ is an $n \times 1$ vector containing the dose variable, first column of $\mathbf{X}$, and η is a random effect with expected value $E(\eta) = 0$ and variance $E(\eta\eta') = \tau^2$, and the random variables η and $\mathbf{e}$ are independent. The τ^2 represents a between-study variance component and quantifies the amount of spread about an overall slope (trend) of the dose variable in the reference category of all covariates specified in $\mathbf{X}$. We estimate the between-study variance using the moment estimator

$$\hat{\tau}^2 = \frac{Q - (n - p)}{\text{tr}(\mathbf{\Sigma}^{-1}) - \text{tr}\{\mathbf{\Sigma}^{-1}\mathbf{X}(\mathbf{X}'\mathbf{\Sigma}^{-1}\mathbf{X})^{-1}\mathbf{X}'\mathbf{\Sigma}^{-1}\}}$$

where tr denotes the trace of a matrix. A revised variance–covariance matrix, $\mathbf{\Sigma}$, is obtained by replacing the matrices $\mathbf{\Sigma}_k = \mathbf{\Sigma}_k + \hat{\tau}^2\mathbf{Z}_k\mathbf{Z}_k'$ in the block diagonal matrix (10). The revised matrix $\mathbf{\Sigma}$ is plugged into the GLS estimators $\mathbf{b}$ and $\mathbf{v}$, defined by (2, 3), and into the Q statistic, defined by (8). To get a fully efficient estimator, this procedure is repeated until the difference between successive estimates of $\hat{\tau}^2$ is less than 10^{-5}. Whenever $\hat{\tau}^2$ is negative, because $Q < n - p$, it is set to zero. The above iterative GLS method is approximately equivalent to first estimating the slope for each study and then pooling the slopes with a random-effects model (DerSimonian and Laird 1986).

3 The glst command

The estimation command `glst` is written for Stata 9.1, and it uses several inline Mata functions (see [M-5] **intro**).

3.1 Syntax of glst

`glst` *depvar dose* [*indepvars*] [*if*] [*in*], <u>se</u>(*stderr*) <u>cov</u>(*n cases*) [[cc|ir|ci]
    <u>pf</u>irst(*id study*) <u>random</u> <u>level</u>(*#*) eform]

where *depvar*, the outcome variable, contains log relative risks; *dose*, a required covariate, contains the exposure levels; and *indepvars* may contain other covariates, such as transformations of *doses* or interaction terms.

3.2 Options

`se`(*stderr*) specifies an estimate of the standard error of *depvar*. `se()` is required.

`cov`(*n cases*) specifies the variables containing the information required to fit the covariances among correlated log relative risks. At each exposure level, according to the study type, *n* is the number of subjects (controls plus cases) for case–control data (cc); or the total person-time for incidence-rate data (ir); or the total number of persons (cases plus noncases) for cumulative incidence data (ci). The variable *cases* contains the number of cases at each exposure level.

`cc` specifies case–control data. It is required for trend estimation of a single study unless the option `pfirst`(*id study*) is specified.

`ir` specifies incidence-rate data. It is required for trend estimation of a single study unless the option `pfirst`(*id study*) is specified.

`ci` specifies cumulative incidence data. It is required for trend estimation of a single study unless the option `pfirst`(*id study*) is specified.

pfirst(*id study*) specifies the pool-first method with multiple summarized studies. The variable *id* is a numeric indicator variable that takes the same value across correlated log relative risks within a study. The variable *study* must take value 1 for case–control, 2 for incidence-rate, and 3 for cumulative incidence study. Within each group of log relative risks, the first observation is assumed to be the referent.

random specifies the iterative generalized least squares method to estimate a random-effects meta-regression model. Between-study variability of the *dose* coefficient is estimated with the moment estimator.

level(*#*) specifies the confidence level, as a percentage, for confidence intervals. The default is level(95) or as set by set level; see [U] **20.7 Specifying the width of confidence intervals**.

eform reports coefficient estimates as exp(*b*) rather than *b*. Standard errors and confidence intervals are similarly transformed.

3.3 Saved results

glst saves in e():

Scalars

e(N)	number of observations	e(df_gf)	goodness-of-fit degrees of
e(chi2)	model χ^2 statistic		freedom
e(ll)	log likelihood	e(chi2_gf)	goodness-of-fit test
e(tau2)	between-study variance τ^2	e(S)	number of studies
e(df_m)	model degrees of freedom		

Macros

e(cmd)	glst	e(properties)	b V
e(depvar)	name of dependent variable		

Matrices

e(b)	coefficient vector	e(V)	variance–covariance matrix of
e(Sigma)	$\widehat{\Sigma}$ matrix		the estimators

Functions

e(sample)	marks estimation sample

4 Examples

4.1 Case–control data: Alcohol and breast cancer risk

Consider the case–control data shown in table 1 on alcohol and breast cancer (Rohan and McMichael 1988). We use the dataset containing the summarized information, and we calculate the standard errors of the log relative risks from the reported 95% confidence intervals using (5).

```
. use cc_ex

. gen double se = (logub - loglb)/(2*invnormal(.975))
```

We fit the log-linear dose–response model (1) to regress the log relative risks on the exposure level. The command `glst` fits the covariances and uses the GLS estimator to provide a correct estimate of the linear trend.

```
. glst logrr dose, se(se) cov(n case) cc
Generalized least-squares regression          Number of obs  =        3
Goodness-of-fit chi2(2)     =    1.93          Model chi2(1)  =     4.83
Prob > chi2                 =  0.3816          Prob > chi2    =   0.0279
```

logrr	Coef.	Std. Err.	z	P>|z|	[95% Conf. Interval]
dose	.0454288	.0206639	2.20	0.028	.0049284 .0859293

The command `glst` stores the fitted covariance matrix of the log relative risks in `e(Sigma)`

```
. matrix list e(Sigma)
symmetric e(Sigma)[3,3]
            c1         c2         c3
r1   .05417235
r2   .01881768  .05627467
r3   .01943145  .02068682  .05632754
```

The exponentiated linear trend for a change of 11 g/d of alcohol level is 1.65 (95% CI = 1.06, 2.57).

```
. lincom dose*11, eform
 ( 1)   11 dose = 0
```

logrr	exp(b)	Std. Err.	z	P>|z|	[95% Conf. Interval]
(1)	1.648255	.3746524	2.20	0.028	1.055709 2.573384

The goodness-of-fit p-value ($Q = 1.93$, $Pr = 0.3816$) is large. Thus this test detected no problems with the fitted model.

4.2   Incidence-rate data: Fiber intake and coronary heart disease

Consider now the incidence-rate data shown in table 2 on long-term intake of dietary fiber and risk of coronary heart disease among women (Wolk et al. 1999). As we did for case–control data, we use the command `glst` to get an efficient estimate of the slope.

```
. use ir_ex

. gen double se = (logub - loglb)/(2*invnormal(.975))
```

```
. glst logrr doser, se(se) cov(n case) ir
Generalized least-squares regression                Number of obs    =        4
Goodness-of-fit chi2(3)      =      0.18             Model chi2(1)    =     3.47
Prob > chi2                  =    0.9809             Prob > chi2      =   0.0626
```

logrr	Coef.	Std. Err.	z	P>|z|	[95% Conf. Interval]	
doser	-.0232086	.0124649	-1.86	0.063	-.0476394	.0012221

```
. lincom doser*10, eform
 ( 1)  10 doser = 0
```

logrr	exp(b)	Std. Err.	z	P>|z|	[95% Conf. Interval]	
(1)	.7928775	.0988316	-1.86	0.063	.6210185	1.012296

For a 10-g/d increase in total fiber intake, the rate of coronary heart disease decreased by 21% (RR = 0.79, 95% CI = 0.62, 1.01). The linear trend estimated with the `glst` command on summarized data is very close to the linear trend estimated on full data (68,782) reported in the abstract of the paper (RR = 0.81, 95% CI = 0.66, 0.99).

4.3   Cumulative incidence data: High-fat dairy food intake and colorectal cancer risk

Finally, let's consider now the cumulative incidence data shown in table 3 on high-fat dairy food intake and colorectal cancer risk (Larsson, Bergkvist, and Wolk 2005).

```
. use ci_ex
. gen double se = (logub - loglb)/(2*invnormal(.975))
. glst logrr dose, se(se) cov(n case) ci
Generalized least-squares regression                Number of obs    =        4
Goodness-of-fit chi2(3)      =      2.56             Model chi2(1)    =    11.84
Prob > chi2                  =    0.4648             Prob > chi2      =   0.0006
```

logrr	Coef.	Std. Err.	z	P>|z|	[95% Conf. Interval]	
dose	-.073636	.0214036	-3.44	0.001	-.1155863	-.0316857

```
. lincom dose*2, eform
 ( 1)  2 dose = 0
```

logrr	exp(b)	Std. Err.	z	P>|z|	[95% Conf. Interval]	
(1)	.8630591	.0369452	-3.44	0.001	.7936024	.9385948

Each increment of two servings per day of high-fat dairy foods corresponded to a 14% reduction in the risk of colorectal cancer (RR = 0.86, 95% CI = 0.79, 0.94). Once again, the linear trend estimated with the `glst` command on summarized data is very close to the linear trend estimated on full data (60,708) reported in the abstract of the paper ($\mathrm{RR} = 0.87$, 95% $\mathrm{CI} = 0.78$, 0.96).

4.4   Meta-analysis: Lactose intake and ovarian cancer risk

Earlier we showed how to estimate a linear trend for a single study. Here we show how to use the command `glst` to estimate a summary linear trend across multiple studies. We consider as a motivating example a meta-analysis of epidemiological studies (six case–control and three cohort studies) investigating the association between lactose intake and ovarian cancer risk (Larsson, Orsini, and Wolk 2006).

Fixed-effects dose–response meta-regression model

We can easily pool trend estimates across studies with the option `pfirst()`, which specifies the variable names identifying the correlated log relative risks and the type of study (case–control or incidence-rate data).

```
. use ma_ex

. glst logrr dose, se(se) cov(n case) pfirst(id study) eform
Fixed-effects dose-response model              Number of studies   =         9
Generalized least-squares regression               Number of obs   =        28
Goodness-of-fit chi2(27)   =    40.25              Model chi2(1)    =      1.11
Prob > chi2                =   0.0486              Prob > chi2      =    0.2925
```

logrr	exb(b)	Std. Err.	z	P>|z|	[95% Conf. Interval]	
dose	1.025822	.0248455	1.05	0.293	.9782636	1.075693

Overall, there is no evidence of association between milk intake (10 g/d) and risk of ovarian cancer ($\mathrm{RR} = 1.03$, 95% $\mathrm{CI} = 0.98$, 1.08). However, the goodness-of-fit test ($Q = 40.25$, $\mathrm{Pr} = 0.0486$) suggests that we should take into account potential sources of heterogeneity. The estimated association of lactose intake with ovarian cancer risk might depend on the study design. Therefore, we create a product (interaction) term between the type of study (1 for incidence-rate and 0 for case–control data) and the dose variable, and we include it in the model. An alternative would be to stratify the meta-analysis by study design.

```
. gen types = study == 2

. gen doseXtypes = dose*types
```

```
. glst logrr dose doseXtypes, se(se) cov(n case) pfirst(id study)
Fixed-effects dose-response model              Number of studies  =        9
Generalized least-squares regression            Number of obs     =       28
Goodness-of-fit chi2(26)    =     30.55          Model chi2(2)     =    10.80
Prob > chi2                 =    0.2453          Prob > chi2       =   0.0045
```

logrr	Coef.	Std. Err.	z	P>|z|	[95% Conf. Interval]	
dose	-.0340478	.0308599	-1.10	0.270	-.094532	.0264365
doseXtypes	.1550466	.0497982	3.11	0.002	.0574439	.2526492

```
. lincom dose + doseXtypes*0, eform
 ( 1)   dose = 0
```

logrr	exp(b)	Std. Err.	z	P>|z|	[95% Conf. Interval]	
(1)	.9665253	.0298269	-1.10	0.270	.9097986	1.026789

```
. lincom dose + doseXtypes*1, eform
 ( 1)   dose + doseXtypes = 0
```

logrr	exp(b)	Std. Err.	z	P>|z|	[95% Conf. Interval]	
(1)	1.128624	.0441106	3.10	0.002	1.045397	1.218476

No association between milk intake and risk of ovarian cancer was found among six case–control studies ($\text{RR} = 0.97$, 95% $\text{CI} = 0.91$, 1.03). A positive association between milk intake and risk of ovarian cancer was found among three cohort studies ($\text{RR} = 1.13$, 95% $\text{CI} = 1.05$, 1.22). A systematic difference in slopes related to study design might result, for instance, from the existence of recall bias in the case–control studies that would not be present in the cohort studies. Now the goodness-of-fit test ($Q = 30.55$, $\text{Pr} = 0.2453$) detects no further problems with the fitted model.

Random-effects dose–response meta-regression model

We can also check residual heterogeneity across linear trend estimates by fitting a random-effects model.

```
. glst logrr dose doseXtypes, se(se) cov(n case) pfirst(id study) random
Random-effects dose-response model             Number of studies  =        9
Iterative Generalized least-squares regression   Number of obs    =       28
Goodness-of-fit chi2(26)    =     28.37          Model chi2(2)     =     7.29
Prob > chi2                 =    0.3407          Prob > chi2       =   0.0261
```

logrr	Coef.	Std. Err.	z	P>|z|	[95% Conf. Interval]	
dose	-.0443064	.0394422	-1.12	0.261	-.1216116	.0329988
doseXtypes	.1654426	.063171	2.62	0.009	.0416297	.2892555

```
Moment-based estimate of between-study variance of the slope: tau2  =    0.0026
```

The trend estimates for case–control and cohort studies are quite close to the previous ones under fixed-effects models. The between-study standard deviation is close to zero ($\hat{\tau} = 0.0026^{1/2} = 0.05$), which implies that the study-specific trends have only a small spread around the average trend (-0.044) for case–control studies. Furthermore, if we model heterogeneity directly with a random-effects model, without considering any effect modifiers, the results of the meta-analysis briefly described above could not be achieved at all.

```
. glst logrr dose, se(se) cov(n case) pfirst(id study) eform random
Random-effects dose-response model            Number of studies    =        9
Iterative Generalized least-squares regression      Number of obs  =       28
Goodness-of-fit chi2(27)    =    32.17          Model chi2(1)       =     0.20
Prob > chi2                 =    0.2259          Prob > chi2         =   0.6519
```

logrr	exb(b)	Std. Err.	z	P>|z|	[95% Conf. Interval]
dose	1.016753	.0374417	0.45	0.652	.9459546 1.092851

```
Moment-based estimate of between-study variance of the slope: tau2  =   0.0059
```

We would simply conclude that, overall, there is no association between lactose intake on ovarian cancer risk (RR $= 1.02$, 95% CI $= 0.95, 1.09$).

5   Empirical comparison of the WLS and GLS estimates

Here we compare and evaluate the uncorrected (WLS) and corrected (GLS) estimates of the linear trend, b, its standard error, se $= \sqrt{v}$, and the heterogeneity statistic, Q. Table 7 summarizes the results for single (sections 4.1–4.3) and multiple studies (section 4.4)

Table 7. Empirical comparison of GLS and WLS estimates

	GLS			WLS			Difference (%)		
	b	se	Q	b	se	Q	b	se	Q
Single study									
Case–control	0.045	0.021	1.93	0.033	0.019	1.72	26.4	9.5	10.5
Incidence-rate	-0.008	0.006	1.61	-0.007	0.004	0.93	14.6	33.7	42.2
Cumulative									
incidence	-0.073	0.021	2.57	-0.098	0.018	2.20	-33.2	15.6	14.1
Multiple studies									
Case–control	-0.034	0.031	24.02	-0.042	0.026	30.48	-23.1	17.2	-26.9
Incidence-rate	0.121	0.039	6.54	0.142	0.033	3.24	-17.0	15.0	50.5
Overall	0.025	0.024	40.25	0.026	0.020	52.90	-3.2	16.4	-31.4

The relative differences, expressed as percentages, between the GLS and WLS estimates are calculated as $(\text{GLS} - \text{WLS})/\text{GLS} \times 100$. The GLS estimates of the linear trend, b, could be higher or lower than the WLS estimates, and the small differences are not surprising because both estimators are consistent (Greenland and Longnecker 1992). The Q statistic based on GLS estimates could be higher or lower than the one based on WLS estimates. In the WLS procedure the off-diagonal elements of Σ, covariances among log relative risks, are set to zeros, whereas in the GLS the covariances are not zeros (see section 2.4). Therefore, the weighting matrix, Σ^{-1}, in the Q statistic depends both on variances and covariances of the log relative risks. As expected, the GLS estimates of the standard errors, se, are always higher than the WLS estimates of the standard errors for single and multiple studies. The underestimation of the standard error of the uncorrected WLS method somewhat overstates the precision of the trend estimate. Further empirical comparisons between the corrected and uncorrected methods can be found in Greenland and Longnecker (1992).

6   Conclusion

We presented a command, glst, to efficiently estimate the trend from summarized epidemiological dose–response data. As shown with several examples, the method can be applied for published case–control, incidence-rate, and cumulative incidence data, from either a single study or multiple studies. In the latter case, the command glst fits fixed-effects and random-effects meta-regression models to allow a better fit of the dose–response relation and the identification of sources of heterogeneity. Adjusting the standard error of the slope for the within-study covariance is just one of the statistical issues arising in the synthesis of information from different studies. Other important issues, not considered in this paper, are the exposure scale, publication bias, and methodologic bias (Berlin, Longnecker, and Greenland 1993; Shi and Copas 2004; Greenland 2005). A limitation of the method proposed by Greenland and Longnecker (1992) is   the assumption that the correlation matrices of the unadjusted and adjusted log relative risks are approximately equal. In future developments of the command, upper and lower bounds of the covariance matrix will be implemented to assess the sensitivity of the GLS estimators, as pointed out by Berrington and Cox (2003).

7   References

Berlin, J. A., M. P. Longnecker, and S. Greenland. 1993. Meta-analysis of epidemiologic dose–response data. *Epidemiology* 4: 218–228.

Berrington, A., and D. R. Cox. 2003. Generalized least squares for the synthesis of correlated information. *Biostatistics* 4: 423–431.

DerSimonian, R., and N. Laird. 1986. Meta-analysis in clinical trials. *Controlled Clinical Trials* 7: 177–188.

Greenland, S. 1987. Quantitative methods in the review of epidemiologic literature. *Epidemiologic Reviews* 9: 1–30.

———. 2005. Multiple-bias modeling for analysis of observational data (with discussion). *Journal of the Royal Statistical Society, Series A* 168: 267–306.

Greenland, S., and M. P. Longnecker. 1992. Methods for trend estimation from summarized dose–reponse data, with applications to meta-analysis. *American Journal of Epidemiology* 135: 1301–1309.

Grizzle, J. E., C. F. Starmer, and G. G. Koch. 1969. Analysis of categorical data by linear models. *Biometrics* 25: 489–504.

Larsson, S. C., L. Bergkvist, and A. Wolk. 2005. High-fat dairy food and conjugated linoleic acid intakes in relation to colorectal cancer incidence in the Swedish Mammography Cohort. *American Journal of Clinical Nutrition* 82: 894–900.

Larsson, S. C., N. Orsini, and A. Wolk. 2006. Milk, milk products and lactose intake and ovarian cancer risk: A meta-analysis of epidemiological studies. *International Journal of Cancer* 118: 431–441.

Rohan, T. E., and A. J. McMichael. 1988. Alcohol consumption and risk of breast cancer. *International Journal of Cancer* 41: 695–699.

Shi, J. Q., and J. B. Copas. 2004. Meta-analysis for trend estimation. *Statistics in Medicine* 23: 3–19.

Wolk, A., J. E. Manson, M. J. Stampfer, G. A. Colditz, F. B. Hu, F. E. Speizer, C. H. Hennekens, and W. C. Willett. 1999. Long-term intake of dietary fiber and decreased risk of coronary heart disease among women. *Journal of the American Medical Association* 281: 1998–2004.

The Stata Journal (2009)
9, Number 1, st0157, pp. 57–69

Meta-analysis with missing data

Ian R. White
MRC Biostatistics Unit
Cambridge, UK
ian.white@mrc-bsu.cam.ac.uk

Julian P. T. Higgins
MRC Biostatistics Unit
Cambridge, UK
julian.higgins@mrc-bsu.cam.ac.uk

Abstract. A new command, `metamiss`, performs meta-analysis with binary outcomes when some or all studies have missing data. Missing values can be imputed as successes, as failures, according to observed event rates, or by a combination of these according to reported reasons for the data being missing. Alternatively, the user can specify the value of, or a prior distribution for, the informative missingness odds ratio.

Keywords: st0157, metamiss, meta-analysis, missing data, informative missingness odds ratio

1 Introduction

Just as missing outcome data present a threat to the validity of any research study, so they present a threat to the validity of any meta-analysis of research studies. Typically, analyses assume that the data are missing completely at random or missing at random (MAR) (Little and Rubin 2002). If the data are not MAR (i.e., they are informatively missing) but are analyzed as if they were missing completely at random or MAR, then nonresponse bias typically occurs. The threat of bias carries over to meta-analysis, where the problem can be compounded by nonresponse bias applied in a similar way in different studies.

Many methods for dealing with missing outcome data require detailed data for each participant. Dealing with missing outcome data in a meta-analysis raises particular problems because limited information is typically available in published reports. Although a meta-analyst would ideally seek any important but unreported data from the authors of the original studies, this approach is not always successful, and it is uncommon to have access to more than group-level summary data at best. We therefore address the meta-analysis of summary data, focusing on the case of an incomplete binary outcome.

A central concept is the informative missingness odds ratio (IMOR), defined as the odds ratio between the missingness, M, and the true outcome, Y, within groups (White, Higgins, and Wood 2008). A value of 1 indicates MAR, while IMOR = 0 means that missing values are all failures, and IMOR = ∞ means that missing values are all successes. We allow the IMOR to differ across groups and across subgroups of individuals defined by reasons for missingness, or to be specified with uncertainty.

We will describe `metamiss` in the context of a meta-analysis of randomized controlled trials comparing an "experimental group" with a "control group", but it could be used in any meta-analysis of two-group comparisons. `metamiss` only prepares the data for each study, and then it calls `metan` to perform the meta-analysis. It allows two main types of methods: imputation methods and Bayesian methods.

First, `metamiss` offers imputation methods as described in Higgins, White, and Wood (2008). Missing values can be imputed as failures or as successes; using the same rate as in the control group, the same rate as in the experimental group, or the same rate as in their own group; or using IMORs. When reasons for missingness are known, a mixture of the methods can be used.

Second, `metamiss` offers Bayesian methods that allow for user-specified uncertainty about the missingness mechanism (Rubin 1977; Forster and Smith 1998; White, Higgins, and Wood 2008). These use the prior $\text{logIMOR}_{ij} \sim N(m_{ij}, s_{ij}^2)$ in group $j = E, C$ of study i, with $\text{corr}(\text{logIMOR}_{iE}, \text{logIMOR}_{iC}) = r$.

The approach of Gamble and Hollis (2005) is also implemented. In this approach, two extreme analyses are performed for each study, regarding all missing values as successes in one group and failures in the other. The two 95% confidence intervals are then combined (together with intermediate values), and a modified standard error is taken as one quarter the width of this combined confidence interval. This method appears to overpenalize studies with missing data (White, Higgins, and Wood 2008), but it is included here for comparison.

2   metamiss command

2.1   Syntax

`metamiss` requires six variables (rE, fE, mE, rC, fC, and mC), which specify the number of successes, failures, and missing values in each randomized group. There are four syntaxes described below.

Simple imputation

`metamiss` *rE fE mE rC fC mC*, *imputation_method* [*imor_option*
    imputation_options meta_options]

where

> *imputation_method* is one of the imputation methods listed in section 2.2, specified without an argument.

> *imor_option* is either `imor(`*# | varname* [*# | varname*]`)` or `logimor(`*# | varname* [*# | varname*]`)` (see section 2.3).

> *imputation_options* are any of the options described in section 2.4.

meta_options are any of the meta-analysis options listed in section 2.6, as well as any valid option for `metan`, including `random`, `by()`, and `xlabel()` (see section 2.6).

Imputation using reasons

`metamiss` *rE fE mE rC fC mC*, *imputation_method1 imputation_method2*
  [*imputation_method3* ...] [*imor_option imputation_options meta_options*]

where

imputation_method1, *imputation_method2*, etc., are any imputation method listed in section 2.2 except `icab` and `icaw`, specified with arguments to indicate numbers of missing values to be imputed by each method.

imor_option, *imputation_options*, and *meta_options* are the same as documented in *Simple Imputation*.

Bayesian analysis using priors

`metamiss` *rE fE mE rC fC mC*, <u>sd</u>`logimor(`*# | varname* [*# | varname*]`)`
  [*imor_option bayes_options meta_options*]

where

imor_option and *meta_options* are the same as documented in *Simple Imputation*.

bayes_options are any of the options described in section 2.5.

Gamble–Hollis analysis

`metamiss` *rE fE mE rC fC mC*, <u>gamblehollis</u> [*meta_options*]

where

`gamblehollis` specifies to use the Gamble–Hollis analysis.

meta_options are the same as documented in *Simple Imputation*.

2.2   imputation_method

For simple imputation, specify one of the following options without arguments. For imputation using reasons, specify two or more of the following options with arguments. The abbreviations ACA, ICA-0, etc., are explained by Higgins, White, and Wood (2008).

`aca` $\big[$ (# | *varname* $\big[$ # | *varname* $\big]$) $\big]$ performs an available cases analysis (ACA).

`ica0` $\big[$ (# | *varname* $\big[$ # | *varname* $\big]$) $\big]$ imputes missing values as zeros (ICA-0).

`ica1` $\big[$ (# | *varname* $\big[$ # | *varname* $\big]$) $\big]$ imputes missing values as ones (ICA-1).

`icab` performs a best-case analysis (ICA-b), which imputes missing values as ones in the experimental group and zeros in the control group—equivalent to `ica0(0 1)` `ica1(1 0)`. If rE and rC count adverse events, not beneficial events, then `icab` will yield a worst-case analysis.

`icaw` performs a worst-case analysis (ICA-w), which imputes missing values as zeros in the experimental group and ones in the control group—equivalent to `ica0(1 0)` `ica1(0 1)`. If rE and rC count beneficial events, not adverse events, then `icaw` will yield a best-case analysis.

`icape` $\big[$ (# | *varname* $\big[$ # | *varname* $\big]$) $\big]$ imputes missing values by using the observed probability in the experimental group (ICA-pE).

`icapc` $\big[$ (# | *varname* $\big[$ # | *varname* $\big]$) $\big]$ imputes missing values by using the observed probability in the control group (ICA-pC).

`icap` $\big[$ (# | *varname* $\big[$ # | *varname* $\big]$) $\big]$ imputes missing values by using the observed probability within groups (ICA-p).

`icaimor` $\big[$ (# | *varname* $\big[$ # | *varname* $\big]$) $\big]$ imputes missing values by using the IMORs specified by `imor()` or `logimor()` within groups (ICA-IMORs).

The default is `icaimor` if `imor()` or `logimor()` is specified; if no IMOR option is specified, the default is `aca`.

Specifying arguments

Used with arguments, these options specify the numbers of missing values to be imputed by each method. For example, `ica0(mfE mfC) icap(mpE mpC)` indicates that `mfE` individuals in group E and `mfC` individuals in group C are imputed using ICA-0, while `mpE` individuals in group E and `mpC` individuals in group C are imputed using ICA-p. If the second argument is omitted, it is taken to be zero. If, for some group, the total over all reasons does not equal the number of missing observations (e.g., if `mfE` + `mpE` does not equal `mE`), then the missing observations are shared between imputation types in the given ratio. If the total over all reasons is zero for some group, then the missing observations are shared between imputation types in the ratio formed by summing overall numbers of individuals for each reason across all studies. If the total is zero for all studies in one or both groups, then an error is returned. Numerical values can also be given: e.g., `ica0(50 50) icap(50 50)` indicates that 50% of missing values in each group are imputed using ICA-0 and the rest are imputed using ICA-p.

2.3   imor_option

imor(*#* | *varname* [*#* | *varname*]) sets the IMORs or (if the Bayesian method is being used) the prior medians of the IMORs. If one value is given, it applies to both groups; if two values are given, they apply to the experimental and control groups, respectively. Both values default to 1. Only one of imor() or logimor() can be specified.

logimor(*#* | *varname* [*#* | *varname*]) does the same as imor() but on the log scale. Thus imor(1 1) is the same as logimor(0 0). Only one of imor() or logimor() can be specified.

2.4   imputation_options

w1 specifies that standard errors be computed, treating the imputed values as if they were observed. This is included for didactic purposes and should not be used in real analyses. Only one of w1, w2, w3, or w4 can be specified.

w2 specifies that standard errors from the ACA be used. This is useful in separating sensitivity to changes in point estimates from sensitivity to changes in standard errors. Only one of w1, w2, w3, or w4 can be specified.

w3 specifies that standard errors be computed by scaling the imputed data down to the number of available cases in each group and treating these data as if they were observed. Only one of w1, w2, w3, or w4 can be specified.

w4, the default, specifies that standard errors be computed algebraically, conditional on the IMORs. Conditioning on the IMORs is not strictly correct for schemes including ICA-pE or ICA-pC, but the conditional standard errors appear to be more realistic than the unconditional standard errors in this setting (Higgins, White, and Wood 2008). Only one of w1, w2, w3, or w4 can be specified.

listnum lists the reason counts for each study implied by the imputation method option.

listall lists the reason counts for each study after scaling to match the number of missing values and imputing missing values for studies with no reasons.

listp lists the imputed probabilities for each study.

2.5   bayes_options

sdlogimor(*#* | *varname* [*#* | *varname*]) sets the prior standard deviation for log IMORs for the experimental and control groups, respectively. Both values default to 0.

corrlogimor(*#* | *varname*) sets the prior correlation between log IMORs in the experimental and control groups. The default is corrlogimor(0).

method(gh | mc | taylor) determines the method used to integrate over the distribution of the IMORs. method(gh) uses two-dimensional Gauss–Hermite quadrature and is the recommended method (and the default). `method(mc)` performs a full Bayesian analysis by sampling directly from the posterior. This is time consuming, so dots display progress, and you can request more than one of the measures `or`, `rr`, and `rd`. `method(taylor)` uses a Taylor-series approximation, as in section 4 of Forster and Smith (1998), and is faster than the default but typically inaccurate for `sdlogimor()` larger than one or two.

`nip(#)` specifies the number of integration points under `method(gh)`. The default is `nip(10)`.

`reps(#)` specifies the number of Monte Carlo draws under `method(mc)`. The default is `reps(100)`.

`missprior(## [##])` and `respprior(##)` apply when `method(mc)` is used, but they are unlikely to be much used. They specify the parameters of the beta priors for $P(M)$ and $P(Y \mid M = 0)$: the parameters for the first group are given by the first two numbers, and the parameters for the second group are given by the next two numbers or are the same as for the first group. The defaults are both beta$(1, 1)$.

`nodots` suppresses the dots that are displayed to mark the number of Monte Carlo draws completed.

2.6 meta_options

`or`, `rr`, and `rd` specify the measures to be analyzed. Usually, only one measure can be specified; the default is `rr`. However, when using `method(mc)`, all three measures can be obtained for no extra effort, so any combination is allowed. When more than one measure is specified, the formal meta-analysis is not performed, but measures and their standard errors are saved (see section 2.7).

`log` has the results reported on the log risk-ratio (RR) or log odds-ratio scale.

`id(varname)` specifies a study identifier for the results table and forest plot.

Most other options allowed with `metan` are also allowed, including `by()`, `random`, and `nograph`.

2.7 Saved results

`metamiss` saves results in the same way as `metan`: `_ES`, `_selogES`, etc. The sample size, `_SS`, excludes the missing values, but an additional variable, `_SSmiss`, gives the total number of missing values. When `method(mc)` is run, the `log` option is assumed for the measures `or` and `rr`, and the following variables are saved for each measure (`logor`, `logrr`, or `rd`): the ACA estimate, `ESTRAW_measure`; the ACA variance, `VARRAW_measure`; the corrected estimate, `ESTSTAR_measure`; and the corrected variance, `VARSTAR_measure`. If these variables already exist, then they are overwritten.

3 Examples

3.1 Data

We apply the above methods to a meta-analysis of randomized controlled trials comparing haloperidol to placebo in the treatment of schizophrenia. A Cochrane review of haloperidol forms the basis of our data (Joy, Adams, and Lawrie 2006). Further details of our analysis are given in Higgins, White, and Wood (2008).

The main data consist of the variables `author` (the author); `r1`, `f1`, and `m1` (the counts of successes, failures, and missing observations in the intervention group); and `r2`, `f2`, and `m2` (the corresponding counts in the control group).

3.2 Available cases analysis

The following analysis illustrates `metamiss` output, but the same results could in fact have been obtained by using `metan r1 f1 r2 f2, fixedi`:

```
. use haloperidol
. metamiss r1 f1 m1 r2 f2 m2, aca id(author) fixed nograph
*********************************************************************
******** METAMISS: meta-analysis allowing for missing data ********
********              Available cases analysis           ********
*********************************************************************
Measure: RR.
Zero cells detected: adding 1/2 to 6 studies.

(Calling metan with options: label(namevar=author) fixed eform nograph ...)
           Study  |    ES   [95% Conf. Interval]    % Weight
-------------------+-------------------------------------------------
Arvanitis         |  1.417    0.891      2.252       18.86
Beasley           |  1.049    0.732      1.504       31.22
Bechelli          |  6.207    1.520     25.353        2.05
Borison           |  7.000    0.400    122.442        0.49
Chouinard         |  3.492    1.113     10.955        3.10
Durost            |  8.684    1.258     59.946        1.09
Garry             |  1.750    0.585      5.238        3.37
Howard            |  2.039    0.670      6.208        3.27
Marder            |  1.357    0.747      2.466       11.37
Nishikawa_82      |  3.000    0.137     65.903        0.42
Nishikawa_84      |  9.200    0.581    145.759        0.53
Reschke           |  3.793    1.058     13.604        2.48
Selman            |  1.484    0.936      2.352       19.11
Serafetinides     |  8.400    0.496    142.271        0.51
Simpson           |  2.353    0.127     43.529        0.48
Spencer           | 11.000    1.671     72.396        1.14
Vichaiya          | 19.000    1.157    311.957        0.52
-------------------+-------------------------------------------------
I-V pooled ES     |  1.567    1.281      1.916      100.00
-------------------+-------------------------------------------------

  Heterogeneity chi-squared =  27.29 (d.f. = 16) p = 0.038
  I-squared (variation in ES attributable to heterogeneity) =  41.4%

  Test of ES=1 : z=   4.37 p = 0.000
```

The effect size (ES) refers to the RR in this output. For brevity, future listings include only the four largest studies: Arvanitis, Beasley, Marder, and Selman, with 2%, 41%, 3%, and 42% missing data, respectively. Interest therefore focuses on changes in inferences for the Beasley and Selman studies.

3.3   Imputation methods

We illustrate imputing all missing values as zeros, using the weighting scheme `w4`, which correctly allows for uncertainty (although in `ica0`, `w1` gives the same answers):

```
. metamiss r1 f1 m1 r2 f2 m2, ica0 w4 id(author) fixed nograph
**********************************************************************
******** METAMISS: meta-analysis allowing for missing data ********
********                  Simple imputation                 ********
**********************************************************************
Measure: RR.
Method: ICA-0 (impute zeros).
Weighting scheme: w4.
Zero cells detected: adding 1/2 to 6 studies.

(Calling metan with options: label(namevar=author) fixed eform nograph ...)
          Study    |    ES    [95% Conf. Interval]    % Weight
-------------------+-------------------------------------------------
Arvanitis          |  1.362     0.854     2.172         24.38
Beasley            |  1.429     0.901     2.266         25.01
   (output omitted)

Marder             |  1.357     0.745     2.473         14.75
   (output omitted)

Selman             |  2.429     1.189     4.960         10.42
   (output omitted)

-------------------+-------------------------------------------------
I-V pooled ES      |  1.898     1.507     2.390        100.00
-------------------+-------------------------------------------------

   Heterogeneity chi-squared =  21.56 (d.f. = 16) p = 0.158
   I-squared (variation in ES attributable to heterogeneity) =  25.8%

   Test of ES=1 : z=   5.45 p = 0.000
```

The Beasley and Selman trials have more missing data in the control group, so imputing failures increases their estimated RR, and the pooled RR also increases.

3.4   Impute using known IMORs

Now we assume that the IMOR is 0.5 in each group, that is, that the odds of success in missing data are half the odds of success in observed data.

```
. metamiss r1 f1 m1 r2 f2 m2, icaimor imor(1/2 1/2) w4 id(author) fixed nograph
*********************************************************************
******** METAMISS: meta-analysis allowing for missing data ********
********                 Simple imputation              ********
*********************************************************************
Measure: RR.
Method: ICA-IMOR (impute using IMORs 1/2 1/2).
Weighting scheme: w4.
Zero cells detected: adding 1/2 to 6 studies.
(Calling metan with options: label(namevar=author) fixed eform nograph ...)
            Study    |    ES    [95% Conf. Interval]    % Weight
--------------------+------------------------------------------------
Arvanitis           | 1.399    0.878     2.227           22.12
Beasley             | 1.120    0.737     1.700           27.47
   (output omitted)
Marder              | 1.358    0.746     2.473           13.34
   (output omitted)
Selman              | 1.743    0.973     3.121           14.11
   (output omitted)
--------------------+------------------------------------------------
I-V pooled ES       | 1.699    1.365     2.115          100.00
--------------------+------------------------------------------------
   Heterogeneity chi-squared =  24.63 (d.f. = 16) p = 0.077
   I-squared (variation in ES attributable to heterogeneity) =  35.0%
   Test of ES=1 : z=  4.75 p = 0.000
```

The assumption is intermediate between ACA and ICA-0, and so is the result.

3.5   Impute using reasons for missingness

Most studies indicated the distribution of reasons for missing outcomes. We assigned imputation methods as follows:

- For reasons such as "lack of efficacy" or "relapse", we imputed failures (ICA-0).

- For reasons such as "positive response", we imputed successes (ICA-1).

- For reasons such as "adverse event", "withdrawal of consent", or "noncompliance", we considered that the patient had not received the intervention, and we imputed according to the control group rate ICA-pC, implicitly assuming lack of selection bias.

- For reasons such as "loss to follow-up", we assumed MAR and imputed according to the group-specific rate ICA-p.

Counts for these four groups are given by the variables df1, ds1, dc1, and dg1 for the intervention group, and df2, ds2, dc2, and dg2 for the control group.

In some trials, the reasons for missingness were given for a different subset of participants, for example, when clinical outcome and dropout were reported for different time points. In such a case, `metamiss` applies the proportion in each reason-group to the missing population in that trial. In trials that did not report any reasons for missingness, the overall proportion of reasons from all other trials is used.

```
. metamiss r1 f1 m1 r2 f2 m2, ica0(df1 df2) ica1(ds1 ds2) icapc(dc1 dc2)
> icap(dg1 dg2) w4 id(author) fixed nograph
******************************************************************
******** METAMISS: meta-analysis allowing for missing data ********
********            Imputation using reasons            ********
******************************************************************
Measure: RR.
Method: ICA-r combining ICA-0 ICA-1 ICA-pC ICA-p.
Weighting scheme: w4.
Zero cells detected: adding 1/2 to 6 studies.

(Calling metan with options: label(namevar=author) fixed eform nograph ...)
          Study    |    ES    [95% Conf. Interval]    % Weight
-------------------+---------------------------------------------
Arvanitis          |  1.381    0.867     2.201         21.37
Beasley            |  1.349    0.892     2.041         27.10
    (output omitted)
Marder             |  1.368    0.751     2.491         12.91
    (output omitted)
Selman             |  1.767    1.037     3.010         16.36
    (output omitted)

-------------------+---------------------------------------------
I-V pooled ES      |  1.785    1.439     2.214        100.00
-------------------+---------------------------------------------

  Heterogeneity chi-squared =  21.86 (d.f. = 16) p = 0.148
  I-squared (variation in ES attributable to heterogeneity) =  26.8%

  Test of ES=1 : z=   5.27 p = 0.000
```

3.6   Impute using uncertain IMORs

Finally, we allow for uncertainty about the IMORs. In the analysis below, we take a $N(0, 4)$ prior for the log IMORs in each group, with the log IMORs in the two groups being a priori uncorrelated.

```
. metamiss r1 f1 m1 r2 f2 m2, sdlogimor(2) logimor(0) w4 id(author) fixed
> nograph
**********************************************************************
******** METAMISS: meta-analysis allowing for missing data ********
********            Bayesian analysis using priors         ********
**********************************************************************
Measure: RR.
Zero cells detected: adding 1/2 to 6 studies.
Priors used:  Group 1: N(0,2^2). Group 2: N(0,2^2). Correlation: 0.
Method: Gauss-Hermite quadrature (10 integration points).

(Calling metan with options: label(namevar=author) fixed eform nograph ...)
             Study    |   ES   [95% Conf. Interval]   % Weight
--------------------+-------------------------------------------------
Arvanitis            | 1.416    0.889    2.257          30.37
Beasley              | 1.085    0.506    2.324          11.36
    (output omitted)

Marder               | 1.350    0.737    2.472          18.04
    (output omitted)

Selman               | 1.596    0.671    3.799           8.77
    (output omitted)

--------------------+-------------------------------------------------
I-V pooled ES        | 1.867    1.444    2.413         100.00
--------------------+-------------------------------------------------

  Heterogeneity chi-squared =  20.93 (d.f. = 16) p = 0.181
  I-squared (variation in ES attributable to heterogeneity) =  23.6%

  Test of ES=1 : z=   4.76 p = 0.000
```

Note how the weight assigned to the Beasley and Selman studies is greatly reduced. Because these studies have estimates below the pooled mean, the pooled mean increases.

4 Details

4.1 Zero cell counts

Like metan, metamiss adds one half to all four cells in a 2×2 table for a particular study if any of those cells contains zero. However, this behavior is modified under methods that impute with certainty (ICA-0, ICA-1, ICA-b, and ICA-w): the certain imputation is performed before metamiss decides whether to add one half. As a result, apparently similar options such as ica1 and logimor(99) differ slightly in the haloperidol data, because the logimor(99) analysis adds one half to six studies with $r2 = 0$, whereas the ica1 analysis does this only for three studies with $r2 + m2 = 0$.

4.2 Formula

For the imputation methods, in a given group of a given study, let r, f, and m be the number of observed successes, failures, and missing observations; let $\widehat{\pi} = r/(r + f)$ be the observed success fraction; and let $N = r + f + m$ be the total count. Let k index reason-groups with counts m_k and IMOR θ_k, so that, for example, a group imputed by ICA-0 has $\theta_k = 0$. Then the estimated success fraction is

$$\widehat{\pi}^* = \frac{1}{N} \left(r + \sum_k \frac{m_k \theta_k \widehat{\pi}}{1 - \widehat{\pi} + \theta_k \widehat{\pi}} \right)$$

with the variance obtained by a Taylor-series expansion (Higgins, White, and Wood 2008).

For the Bayesian methods, let δ_j be the log IMOR in group j. Then

$$\widehat{\pi}_j^*(\delta_j) = \frac{1}{N_j} \left(r_j + \frac{m_j e^{\delta_j} \widehat{\pi}_j}{1 - \widehat{\pi}_j + e^{\delta_j} \widehat{\pi}_j} \right)$$

and, for example, the log risk-ratio is obtained by finding the expectation of

$$\log \widehat{\pi}_E^*(\delta_E) - \log \widehat{\pi}_C^*(\delta_C)$$

over the prior $p(\delta_E, \delta_C)$ by numerical integration. The variance is obtained by combining the variance conditional on $p(\delta_E, \delta_C)$ with the variance over $p(\delta_E, \delta_C)$ (White, Higgins, and Wood 2008).

5 Discussion

We believe that ACA is a suitable starting point for a sensitivity analysis that might encompass, for example, `imor(1/2 1/2)`, `imor(1/2 2)`, `sdlogimor(2) corrlogimor(1)`, and `sdlogimor(2) corrlogimor(0)` (Higgins, White, and Wood 2008; White, Higgins, and Wood 2008). However, a "best" analysis might use reasons for missingness together with subject matter knowledge to assign suitable IMORs. Future work will explore how to integrate the two approaches.

6 References

Forster, J. J., and P. W. F. Smith. 1998. Model-based inference for categorical survey data subject to non-ignorable non-response. *Journal of the Royal Statistical Society, Series B* 60: 57–70.

Gamble, C., and S. Hollis. 2005. Uncertainty method improved on best–worst case analysis in a binary meta-analysis. *Journal of Clinical Epidemiology* 58: 579–588.

Higgins, J. P. T., I. R. White, and A. M. Wood. 2008. Imputation methods for missing outcome data in meta-analysis of clinical trials. *Clinical Trials* 5: 225–239.

Joy, C. B., C. E. Adams, and S. M. Lawrie. 2006. Haloperidol versus placebo for schizophrenia. *Cochrane Database of Systematic Reviews* 4: CD003082.

Little, R. J. A., and D. B. Rubin. 2002. *Statistical Analysis with Missing Data*. 2nd ed. Hoboken, NJ: Wiley.

Rubin, D. B. 1977. Formalizing subjective notions about the effect of nonrespondents in sample surveys. *Journal of the American Statistical Association* 72: 538–543.

White, I. R., J. P. T. Higgins, and A. M. Wood. 2008. Allowing for uncertainty due to missing data in meta-analysis - Part 1: Two-stage methods. *Statistics in Medicine* 27: 711–727.

The Stata Journal (2015)
15, Number 3, st0398, forthcoming

Fitting fixed- and random-effects meta-analysis models using structural equation modeling with the sem and gsem commands

Tom M. Palmer
Department of Mathematics and Statistics
Lancaster University
Lancaster, UK
t.palmer1@lancaster.ac.uk

Jonathan A. C. Sterne
School of Social and Community Medicine
University of Bristol
Bristol, UK

Abstract. In this article, we demonstrate how to fit fixed- and random-effects meta-analysis, meta-regression, and multivariate outcome meta-analysis models under the structural equation modeling framework using the `sem` and `gsem` commands. While all of these models can be fit using existing user-written commands, formulating the models in the structural equation modeling framework provides deeper insight into how they work. Further, the heterogeneity test for meta-regression and multivariate meta-analysis is readily available from this output.

Keywords: st0398, gsem, meta-analysis, meta-regression, metan, metareg, mvmeta, sem, structural equation modeling

1 Introduction

Stata 12 introduced the `sem` command for structural equation modeling (SEM). Stata 13 introduced the `gsem` command for generalized SEM, including SEMs with latent variables (also known as random effects in the multilevel modeling literature). These commands allow the user to place constraints on the variances of the error terms of the outcome variables in the models, which means that for the first time, it is possible to fit random-effects meta-analysis using an official Stata command (the official Stata command `vwls`, which implements variance-weighted least squares, can fit fixed-effects meta-analysis and meta-regression).

The user-written Stata commands for meta-analysis that have been published in the *Stata Journal* were summarized by Sterne (2009). The `metan` command is the main command that can implement univariate outcome fixed- and random-effects meta-analysis models, and it can also produce forest plots (Bradburn, Deeks, and Altman 1998; Harris et al. 2008). The `metareg` command implements random-effects meta-regression (Sharp 1998; Harbord and Higgins 2008). The `mvmeta` command implements multivariate fixed- and random-effects meta-analysis and meta-regression, including maximum likelihood, restricted maximum likelihood, and method of moments estimation for random-effects models (White 2009, 2011).

Performing meta-analysis in the SEM framework has been discussed in several articles (for example, see Cheung [2008, 2010, 2013c]), and these ideas have been implemented in the `metaSEM` package for R (Cheung 2013a; R Core Team 2014).

We begin this article by showing how to perform a fixed-effects meta-analysis using the `sem` command and then explaining how to do random-effects meta-analysis using `gsem`. We then demonstrate a fixed- and random-effects meta-regression. We also show how to fit multivariate outcome meta-analysis models with both zero and nonzero within-study covariances.

For readers unfamiliar with standard meta-analysis concepts, we recommend Bradburn, Deeks, and Altman (1998), Sutton et al. (2000), Harris et al. (2008), and Borenstein et al. (2009).

2   Univariate outcome meta-analysis models

2.1   Fixed-effects meta-analysis

The fixed-effects meta-analysis model is defined as

$$y_i \sim N(\theta, \ \sigma_i^2)$$

where subscript i denotes a study, y_i denotes the study outcomes, σ_i^2 denotes the square of the standard error of the study outcomes, and θ denotes the summary estimate.

Below, we give an example of how to fit this model using the user-written command `metan` in an example meta-analysis by Turner et al. (2000) that consists of nine clinical trials examining the effect of taking diuretics during pregnancy on the risk of pre-eclampsia. The dataset is available from the Stata 13 website and can be loaded into Stata 13 with the command `webuse diuretics`. The dataset includes the estimated log odds-ratio for the association between diuretics and pre-eclampsia from each study (variable `or`, which we rename `logor` for clarity) and the corresponding estimated variance (variable `varor`, which we rename `varlogor` for clarity). It also includes the variable `std` with the square root of the variances (the estimated standard errors of the log odds-ratios, which we rename `selogor`).

We can install the `metan` command from the Statistical Software Components website (using `ssc install metan`). We then load the data and run the fixed-effects meta-analysis.

```
. webuse diuretics
(Meta analysis of clinincal trials studying diuretics and pre-eclampsia)

. rename or logor

. rename varor varlogor

. rename std selogor

. metan logor selogor, fixed nograph
            Study    |     ES    [95% Conf. Interval]    % Weight
---------------------+----------------------------------------------------
1                    |  0.040    -0.744     0.824          3.99
2                    | -0.920    -1.599    -0.241          5.32
3                    | -1.120    -1.952    -0.288          3.55
4                    | -1.470    -2.544    -0.396          2.13
5                    | -1.390    -2.040    -0.740          5.80
6                    | -0.300    -0.496    -0.104         63.84
7                    | -0.260    -0.939     0.419          5.32
8                    |  1.090    -0.538     2.718          0.93
9                    |  0.140    -0.379     0.659          9.12
---------------------+----------------------------------------------------
I-V pooled ES        | -0.382    -0.538    -0.225        100.00
---------------------+----------------------------------------------------

    Heterogeneity chi-squared =   27.56 (d.f. = 8) p = 0.001
    I-squared (variation in ES attributable to heterogeneity) =  71.0%

    Test of ES=0 : z=   4.78 p = 0.000
```

To perform a fixed-effects meta-analysis using the sem command, we first generate a variable containing the inverse of the variance of the log odds-ratios, which is used to weight the analysis by specifying importance weights with [iw=weight]. We specify a simple linear model in which the log odds-ratios (logor) depend only on a constant with the syntax (logor <-). Note that Stata includes the constant by default (we could also explicitly include the constant by writing _cons to the right of the arrow).

It is important to include the constraint that the variance of the error term of the log odds-ratios is equal to 1; we do this using the option variance(e.logor@1). This ensures that the standard error of the summary fixed-effects estimate is correct (Cheung 2008). We can understand this by expressing the model in matrix notation,

$$\mathbf{Y} \sim N(\mathbf{X}\theta, \sigma^2 \mathbf{W}^{-1})$$

where $\mathbf{Y}$ is the vector of log odds-ratios, $\mathbf{X}$ is the vector of 1s for the constant, $\mathbf{W}$ is a matrix with the inverse variance weights on its diagonal and 0 everywhere else, and w_i is the inverse variance weight for each study.

The weighted least-squares summary estimate for N studies is $\widehat{\theta} = (\mathbf{X}'\mathbf{W}\mathbf{X})^{-1}$ $\mathbf{X}'\mathbf{W}\mathbf{Y} = (\sum_{i=1}^{N} w_i y_i)/(\sum_{i=1}^{N} w_i)$, and its associated variance is $\sigma^2(\mathbf{X}'\mathbf{W}\mathbf{X})^{-1}$. With the constraint that $\sigma^2 = 1$, the variance of the summary estimate is simply $(\mathbf{X}'\mathbf{W}\mathbf{X})^{-1}$, which equals $1/\sum_{i=1}^{N} w_i = 1/\sum_{i=1}^{N}(1/\sigma_i^2)$, which is the required form of the variance of the summary estimate.

To reduce the amount of output, we additionally specify the nodescribe option to suppress the variable classification table, the nocnsreport option to suppress the reporting of the constraints, and the nolog option to suppress the log of the maximum likelihood iterations. In Stata 13, one cannot use `gsem` rather than `sem`, because `gsem` does not allow weights; however in Stata 14, one can use `gsem` because it does allow weights.

```
. generate double weight = 1/varlogor
. sem (logor <- ) [iw=weight], variance(e.logor@1) nodescribe nocnsreport nolog
Structural equation model                         Number of obs     =           9
Estimation method  = ml
Log likelihood      = -157.71614
```

	Coef.	OIM Std. Err.	z	P>|z|	[95% Conf.	Interval]
Structural						
logor <-						
_cons	-.3815467	.0799025	-4.78	0.000	-.5381527	-.2249406
var(e.logor)	1	(constrained)				

```
LR test of model vs. saturated: chi2(1)     =     143.07, Prob > chi2 = 0.0000
```

The summary log odds-ratio is -0.3815467, which corresponds to the coefficient for the constant (`_cons`). The estimate and 95% confidence interval (CI) are identical to those displayed in the `metan` output shown above. We would generally exponentiate this summary estimate and its CI limits to report the fixed-effects estimate on the odds-ratio scale, using the syntax `lincom _b[logor:_cons], eform`.

We can also use `sem` to perform a test for between-study heterogeneity (more variability between the results of the different studies than is expected under the fixed-effects assumption). To do this, we scale the outcomes by the inverse of the standard errors and regress this transformed outcome variable on a constant that is scaled in the same way (Kalaian and Raudenbush 1996; Cheung 2008, 2013c). To obtain the heterogeneity test, we leave the variance of the error term of the outcome unconstrained. The test statistic (Q) is given by the estimated variance of the error term of the scaled outcomes multiplied by the number of studies, and it is then compared with a χ^2 critical value with degrees of freedom (d.f.) equal to the number of studies minus one. From this output, we can also calculate the I^2 statistic, which is given by $(Q - \mathrm{d.f.})/Q$ (Higgins and Thompson 2002).

```
. generate double invselogor = 1/selogor
. generate double logortr = logor*invselogor
. sem (logortr <- invselogor, noconstant), noheader nodescribe nocnsreport nolog
```

	Coef.	OIM Std. Err.	z	P>|z|	[95% Conf.	Interval]
Structural						
logortr <-						
invselogor	-.3815467	.1398305	-2.73	0.006	-.6556094	-.107484
_cons	0	(constrained)				
var(e.logortr)	3.062546	1.443698			1.215689	7.715122

```
LR test of model vs. saturated: chi2(1)     =      0.61, Prob > chi2 = 0.4344
```

```
. local Q = _b[var(e.logortr):_cons]*e(N)

. local df = e(N) - 1

. display "Het. test statistic = " `Q´
Het. test statistic = 27.562912

. display "Het. test p-value = " chi2tail(`df´, `Q´)
Het. test p-value = .00056491

. display "I-squared = " (`Q´ - `df´)/`Q´
I-squared = .7097549
```

The test statistic, p-value, and I^2 statistic correspond to those reported at the bottom of the `metan` output shown above.

We could also specify these models using the SEM builder.[1] SEM builder path diagrams for these fixed-effects models are shown in figure 1. On these diagrams, observed variables are shown in rectangular boxes, and error terms for outcome (endogenous) variables are in circles. The constraint that the variance of the error term of the outcomes is 1 is shown by the 1 to the right of the ε_1 nodes. It is not possible to show the inverse variance weights in the diagram in figures 1(a) and 1(b). In figures 1(c) and 1(d), the omission of the constant is represented by the 0 in the bottom right of the box for the transformed outcome variable `logortr`. After estimation, the coefficients are displayed on the diagram, and the mean and variance of exogenous variables, such as `invselogor` in figure 1(d), are also displayed.

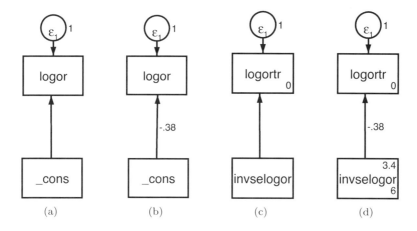

(a)   (b)   (c)   (d)

Figure 1. Stata SEM Builder path diagrams for fixed-effects meta-analysis. Diagrams (a) and (b) use untransformed variables; (c) and (d) use transformed variables. Diagrams (b) and (d) are postestimation.

---

1. To get to the SEM Builder in Stata, select the following menus: **Statistics > SEM (structural equation modeling) > Model building and estimation**.

2.2   Random-effects meta-analysis

The random-effects meta-analysis model is defined as

$$y_i \sim N(\theta + \nu_i, \ \sigma_i^2)$$
$$\nu_i \sim N(0, \ \tau^2)$$

where ν_i denotes the random effects (the differences between the intervention effects in study i and the overall mean intervention effect).

In this model, τ^2 is commonly referred to as the between-study variance. To fit this model, we assume that the estimated values of the study variances are equal to their true values (σ_i^2). We first perform a random-effects meta-analysis using `metan` on the same example dataset, using the syntax `metan logor selogor, random`. The summary log odds-ratio is -0.52 (95% CI $[-0.91, -0.12]$) with I^2 (Higgins and Thompson 2002; Higgins et al. 2003) and τ^2 estimated to be 71% and 0.22, respectively.

Example 6 in the *Stata 13 Multilevel Mixed-Effects Reference Manual* provides the syntax to perform the random-effects meta-analysis with the `meglm` command (Stata-Corp 2013). We provide the equivalent `gsem` syntax below.[2] In this syntax, the latent variables specify the within-study effects; therefore, the model is specified with one random effect per study. In this case, we do not constrain the variance of the error term of the outcome, `var(e.logor)`, because this represents the between-study variance τ^2.

The code below begins by generating the indicator variables (variables `tr1`–`tr9`) using the `generate()` option of the `tabulate` command. We define one latent variable (random effect) for each study. The latent variables are specified by variable names starting with a capital letter, which importantly do not already exist in our dataset. We use latent variable names `M1`–`M9`. Each latent variable interacts with the indicator variable for the corresponding study via the `#` symbol. The coefficients for each latent variable by indicator interaction are constrained to 1 by `@1` (although Stata would impose this constraint by default). The variance of each latent variable is constrained to the variance of the corresponding study's log odds-ratio, and the latent variables are constrained to be uncorrelated. This constraint is achieved by putting the sample variances along the diagonal of a matrix, which we call `f`, with all other entries equal to 0 by using the `mkmat` and `matrix` commands and then specifying `covstructure(_LEx, fixed(f))` in `gsem`. Here `_LEx` stands for latent exogenous variables (our nine latent variables). We additionally specify the `intmethod(laplace)` option to use the Laplace approximation to speed up the numerical integration.

---

2. Note that `meglm` and `gsem` use the same underlying adaptive quadrature routines to integrate over the random effects and, hence, their output is identical.

```
. mkmat varlogor, matrix(f)

. matrix f = diag(f)

. quietly tabulate trial, gen(tr)

. gsem (logor <- M1#c.tr1@1 M2#c.tr2@1 M3#c.tr3@1
>            M4#c.tr4@1 M5#c.tr5@1 M6#c.tr6@1
>            M7#c.tr7@1 M8#c.tr8@1 M9#c.tr9@1),
>            covstructure(_LEx, fixed(f)) intmethod(laplace) nocnsreport nolog
Generalized structural equation model          Number of obs     =          9
Response       : logor
Family         : Gaussian
Link           : identity
Log likelihood = -9.4552759
```

	Coef.	Std. Err.	z	P>|z|	[95% Conf. Interval]
logor <-					
c.tr1#c.M1	1	(constrained)			
c.tr2#c.M2	1	(constrained)			
c.tr3#c.M3	1	(constrained)			
c.tr4#c.M4	1	(constrained)			
c.tr5#c.M5	1	(constrained)			
c.tr6#c.M6	1	(constrained)			
c.tr7#c.M7	1	(constrained)			
c.tr8#c.M8	1	(constrained)			
c.tr9#c.M9	1	(constrained)			
_cons	-.5166151	.2059448	-2.51	0.012	-.9202594 -.1129707
var(M1)	.16	(constrained)			
var(M2)	.12	(constrained)			
var(M3)	.18	(constrained)			
var(M4)	.3	(constrained)			
var(M5)	.11	(constrained)			
var(M6)	.01	(constrained)			
var(M7)	.12	(constrained)			
var(M8)	.69	(constrained)			
var(M9)	.07	(constrained)			
var(e.logor)	.2377469	.1950926			.0476023 1.187413

Our summary log odds-ratio of -0.52 is the same as reported by metan. However, our estimate for var(e.logor), which is our estimate of τ^2, at 0.2377 is slightly larger than that given by metan. This is because metan reports a method of moments estimate of τ^2, whereas our gsem output gives a maximum likelihood estimate (Harris et al. 2008; DerSimonian and Laird 1986).

Harris et al. (2008) explain that when we additionally specify the `rfdist` option of `metan`, we obtain the 95% prediction interval for the summary estimate on the forest plot, which here is -1.72 to 0.69. We can obtain the prediction interval from our `gsem` output as shown below. This interval is slightly wider than the interval obtained from `metan` because of the larger estimate of τ^2.

```
. local setotal = sqrt(_se[logor:_cons]^2 + _b[var(e.logor):_cons])
. local pilow = _b[logor:_cons] - invt(e(N) - 2, .975)*`setotal'
. local piupp = _b[logor:_cons] + invt(e(N) - 2, .975)*`setotal'
. display "95% Prediction interval:", `pilow', `piupp'
95% Prediction interval: -1.7682144 .73498424
```

The *Stata Multilevel Mixed-Effects Reference Manual* shows that it is also possible to fit the random-effects meta-analysis model by using a single random effect (StataCorp 2013). In this syntax, one latent variable is used to incorporate the random effects. We do not need to constrain the variance of the error term of the outcome (which represents τ^2) because the variance of the latent variable is constrained to 1. The equivalent `gsem` code is

```
. gsem (logor <- ibn.trial#c.selogor#c.M@1), variance(M@1) nolog nocnsreport
  (output omitted )
```

We omit the output because it gives the same estimates as the previous `gsem` model. In this formulation, we specify a single latent variable at the trial level, which we call M, scaled by the observed standard errors of the log odds-ratios from each study. The variance of this latent variable is constrained to 1 using `variance(M@1)`. Because `gsem` allows the use of Stata's factor-variable notation, we can include indicator variables for each study with the `i.trial` notation, and we specify `ibn.trial` to assign no baseline level. This prevents the indicator variable for trial 1 from being omitted from the model. Because the individual studies are already on separate rows in the dataset, we could omit `ibn.trial#` from the interaction and this code would still work.

In this code, there is a three-way interaction between the indicator variables, the standard errors, and the latent variable. In these interactions, it is also necessary to specify that the standard-error variable (`selogor`) is continuous by using `c.selogor`. Otherwise, Stata will treat it as a factor variable and will split it into indicator variables for its different values. By default, Stata assumes that the latent variable (M) is continuous and includes it in the model as `c.M`.

We can also fit the random-effects model using the same inverse standard-error transformation that we used to fit the heterogeneity test under the fixed-effects model. The scaling is applied to all variables in the model, including the latent variable. In this case, we do need to constrain the variance of the error term of the outcome to 1 to obtain correct standard errors because, in this syntax, the variance of the latent variable represents τ^2. We specify the `latent(M)` option so that Stata knows to treat only M as a latent variable (in case there are any other variables in the dataset whose names start with a capital letter). Again, we omit the output because the estimates are the same as those from the first `gsem` model.

```
. gsem (logortr <- invselogor c.invselogor#c.M@1, noconstant),
>    variance(e.logortr@1) latent(M) nolog nocnsreport
    (output omitted)
```

Restricted maximum-likelihood (REML) estimates of random-effects models are usually preferred because maximum likelihood estimates of variance components are often negatively biased (Viechtbauer 2005; Prevost et al. 2007). Cheung (2015a, 2013b) claims that it is possible to obtain REML estimates of variance components by using the maximum likelihood software OpenMx (Boker et al. 2011) with the approach of Patterson and Thompson (1971) and Harville (1977). These REML estimates of the variance components can be used to obtain REML summary estimates by fitting a model with the variance components constrained at the values of the REML estimates. Unfortunately, it appears that currently it is not possible to implement REML estimation of variance components in `gsem` because it is not possible to constrain the variance components in the required form.

If we are concerned that the numerical integration routines in `gsem` have converged in the wrong place, we can compare our estimates with the maximum likelihood estimates from `mvmeta` (White 2009, 2011). The `mvmeta` command does not perform numerical integration but instead maximizes the marginal likelihood of the random-effects model. The `mvmeta` syntax is given below, and in this case, the estimates are the same, so we omit the output. We can also obtain REML estimates from `mvmeta` by specifying its `reml` option (which is its default method) instead of the `ml` option.

```
. generate b_logor_1 = logor
. generate V_logor_1_logor_1 = varlogor
. mvmeta b V, vars(b_logor_1) nolog ml i2 print(bscov)
. mvmeta b V, vars(b_logor_1) nolog reml i2 print(bscov)
    (output omitted)
```

As with the fixed-effects models, we can define and fit these models by using Stata's SEM Builder to make their path diagrams, which we show in figure 2. The diagrams show the latent variable M in an oval and show an interaction between two variables (namely, between `selogor` and M and between `invselogor` and M) using one path going into another path. Figure 2(a) shows the diagram using untransformed variables. The 0 on the path from `selogor` to `logor` shows that the main effect of `selogor` is omitted from the model. Figure 2(b) also shows the path diagram using the untransformed variables, except it is postestimation. Here we see the summary estimate on the path from the constant (`_cons`) to the outcome (`logor`), as well as the estimate of τ^2 as the variance of the ε_1 node. Figures 2(c) and 2(d) show diagrams using the transformed variables; figure 2(d) shows the diagram postestimation, and we see the summary estimate on the path for the main effect of `invselogor` as well as the estimate of τ^2 as the variance of M.

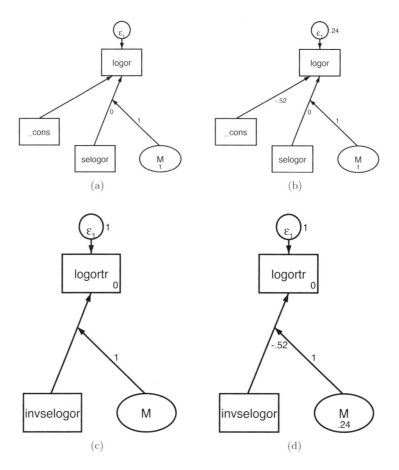

Figure 2. Stata SEM Builder path diagrams for random-effects meta-analysis. Diagrams (a) and (b) use untransformed variables; (c) and (d) use transformed variables. Diagrams (b) and (d) are postestimation.

3 Univariate outcome meta-regression models

3.1 Fixed-effects meta-regression

Fixed-effects meta-regression is generally not recommended in practice because it assumes that the heterogeneity is explained by the covariates in the model. It, therefore, tends to give standard errors that are too small when there is moderate or large heterogeneity (Thompson and Sharp 1999; Higgins and Thompson 2004; Harbord and Higgins 2008). However, we present the code here because it is instructive for fitting the random-effects version and because the heterogeneity test can be derived from it.

The fixed-effects meta-regression model is given by

$$y_i \sim N(\mathbf{X}_i\boldsymbol{\beta}, \ \sigma_i^2)$$

where $\mathbf{X}$ represents an $N \times (c + 1)$ matrix—here N is the number of studies and c is the number of covariates (with the first column a constant and the other columns study-level covariates)—and $\boldsymbol{\beta}$ is the vector of fixed-effects parameters to be estimated.

In this section, we use the same example meta-analysis as used by Harbord and Higgins (2008), which was taken from table 1 of Thompson and Sharp (1999). The data are from 28 randomized controlled trials of cholesterol-lowering interventions for reducing the risk of ischaemic heart disease. The `cholesterol.dta` dataset can be downloaded in Stata using the command `net get sbe23_1`. The measure of effect size is again the log odds-ratio (variable `logor`). The meta-regression estimates the change in the summary estimate per unit change in cholesterol reduction observed in the trial (quantified in the variable `cholreduc`).

The `sem` code for fixed-effects meta-regression is below.

```
. use cholesterol, clear
(Serum cholesterol reduction & IHD)

. generate weight = 1/varlogor

. sem (logor <- cholreduc) [iweight=weight], variance(e.logor@1)
> nodescribe nolog nocnsreport
Structural equation model                      Number of obs      =          28
Estimation method  = ml
Log likelihood     = -837.19264
```

	Coef.	OIM Std. Err.	z	P>|z|	[95% Conf. Interval]	
Structural						
logor <-						
cholreduc	-.4752451	.1382083	-3.44	0.001	-.7461284	-.2043617
_cons	.1207613	.0972033	1.24	0.214	-.0697538	.3112763
var(e.logor)	1	(constrained)				

```
LR test of model vs. saturated: chi2(1)    =   1945.33, Prob > chi2 = 0.0000
```

For every unit increase in cholesterol reduction, the log odds-ratio is estimated to increase by -0.475 (95% CI $[-0.746, -0.204]$).

The estimate for the constant (`_cons`) is the fixed-effects summary estimate for a trial with zero cholesterol reduction (for comparison, the fixed-effects summary estimate assuming no between-trial heterogeneity is -0.193 (95% CI $[-0.259, -0.127]$) and the heterogeneity test p-value is 0.005). We can perform the heterogeneity test for this fixed-effects meta-regression by scaling the variables in the model by the inverse of the standard errors and then removing the constraint on the variance of the error term of the outcome variable.

```
. generate double invselogor = 1/sqrt(varlogor)

. generate double logortr = logor*invselogor

. generate double cholreductr = cholreduc*invselogor

. sem (logortr <- cholreductr invselogor, noconstant),
>        nodescribe nolog nocnsreport
Structural equation model                        Number of obs    =        28
Estimation method  = ml
Log likelihood     = -164.50782
```

		OIM					
	Coef.	Std. Err.	z	P>|z|	[95% Conf. Interval]	
Structural						
logortr <-						
cholreductr	-.4752451	.1607242	-2.96	0.003	-.7902588	-.1602313
invselogor	.1207613	.113039	1.07	0.285	-.1007911	.3423136
_cons	0	(constrained)				
var(e.logortr)	1.352366	.3614347			.8009426	2.283428

```
LR test of model vs. saturated: chi2(1)    =      0.01, Prob > chi2 = 0.9230
. local Q = _b[var(e.logortr):_cons]*e(N)

. local df = e(N) - 2

. display "Het. test statistic = " `Q´
Het. test statistic = 37.866258
. display "Het. test p-value = " chi2tail(`df´, `Q´)
Het. test p-value = .06231403
```

The estimated variance of the transformed log odds-ratios was 1.35, which led to the test statistic of 37.9. Including the covariate for cholesterol explained some of the between-trial heterogeneity.

3.2 Random-effects meta-regression

The random-effects meta-regression model is defined as

$$y_i \sim N(\mathbf{X}_i\boldsymbol{\beta} + \nu_i, \ \sigma_i^2)$$
$$\nu_i \sim N(0, \ \tau^2)$$

where, again, ν_i represents the random effects.

We start by fitting a random-effects meta-regression model using the user-written command `metareg` (Harbord and Higgins 2008).

```
. metareg logor cholreduc, wsse(selogor)
```

Meta-regression	Number of obs	=	28
REML estimate of between-study variance	tau2	=	.0097
% residual variation due to heterogeneity	I-squared_res	=	31.34%
Proportion of between-study variance explained	Adj R-squared	=	69.02%
With Knapp-Hartung modification			

logor	Coef.	Std. Err.	t	P>|t|	[95% Conf. Interval]	
cholreduc	-.5056849	.1834858	-2.76	0.011	-.8828453	-.1285244
_cons	.1467225	.137463	1.07	0.296	-.1358367	.4292816

For every unit increase in cholesterol reduction, the log odds-ratio is estimated to increase by -0.506 (95% CI $[-0.883, -0.129]$). The estimated log odds-ratio for a trial with zero cholesterol reduction is 0.147 (95% CI $[-0.136, 0.429]$). The random-effects meta-analysis summary (method of moments) estimate from `metan` (without adjustment) was -0.213 (95% CI $[-0.322, -0.103]$), indicating that, on average, the trials led to a reduction in the odds of ischaemic heart disease.

We can fit the random-effects meta-regression model with `gsem` by including the `cholreduc` variable in the second random-effects meta-analysis model syntax from section 2.2. In this dataset, the studies are identified by the variable `id`. Because there are no rows of repeated study results in the dataset (each study is on a separate row), we can omit `ibn.id#` from the interaction. In this example, the default estimation settings make it difficult for the numerical integration routines to converge; therefore, to help convergence, we specify a smaller number of integration points with `intpoints(6)`.

```
. gsem (logor <- c.selogor#c.M@1 cholreduc), variance(M@1) intpoints(6)
> nolog nocnsreport
```

Generalized structural equation model	Number of obs	=	28
Response : logor			
Family : Gaussian			
Link : identity			
Log likelihood = -10.593922			

	Coef.	Std. Err.	z	P>|z|	[95% Conf. Interval]	
logor <-						
cholreduc	-.4756347	.1393961	-3.41	0.001	-.748846	-.2024234
c.selogor#c.M	1	(constrained)				
_cons	.121052	.0998135	1.21	0.225	-.0745789	.3166829
var(M)	1	(constrained)				
var(e.logor)	2.64e-10	2.29e-07			.	.

The estimated change in the log odds-ratio per unit change in cholesterol reduction is -0.476 (95% CI $[-0.749, -0.202]$), which is slightly smaller in magnitude than the estimate from `metareg`. We think this difference is caused by `gsem`'s smaller estimate of τ^2 (the `var(e.logor)` parameter), which is a result of `gsem`'s use of maximum likelihood estimation (whereas `metareg` uses REML estimation). For comparison, we fit the model using the inverse standard-error transformation with the `metareg` estimate of τ^2 as an initial value.

```
. gsem (logortr <- c.invselogor#c.M@1 invselogor cholreductr, noconstant),
> variance(e.logortr@1 (M, init(.0097))) latent(M)
> nolog nocnsreport
Generalized structural equation model          Number of obs      =         28
Response        : logortr
Family          : Gaussian
Link            : identity
Log likelihood = -44.663427
```

	Coef.	Std. Err.	z	P>|z|	[95% Conf.	Interval]
logortr <-						
invselogor	.1207613	.0972033	1.24	0.214	-.0697538	.3112763
cholreductr	-.4752451	.1382083	-3.44	0.001	-.7461284	-.2043617
c.invselogor#c.M	1	(constrained)				
_cons	0	(omitted)				
var(M)	1.89e-34	6.40e-18			.	.
var(e.logortr)	1	(constrained)				

These parameter estimates are very similar to those from our previous model. We can compare these estimates with maximum likelihood estimates from `mvmeta` by using the following syntax. The `mvmeta` estimates are virtually identical, so we omit the output.

```
. generate double b_logor_1 = logor
. generate double V_logor_1_logor_1 = varlogor
. mvmeta b V cholreduc, vars(b_logor_1) ml nolog print(bscov)
  (output omitted)
```

We could also derive prediction intervals for the estimates, as described in section 2.2.

4   Multivariate outcome meta-analysis with zero within-study covariances

In the next two sections, we split multivariate outcome meta-analysis models into those with zero and those with nonzero within-study covariances. To use the multivariate outcomes on their natural scale and include a different within-study covariance matrix for each study would require `sem` and `gsem` to allow the use of so-called definition variables (Mehta and Neale 2005; Cheung 2013c). However, to our knowledge, only the OpenMx package (Boker et al. 2011) has this feature. The use of OpenMx for meta-analysis is automated in the `metaSEM` package for R (Cheung 2013a).

To include the study-specific within-study covariance matrices, we must transform the within-study outcomes by using a Cholesky decomposition of the inverse of the within-study covariance matrices to make the outcomes independent of one another, as per Cheung (2013c). This is the multivariate equivalent of scaling the variables in the univariate outcome model by the inverse of the standard errors, as we demonstrated in the previous sections.

4.1 Fixed-effects models

The multivariate fixed-effects model is defined below for a two outcome meta-analysis, where MVN stands for multivariate normal distribution, $\mathbf{V}_i$ represents the within-study variance–covariance matrix for study i, $\boldsymbol{\Sigma}$ represents the within-study block-diagonal covariance matrix, $\boldsymbol{\theta}$ is the vector of parameters to be estimated, and $\mathbf{Y}$ is the vector of study outcomes with the study-specific outcomes stacked on top of one another (y_{11} and y_{12} are outcomes 1 and 2 for study 1).

$$\mathbf{Y} = \begin{bmatrix} y_{11} \\ y_{12} \\ \vdots \\ y_{N1} \\ y_{N2} \end{bmatrix}, \; \boldsymbol{\theta} = \begin{bmatrix} \theta_1 \\ \theta_2 \end{bmatrix}, \; \mathbf{V}_i = \begin{bmatrix} \sigma_{i,11}^2 & \sigma_{i,12} \\ \sigma_{i,12} & \sigma_{i,22}^2 \end{bmatrix}, \; \boldsymbol{\Sigma} = \begin{bmatrix} \mathbf{V}_1 & 0 & 0 \\ 0 & \ddots & 0 \\ 0 & 0 & \mathbf{V}_N \end{bmatrix}$$

The model is then defined as

$$\mathbf{Y} \sim \text{MVN}(\boldsymbol{\theta}, \; \boldsymbol{\Sigma})$$

In this section, we assume that all within-covariance terms ($\sigma_{i,12}$) are 0, which means that we do not have to use the Cholesky decomposition transformation. We use the example meta-analysis from Riley et al. (2007), which consists of 10 studies from the systematic review of Glas et al. (2003) investigating the sensitivity (the true-positive rate) and specificity (the true-negative rate) of tumor markers used for diagnosing primary bladder cancer. The example data are for the telomerase marker. The `telomerase.dta` dataset is available as an ancillary file from a *Stata Journal* article for `mvmeta` by using the command `net get st0156`. This example has zero within-study covariances (and, therefore, zero correlations) because the sensitivity and specificity outcomes are independent of one another. The `metandi` and `midas` commands use an alternative hierarchical logistic regression modeling approach to meta-analyzing such data (Harbord and Whiting 2009; Dwamena 2007).

Each study reports data on two outcome measures: the logit, $\log\{p/(1-p)\}$, of the sensitivity (`y1`) and the logit of the specificity (`y2`), which we will meta-analyze jointly. The standard errors are in the variables `s1` and `s2`. Because `sem` can use only one `iweight` variable, we follow Cheung (2013c) in fitting the multivariate model as a univariate model by stacking the multiple outcomes in one variable (`y`) using `reshape long`. We also create a stacked variable of inverse variance weights (`weight`) and an indicator variable for the second outcome (`y2cons`).

As with the previous two fixed-effects model syntaxes using `iweights`, it is important to constrain the variance of the error term of the outcomes to 1 with `variance(e.y@1)` to obtain correct standard errors for our summary estimates. The summary estimate for logit sensitivity is given by the estimate for _cons. We then use `lincom` to obtain the summary estimate for logit specificity. This is given by adding the coefficients for _cons and y2cons. To fit an equivalent model, we could include an additional indicator variable for the logit sensitivity and omit the constant, which would avoid the need to use `lincom`. However, `sem` could not find initial values for this model.

```
. use telomerase, clear
(Riley's telomerase data)
. reshape long y s, i(study) j(outcome)
(note: j = 1 2)
```

Data		wide	->	long
Number of obs.		10	->	20
Number of variables		5	->	4
j variable (2 values)			->	outcome
xij variables:				
		y1 y2	->	y
		s1 s2	->	s

```
. generate byte y2cons = (outcome == 2)
. generate weight = 1/(s^2)
. sem (y <- y2cons) [iw=weight], variance(e.y@1) nocaps nodescribe nolog
> nocnsreport
Structural equation model                        Number of obs    =        20
Estimation method  = ml
Log likelihood     = -210.20211
```

	Coef.	OIM Std. Err.	z	P>|z|	[95% Conf. Interval]
Structural					
y <-					
y2cons	.0834338	.2104572	0.40	0.692	-.3290547	.4959223
_cons	1.126318	.1177527	9.57	0.000	.8955267	1.357109
var(e.y)	1	(constrained)			

```
LR test of model vs. saturated: chi2(1)    =    1.04, Prob > chi2 = 0.3068
. lincom [y]_cons + [y]y2cons
 ( 1)  [y]y2cons + [y]_cons = 0
```

	Coef.	Std. Err.	z	P>|z|	[95% Conf. Interval]
(1)	1.209752	.174432	6.94	0.000	.867871	1.551632

When back-transformed, these estimates give summary estimates of sensitivity and specificity of 0.76 (95% CI [0.71, 0.80]) and 0.77 (95% CI [0.71, 0.83]), respectively. This code can easily be extended to fit multivariate fixed-effects meta-regression models by including covariates, typically interacted with y2cons, as before.

The heterogeneity test can again be performed by scaling the variables in the model by the inverse of the standard errors of the outcomes and removing the constraint on the variance of the error term of the outcome. In this case, the d.f. is given by the number of observed outcomes minus two (because there are two outcomes).

```
. generate double invse = 1/s
. generate double ytr = y*invse
. generate double y2constr = y2cons*invse
. quietly sem (ytr <- y2constr invse, noconstant), nocaps nodescribe nolog
> nocnsreport
. local Q = _b[var(e.ytr):_cons]*e(N)
. local df = e(N) - 2
. display "Het. test statistic = " `Q'
Het. test statistic = 90.865377
. display "Het. test p-value = " chi2tail(`df', `Q')
Het. test p-value = 1.009e-11
```

The heterogeneity test shows very strong evidence against homogeneity, so we would fit a random-effects model as shown in the next section.

4.2   Random-effects models

The multivariate random-effects meta-analysis model is defined as

$$\mathbf{Y} \sim \text{MVN}(\boldsymbol{\theta} + \boldsymbol{\nu}, \ \boldsymbol{\Sigma})$$

$$\boldsymbol{\nu} \sim N(\mathbf{0}, \ \mathbf{T}^2), \text{ where for a two-outcome model, } \mathbf{T}^2 = \begin{bmatrix} \tau_1^2 & \tau_{12} \\ \tau_{12} & \tau_2^2 \end{bmatrix}$$

We give the gsem code for the model below. We start by reloading the dataset, which is in wide format. This means that we fit the multivariate model using gsem's multiple-equation syntax, with each equation being defined within parentheses. Because the variables are on their natural scale, this syntax uses the latent variables M1 and M2 to define the within-study variances. Constraining the within-study variances to the observed values for each study is achieved by scaling the latent variables by the standard errors in the c.s1#c.M1 and c.s2#c.M2 interactions and constraining the variances of the latent variables to 1. Because each study is on a separate row, we can omit ibn.study# from the interaction terms (study is the variable identifying the studies). In this syntax, the variances of the error terms of the outcomes define the between-study variance components. In the covariance option, we set the covariance between the latent variables to 0 with M1*M2@0, which represents the zero within-study correlations. In this option, we also allow a between-study covariance with e.y1*e.y2. We use intmethod(laplace) to help the numerical integration converge.

```
. quietly use telomerase, clear

. gsem (y1 <- c.s1#c.M1@1) (y2 <- c.s2#c.M2@1),
>       covariance(M1@1 M2@1 M1*M2@0 e.y1*e.y2) latent(M1 M2)
>       intmethod(laplace) nolog nocnsreport
Generalized structural equation model           Number of obs      =          10
Response       : y1
Family         : Gaussian
Link           : identity
Response       : y2
Family         : Gaussian
Link           : identity
Log likelihood = -25.598741
```

	Coef.	Std. Err.	z	P>|z|	[95% Conf. Interval]
y1 <-					
c.s1#c.M1	1	(constrained)			
_cons	1.157863	.1796031	6.45	0.000	.8058472 1.509879
y2 <-					
c.s2#c.M2	1	(constrained)			
_cons	2.020725	.5340579	3.78	0.000	.9739904 3.067459
var(M1)	1	(constrained)			
var(M2)	1	(constrained)			
var(e.y1)	.1680775	.1242589			.0394658 .7158118
var(e.y2)	2.247177	1.231404			.7677099 6.57775
cov(e.y2,e.y1)	-.3975715	.3042327	-1.31	0.191	-.9938566 .1987135

These estimates correspond to summary estimates of sensitivity and specificity of 0.76 (95% CI [0.70, 0.82]) and 0.88 (95% CI [0.77, 0.99]), respectively, which are very similar to those originally reported by Glas et al. (2003). We could also fit this model using the transformed variables. In this case, the variance components of the latent variables represent the between-study variance components, and therefore, we constrain the variances of the error terms of the outcomes to be 1 to obtain correct standard errors on our summary estimates.

```
. generate double y1tr = y1/s1

. generate double invs1 = 1/s1

. generate double y2tr = y2/s2

. generate double invs2 = 1/s2

. gsem (y1tr <- c.invs1#c.M1@1 invs1, noconstant)
>     (y2tr <- c.invs2#c.M2@1 invs2, noconstant),
>     cov(e.y1tr@1 e.y2tr@1 e.y1tr*e.y2tr@0)
>     latent(M1 M2) nolog nocnsreport
  (output omitted)
```

We could also derive prediction intervals for the estimates, as described in section 2.2. We can also check our summary estimates with maximum likelihood estimates from mvmeta using the code below. We find some minor differences between the estimates of these three models at the second decimal place, but these lead to almost identical summary estimates of sensitivity and specificity.

```
. generate double S11=s1^2
. generate double S22=s2^2
. mvmeta y S, vars(y1 y2) wscorr(0) print(bscov) nolog ml
  (output omitted)
```

5 Multivariate outcome meta-analysis with nonzero within-study covariances

In this section, we use the example Fibrinogen Studies Collaboration (2004) meta-analysis provided by White (2009), which is also available using the command net get st0156 in Stata. This meta-analysis compares the incidence of coronary heart disease across 31 studies (Fibrinogen Studies Collaboration 2005). In the original analysis, there were 10 groups, 5 of which were presented by White (2009). We use only two of the comparisons as outcomes (variables b_Ifg_2 and b_Ifg_5, which are on the log hazard-ratio scale, and their variances and covariances in variables V_Ifg_2_Ifg_2, V_Ifg_5_Ifg_5, and V_Ifg_2_Ifg_5), but the code easily generalizes to more outcomes.

Because we cannot incorporate different study-specific within-study covariance matrices when using the outcome variables in their natural scales in sem and gsem, we must transform the outcomes by using the multivariate equivalent of scaling them by the inverse of their standard errors. We do this by multiplying the outcomes and study indicators by the Cholesky decomposition of the inverse of the within-study covariance matrices, as described by Cheung (2013c). For each study i, we calculate $\mathbf{W}_i^{1/2} = \mathbf{V}_i^{-1/2}$, where a matrix to the power $1/2$ denotes the Cholesky decomposition.

5.1 Fixed-effects models

The fixed-effects model using the transformed variables, which we denote with $*$, is defined as

$$\mathbf{W}^{1/2}\mathbf{Y} \sim \text{MVN}\left\{\mathbf{W}^{1/2}\mathbf{X}\boldsymbol{\beta},\ \mathbf{W}^{1/2}\boldsymbol{\Sigma}\left(\mathbf{W}^{1/2}\right)'\right\}$$

$$\mathbf{Y}^* \sim \text{MVN}(\mathbf{X}^*\boldsymbol{\beta},\ \mathbf{I})$$

We fit the model using the code below. We start by creating the $\mathbf{V}_i$ matrices in Stata and then pass them to Mata to perform the inverse and Cholesky decomposition. We create stacked $\mathbf{Y}_i^*$ vectors and $\mathbf{X}_i^*$ matrices in Mata and then combine them, and then we pass these to Stata using getmata. In the calculations, we multiply by the transpose of $\mathbf{W}_i^{1/2}$ because Stata's and Mata's cholesky() functions return the lower triangular decomposition, whereas we require the upper triangular decomposition.

In the `sem` syntax, we must constrain the variance of the error term of the outcome to 1 to obtain correct standard errors on the summary estimates. The summary estimates for the two outcome variables are given by the coefficients of the `xstarstack#` variables. Because these are on the log hazard-ratio scale, we use `lincom` with its `eform` option to show the estimates on the hazard-ratio scale.

```
. use fscstage1, clear
. forvalues i=1/31 {
  2.          matrix V`i´ = (V_Ifg_2_Ifg_2[`i´], V_Ifg_2_Ifg_5[`i´] \
  >                 V_Ifg_2_Ifg_5[`i´], V_Ifg_5_Ifg_5[`i´])
  3.          mata V`i´ = st_matrix("V`i´")
  4.          mata invV`i´ = invsym(V`i´)
  5.          mata W`i´ = cholesky(invV`i´)
  6.          matrix y`i´ = (b_Ifg_2[`i´] \ b_Ifg_5[`i´])
  7.          mata y`i´ = st_matrix("y`i´")
  8.          mata ystar`i´ = W`i´´*y`i´
  9.          mata x`i´ = I(2)
 10.          mata xstar`i´ = W`i´´*x`i´
 11.          if `i´ == 1 {
 12.              mata ystarstack = ystar1
 13.              mata xstarstack = xstar1
 14.          }
 15.          else {
 16.              mata ystarstack = (ystarstack \ ystar`i´)
 17.              mata xstarstack = (xstarstack \ xstar`i´)
 18.          }
 19. }
. clear
. getmata ystarstack (xstarstack*)=xstarstack
. sem (ystarstack <- xstarstack1 xstarstack2, noconstant),
>         variance(e.ystarstack@1) nocapslatent nolog nocnsreport nodescribe
Structural equation model                   Number of obs      =          62
Estimation method  = ml
Log likelihood     = -384.49772
```

	Coef.	OIM Std. Err.	z	P>|z|	[95% Conf. Interval]	
Structural						
ystarstack <-						
xstarstack1	.2042387	.0529888	3.85	0.000	.1003826	.3080947
xstarstack2	.8639001	.0536208	16.11	0.000	.7588052	.968995
_cons	0	(constrained)				
var(e.ystarstack)	1	(constrained)				

```
LR test of model vs. saturated: chi2(2)   =     15.87, Prob > chi2 = 0.0004
. lincom _b[ystarstack:xstarstack1], eform
 ( 1)  [ystarstack]xstarstack1 = 0
```

	exp(b)	Std. Err.	z	P>|z|	[95% Conf. Interval]	
(1)	1.226591	.0649955	3.85	0.000	1.105594	1.36083

```
. lincom _b[ystarstack:xstarstack2], eform
 ( 1)  [ystarstack]xstarstack2 = 0
```

	exp(b)	Std. Err.	z	P>|z|	[95% Conf.	Interval]
(1)	2.372395	.1272098	16.11	0.000	2.135723	2.635295

We can then perform the heterogeneity test by fitting the fixed-effects model with the variance of the error term of the outcome variable, unconstrained (Cheung 2013c).

```
. quietly sem (ystarstack <- xstarstack1 xstarstack2, noconstant), nocapslatent
. display "var(e.ystarstack) = " _b[var(e.ystarstack):_cons]
var(e.ystarstack) = 1.8483607
. local Q = _b[var(e.ystarstack):_cons]*e(N)
. local df = e(N) - 2
. display "Het. test statistic = " `Q´
Het. test statistic = 114.59836
. display "Het. test p-value = " chi2tail(`df´, `Q´)
Het. test p-value = .00002803
```

We can also specify this model in **sem** with the dataset in wide format by using two equations. In this code, the relevant parameters are constrained to be equal using the **@c#** syntax (parameters with the same **c#** constraint are estimated to be equal). We can omit the **xstarstack12** variable because all of its values are 0 (Cheung 2015b). We omit the output because the estimates are the same.

```
. generate study = round(_n/2)
. generate outcome = (mod(_n,2)==0) + 1
. reshape wide ystarstack xstarstack1 xstarstack2, i(study) j(outcome)
. assert xstarstack12 == 0
. sem (ystarstack1 <- xstarstack11 xstarstack21@c1)
>     (ystarstack2 <- xstarstack22@c1), noconstant
>     variance(e.ystarstack1@1 e.ystarstack2@1) nocaps nolog nocnsreport nodescribe
  (output omitted )
```

In this syntax, if we removed the constraints from the variances of the error terms of **ystarstack1** and **ystarstack2**, we could perform the heterogeneity test for each outcome separately.

5.2 Random-effects models

The random-effects model using the transformed variables is defined as

$$\mathbf{W}^{1/2}\mathbf{Y} \sim \text{MVN}\left\{\mathbf{W}^{1/2}\mathbf{X}\boldsymbol{\beta} + \mathbf{W}^{1/2}\boldsymbol{\nu}, \ \mathbf{W}^{1/2}\boldsymbol{\Sigma}\left(\mathbf{W}^{1/2}\right)'\right\}$$

$$\boldsymbol{\nu} \sim N\left(\mathbf{0}, \ \mathbf{T}^2\right)$$

$$\text{which implies} \quad \mathbf{Y}^* \sim \text{MVN}\left\{\mathbf{X}^*\boldsymbol{\beta}, \ \mathbf{I} + \mathbf{W}^{1/2}\mathbf{T}^2\left(\mathbf{W}^{1/2}\right)'\right\}$$

Reverting to the long form of the dataset and continuing the code above, as per Cheung (2013c), we need to introduce two latent variables (because there are two outcomes) for the random effects. Using the long format of the data, we specify the latent variables at the study level using the M1[study] syntax. As with our previous analyses using the transformed scale, we must also scale the random effects.

```
. quietly reshape long

. gsem (ystarstack <- c.xstarstack1#c.M1[study]@1
> c.xstarstack2#c.M2[study]@1 xstarstack1 xstarstack2, nocons),
> latent(M1 M2) nocnsreport nolog
> cov(e.ystarstack@1 M1[study]*M2[study])
Generalized structural equation model          Number of obs     =         62
Response      : ystarstack
Family        : Gaussian
Link          : identity
Log likelihood = -101.66433
```

	Coef.	Std. Err.	z	P>|z|	[95% Conf. Interval]	
ystarstack <-						
xstarstack1	.1875603	.0690866	2.71	0.007	.0521531	.3229675
xstarstack2	.8585811	.0887304	9.68	0.000	.6846728	1.032489
c. xstarstack1# M1[study]	1	(constrained)				
c. xstarstack2# M2[study]	1	(constrained)				
_cons	0	(omitted)				
var(M1[study])	.0221546	.0324089			.0012597	.3896245
var(M2[study])	.09458	.0614174			.0264883	.3377098
cov(M2[study], M1[study])	.0272542	.0382754	0.71	0.476	-.0477642	.1022726
var(e.ystar~k)	1	(constrained)				

For comparison, we can also fit this model using the wide-format data and two equations. We omit the output as the estimates are identical.

```
. quietly reshape wide
. gsem (ystarstack1 <- c.xstarstack11#c.M1@1 c.xstarstack21#c.M2@1
>               xstarstack11 xstarstack21@c1, nocons)
>         (ystarstack2 <- c.xstarstack22#c.M2@1
>               xstarstack22@c1, nocons),
>         cov(e.ystarstack1@1 e.ystarstack2@1) latent(M1 M2)
>         collinear nocnsreport nolog
  (output omitted)
```

We also compared our estimates with maximum likelihood estimates from `mvmeta`.

```
. quietly use fscstage1, clear
. mvmeta b V, vars(b_Ifg_2 b_Ifg_5) print(bscov) ml nolog
Note: using method ml
Note: using variables b_Ifg_2 b_Ifg_5
Note: 31 observations on 2 variables
Note: variance-covariance matrix is unstructured

Multivariate meta-analysis
Variance-covariance matrix = unstructured
Method = ml                                     Number of dimensions   =      2
Log likelihood = -64.533812                     Number of observations =     31
```

	Coef.	Std. Err.	z	P>|z|	[95% Conf. Interval]	
Overall_mean						
b_Ifg_2	.1875603	.0690844	2.71	0.007	.0521574	.3229631
b_Ifg_5	.8585811	.0887347	9.68	0.000	.6846644	1.032498

```
Estimated between-studies covariance matrix Sigma:
          b_Ifg_2    b_Ifg_5
b_Ifg_2   .02215463
b_Ifg_5   .0272542   .09457994
```

Our summary estimates using `gsem` and the `mvmeta` summary estimates are essentially identical. We could also derive prediction intervals for the estimates as described in section 2.2.

6   Conclusion

In this article, we have shown how to fit fixed- and random-effects meta-analysis, meta-regression, and multivariate outcome meta-analysis models under the SEM framework using the `sem` and `gsem` commands. This is possible because `sem` and `gsem` allow the user to constrain the variance of the error term of the outcome variables and the variance of the latent variables in the model while placing additional constraints on other parameters as required.

Our multivariate meta-analysis code easily extends to incorporate multivariate meta-regression models. Additionally, because `gsem` can fit random-effects models with many different link functions, it can be used to fit individual patient data meta-analysis models and meta-analysis models of diagnostic accuracy studies as per `metandi` and `midas`, which use hierarchical logistic regression (Harbord and Whiting 2009; Dwamena 2007). Because `sem` and `gsem` can fit multivariate normal meta-analysis models, they can also fit network meta-analysis models, including consistency and inconsistency models, using the same approaches developed for use with `mvmeta` by Higgins et al. (2012) and White et al. (2012).

We found that for multivariate outcome models with nonzero within-study covariances, we need to transform the outcome variables using the Cholesky decomposition of the inverse of their within-study covariance matrices (Cheung 2013c). This is because `sem` and `gsem` do not allow the use of definition variables as per the OpenMx package (Boker et al. 2011).

In the random-effects models, the estimate of the between-study variance (τ^2) that we obtain from `gsem` is arguably suboptimal because it is a maximum likelihood estimate rather than an REML estimate. Maximum likelihood estimates of variance components have been shown to be negatively biased (Viechtbauer 2005; Prevost et al. 2007). Cheung (2013b, 2015a) demonstrated that REML estimates of τ^2 can be obtained using maximum likelihood estimation in OpenMx. However, this is not currently possible using `gsem` because constraints cannot be placed on the variances of the random effects in such a flexible manner. REML estimates of τ^2 and summary estimates can be obtained using `mvmeta`.

When performing estimation including latent variables, `gsem` performs numerical integration, also referred to as quadrature, which can prove problematic. In such cases, we found that it is very helpful to compare estimates with maximum likelihood estimates from `mvmeta` (White 2009, 2011). Maximum likelihood estimates of random-effects variance components from `mvmeta` can be used as initial values in `gsem`. It is also helpful to fit the fixed-effects model first, confirming the estimates with the fixed estimates from `mvmeta`, and then build the code to the random-effects model second. The `gsem` command also has several integration methods and options available. We recommend starting with the default mean-and-variance adaptive Gauss–Hermite quadrature (`mvaghermite`) integration method, and then experimenting with the mode-and-curvature adaptive Gauss–Hermite quadrature (`mcaghermite`) and nonadaptive Gauss–Hermite quadrature (`ghermite`) methods after that. Finally, we recommend trying the Laplace method, which performs the Laplacian approximation instead of quadrature. It can also be helpful to experiment with the number of integration points, but we would not recommend reducing this to fewer than three points.

One advantage of our approach is that it is straightforward to perform the heterogeneity test after fitting the fixed-effects meta-analysis, meta-regression, and multivariate outcome meta-analysis models by leaving the variance of the error term of the transformed outcomes unconstrained. This test, although simple, does not appear to be implemented for multivariate outcome meta-analysis models in the existing user-written commands, although `mvmeta` reports I^2 statistics for these models.

7 Acknowledgments

We thank Dr. Ian White (MRC Biostatistics Unit), associate professor Mike Cheung (National University of Singapore), Rebecca Pope and the `gsem` development team from StataCorp, and an anonymous reviewer for very helpful comments.

8 References

Boker, S., M. C. Neale, H. Maes, M. Wilde, M. Spiegel, T. Brick, J. Spies, R. Estabrook, S. Kenny, T. Bates, P. Mehta, and J. Fox. 2011. OpenMx: An open source extended structural equation modeling framework. *Psychometrika* 76: 306–317.

Borenstein, M., L. V. Hedges, J. P. T. Higgins, and H. R. Rothstein. 2009. *Introduction to Meta-Analysis*. Chichester, UK: Wiley.

Bradburn, M. J., J. J. Deeks, and D. G. Altman. 1998. sbe24: metan—an alternative meta-analysis command. *Stata Technical Bulletin* 44: 4–15. Reprinted in *Stata Technical Bulletin Reprints*, vol. 8, pp. 86–100. College Station, TX: Stata Press. (Reprinted in this collection on pp. 3–28.)

Cheung, M. W.-L. 2008. A model for integrating fixed-, random-, and mixed-effects meta-analyses into structural equation modeling. *Pyschological Methods* 13: 182–202.

———. 2010. Fixed-effects meta-analyses as multiple-group structural equation models. *Structural Equation Modeling: A Multidisciplinary Journal* 17: 481–509.

———. 2013a. The metaSEM package. http://courses.nus.edu.sg/course/psycwlm/Internet/metaSEM/.

———. 2013b. Implementing restricted maximum likelihood estimation in structural equation models. *Structural Equation Modeling: A Multidisciplinary Journal* 20: 157–167.

———. 2013c. Multivariate meta-analysis as structural equation models. *Structural Equation Modeling: A Multidisciplinary Journal* 20: 429–454.

———. 2015a. Advanced topics in SEM-based meta-analysis. In *Meta-Analysis: A Structural Equation Modeling Approach*, chap. 8. Chichester, UK: Wiley.

———. 2015b. Conducting meta-analysis with Mplus. In *Meta-Analysis: A Structural Equation Modeling Approach*, chap. 9. Chichester, UK: Wiley.

DerSimonian, R., and N. Laird. 1986. Meta-analysis in clinical trials. *Controlled Clinical Trials* 7: 177–188.

Dwamena, B. 2007. midas: Stata module for meta-analytical integration of diagnostic test accuracy studies. Statistical Software Components S456880, Department of Economics, Boston College. https://ideas.repec.org/c/boc/bocode/s456880.html.

Fibrinogen Studies Collaboration. 2004. Collaborative meta-analysis of prospective studies of plasma fibrinogen and cardiovascular disease. *European Journal of Cardiovascular Prevention and Rehabilitation* 11: 9–17.

———. 2005. Plasma fibrinogen level and the risk of major cardiovascular diseases and nonvascular mortality: An individual participant meta-analysis. *Journal of the American Medical Association* 294: 1799–1809.

Glas, A. S., D. Roos, M. Deutekom, A. H. Zwinderman, P. M. Bossuyt, and K. H. Kurth. 2003. Tumor markers in the diagnosis of primary bladder cancer: A systematic review. *Journal of Urology* 169: 1975–1982.

Harbord, R. M., and J. P. T. Higgins. 2008. Meta-regression in Stata. *Stata Journal* 8: 493–519. (Reprinted in this collection on pp. 85–111.)

Harbord, R. M., and P. Whiting. 2009. metandi: Meta-analysis of diagnostic accuracy using hierarchical logistic regression. *Stata Journal* 9: 211–229. (Reprinted in this collection on pp. 213–231.)

Harris, R. J., M. J. Bradburn, J. J. Deeks, R. M. Harbord, D. G. Altman, and J. A. C. Sterne. 2008. metan: Fixed- and random-effects meta-analysis. *Stata Journal* 8: 3–28. (Reprinted in this collection on pp. 29–54.)

Harville, D. A. 1977. Maximum likelihood approaches to variance component estimation and to related problems. *Journal of the American Statistical Association* 72: 320–338.

Higgins, J. P. T., D. Jackson, J. K. Barrett, G. Lu, A. E. Ades, and I. R. White. 2012. Consistency and inconsistency in network meta-analysis: Concepts and models for multi-arm studies. *Research Synthesis Methods* 3: 98–110.

Higgins, J. P. T., and S. G. Thompson. 2002. Quantifying heterogeneity in a meta-analysis. *Statistics in Medicine* 21: 1539–1558.

———. 2004. Controlling the risk of spurious findings from meta-regression. *Statistics in Medicine* 23: 1663–1682.

Higgins, J. P. T., S. G. Thompson, J. J. Deeks, and D. G. Altman. 2003. Measuring inconsistency in meta-analyses. *British Medical Journal* 327: 557–560.

Kalaian, H. A., and S. W. Raudenbush. 1996. A multivariate mixed linear model for meta-analysis. *Psychological Methods* 1: 227–235.

Mehta, P., and M. C. Neale. 2005. People are variables too: Multilevel structural equations modeling. *Psychological Methods* 10: 259–284.

Patterson, H. D., and R. Thompson. 1971. Recovery of inter-block information when block sizes are unequal. *Biometrika* 58: 545–554.

Prevost, A. T., D. Mason, S. Griffin, A.-L. Kinmonth, S. Sutton, and D. J. Spiegelhalter. 2007. Allowing for correlations between correlations in random-effects meta-analysis of correlation matrices. *Psychological Methods* 12: 434–450.

R Core Team. 2014. *R: A Language and Environment for Statistical Computing*. R Foundation for Statistical Computing, Vienna, Austria. http://www.R-project.org.

Riley, R. D., K. R. Abrams, A. J. Sutton, P. C. Lambert, and J. R. Thompson. 2007. Bivariate random-effects meta-analysis and the estimation of between-study correlation. *BMC Medical Research Methodology* 7: 3.

Sharp, S. 1998. sbe23: Meta-analysis regression. *Stata Technical Bulletin* 42: 16–22. Reprinted in *Stata Technical Bulletin Reprints*, vol. 7, pp. 148–155. College Station, TX: Stata Press. (Reprinted in this collection on pp. 112–120.)

StataCorp. 2013. *Stata 13 Multilevel Mixed-Effects Reference Manual*. College Station, TX: Stata Press.

Sterne, J. A. C., ed. 2009. *Meta-Analysis: An Updated Collection from the Stata Journal*. College Station, TX: Stata Press.

Sutton, A. J., K. R. Abrams, D. R. Jones, T. A. Sheldon, and F. Song. 2000. *Methods for Meta-Analysis in Medical Research*. Chichester, UK: Wiley.

Thompson, S. G., and S. Sharp. 1999. Explaining heterogeneity in meta-analysis: A comparison of methods. *Statistics in Medicine* 18: 2693–2708.

Turner, R. M., R. Z. Omar, M. Yang, H. Goldstein, and S. G. Thompson. 2000. A multilevel model framework for meta-analysis of clinical trials with binary outcomes. *Statistics in Medicine* 19: 3417–3432.

Viechtbauer, W. 2005. Bias and efficiency of meta-analytic variance estimators in the random-effects model. *Journal of Educational and Behavioral Statistics* 30: 261–293.

White, I. R. 2009. Multivariate random-effects meta-analysis. *Stata Journal* 9: 40–56. (Reprinted in this collection on pp. 232–248.)

———. 2011. Multivariate random-effects meta-regression: Updates to mvmeta. *Stata Journal* 11: 255–270. (Reprinted in this collection on pp. 249–264.)

White, I. R., J. K. Barrett, D. Jackson, and J. P. T. Higgins. 2012. Consistency and inconsistency in network meta-analysis: Model estimation using multivariate meta-regression. *Research Synthesis Methods* 3: 111–125.

The Stata Journal (2013)
13, Number 1, st0284, pp. 577–591

462

Trial sequential boundaries for cumulative meta-analyses

Branko Miladinovic
Center for Evidence-Based Medicine and Health Outcomes Research
University of South Florida
Tampa, FL
bmiladin@health.usf.edu

Iztok Hozo
Department of Mathematics
Indiana University Northwest
Gary, IN

Benjamin Djulbegovic
Center for Evidence-Based Medicine and Health Outcomes Research
University of South Florida
Tampa, FL

Abstract. We present a new command, `metacumbounds`, for the estimation of trial sequential monitoring boundaries in cumulative meta-analyses. The approach is based on the Lan–DeMets method for estimating group sequential boundaries in individual randomized controlled trials by using the package `ldbounds` in R statistical software. Through Stata's `metan` command, `metacumbounds` plots the Lan–DeMets bounds, z-values, and p-values obtained from both fixed and random-effects cumulative meta-analyses. The analysis can be performed with count data or on the hazard scale for time-to-event data.

Keywords: st0284, metacumbounds, trial sequential analysis, cumulative meta-analysis, information size, Lan–DeMets bounds, monitoring boundary, cumulative z score, heterogeneity

1   Introduction

Randomized controlled trials (RCTs) are the gold standard for making causal inferences regarding treatment effects. Meta-analyses of RCTs increase both the power and the precision of estimated treatment effects. However, there is a risk that a meta-analysis may report false positive results, that is, report a treatment effect when in reality there is none. This is especially true when the pooled estimates are updated with the publication of a new trial in cumulative meta-analyses. A small RCT may result in chance findings and overestimation. To avoid false conclusions, Pogue and Yusuf (1997, 1998) advocated constructing Lan–DeMets trial sequential monitoring boundaries for cumulative meta-analysis. This is analogous to constructing interim treatment sequential monitoring boundaries in a single RCT, where a trial would be terminated if the cumulative z curve crossed the discrete sequential boundary and a treatment larger than expected occurred. They calculated the optimal information size based on the assumption that participants originated from a single trial.

More recently, Wetterslev et al. (2008) adjusted the method for heterogeneity and labeled it trial sequential analysis (TSA). Their approach accounted for bias and observed heterogeneity in a retrospective cumulative meta-analysis. We implement TSA in Stata under the command `metacumbounds` and with the `ldbounds` package in open-source R statistical software, which calculates bounds by using the Lan–DeMets α spending function approach. `metacumbounds` is the first widely available package to construct monitoring bounds for cumulative meta-analysis for both count data and information in the form of hazard ratios for time-to-event data. Analyzing time-to-event data on the count scale leads to the loss of valuable information, decreases the power, and should be avoided. Tierney et al. (2007) discuss methods for extracting hazard ratios from published data. The option to construct monitoring bounds for cumulative meta-analysis on the hazard scale has not been available in the domain of public software and, to our knowledge, is presented here for the first time. In section 2, we discuss the methodology behind TSA. In section 3, we describe how to install R and the packages needed to implement `metacumbounds`. In section 4, we present the command `metacumbounds`, and in section 5, the command is illustrated with two examples from published literature.

2 Methods

Group sequential analysis for individual RCTs was introduced by Armitage (1969) and Pocock (1977). Gordon Lan and DeMets (1983) made the methods for controlling the type I error when interim analyses are conducted more flexible by introducing the z curve and α spending function, which produce either the O'Brien–Fleming or the Pocock type boundaries. Under this method, the progress of a single RCT is measured over time, and the trial is terminated early if the cumulative z curve crosses a discrete sequential boundary. The boundary depends on the number of decision times and the rate at which the prespecified type I error α is spent, independent of the number of future decision times. The probability of terminating a trial early at time t_i is calculated as the proportion of α that should be spent at t_i minus the α already used in the past. We use five different spending functions (DeMets and Gordon Lan 1994):

(i) O'Brien–Fleming spending function

$$\alpha(t) = \begin{cases} 0, & t = 0 \\ 2 - 2\Phi(\frac{Z_{\frac{\alpha}{2}}}{\sqrt{t}}), & 0 < t \leq 1 \end{cases}$$

(ii) Pocock spending function

$$\alpha(t) = \begin{cases} 0, & t = 0 \\ \alpha \ln\{1 + (e - 1)t\}, & 0 < t \leq 1 \end{cases}$$

(iii) Alpha × time

$$\alpha(t) = \begin{cases} 0, & t = 0 \\ \alpha t, & 0 < t \le 1 \end{cases}$$

(iv) Alpha × time$^{1.5}$

$$\alpha(t) = \begin{cases} 0, & t = 0 \\ \alpha t^{1.5}, & 0 < t \le 1 \end{cases}$$

(v) Alpha × time^2

$$\alpha(t) = \begin{cases} 0, & t = 0 \\ \alpha t^2, & 0 < t \le 1 \end{cases}$$

Pogue and Yusuf (1997) extended the methodology to cumulative meta-analysis, where its progress is monitored as the relevant information is accrued over time. The total number of observed patients in the cumulative meta-analysis is defined as the accrued information size (AIS). Assuming that the information size (that is, the sample size) needed is at least equal to the sample size required in an individual RCT, given the prespecified type I error α and power $(1 - \beta)$, then the required a priori anticipated information size (APIS) based on a prespecified intervention effect is defined as

$$\text{APIS} = \frac{4\nu}{\mu^2}(Z_{\frac{\alpha}{2}} + Z_\beta)^2$$

Here μ is the intervention effect and ν its variance, assuming equal size between the intervention and control groups. For count data and the event rates in the control and experimental groups p_c and p_e, $\mu = p_c - p_e$ and $\nu = p^*(1 - p^*)$, where $p^* = (p_c + p_e)/2$. The a priori relative risk reduction (RRR) is defined as $\text{RRR} = 1 - p_e/p_c$.

If we use the results of Lachin and Foulkes (1986), the required APIS for time-to-event data and assumed hazard ratio HR$_0$, expected censoring rate w (that is, loss to follow-up), and average survival rate across studies S is given by

$$\text{APIS} = \frac{(Z_{\frac{\alpha}{2}} + Z_\beta)^2}{(1 - w)(1 - S)}\left(\frac{\text{HR}_0 + 1}{\text{HR}_0 - 1}\right)^2$$

Individual RCTs may be biased. It is well accepted that trials with a high risk of bias due to inadequate randomization sequence generation, intention-to-treat analysis, allocation concealment, masking, or reported incomplete outcome data may overestimate intervention effects. RRR and low-bias information size (LBIS) are thus calculated by applying the intervention effects from low-bias trials only. Combining trials as if participants came from one mega-trial may bias the results because of heterogeneity. To account for uncertainty induced by heterogeneity, we must adjust (multiply) information size by $1/(1 - I^2)$ to calculate the low-bias heterogeneity-adjusted information size (LBHIS). Note that I^2 is heterogeneity defined as

$$I^2 = \frac{(Q - k + 1)}{Q}$$

and Q is Cochran's homogeneity statistic. Once the information size is calculated, while the new trials are published and meta-analyses are updated, the monitoring bounds can be updated over time as well. Brok et al. (2008) present a set of examples of two-sided TSA for four different cumulative z curves (see figure 1).

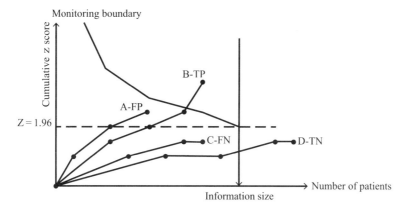

Figure 1. Examples of the upper half of two-sided TSA

(A) Crossing of $Z = 1.96$ provides a significant result but a spurious effect because the z curve does not cross the monitoring boundary. This is a false positive result.

(B) Crossing of the monitoring boundary before reaching the information size provides for firm evidence of effect. This is a true positive result.

(C) z curve not crossing $Z = 1.96$ indicates absence of evidence; that is, the meta-analysis included fewer patients than the required information size. This is a false negative result.

(D) Lack of predefined effect even though the information size is reached. This is a true negative result.

The monitoring boundary typically moves right and down over time. However, it may move right and up if the event rate decreases, intervention effect increases, or heterogeneity increases. In the context of LBIS and LBHIS, crossing of the monitoring bounds before the information size is reached indicates that high-bias risk trials find a larger intervention effect compared with low-bias risk trials.

3   R statistical software

R statistical software is an open-source package that may be downloaded free of charge at http://www.r-project.org. To use `metacumbounds`, after installing R, the user needs to install the R packages `foreign` (to read and write Stata data files) and `ldbounds` (to compute group sequential bounds by using the Lan–Demets method with either the O'Brien–Fleming or the Pocock spending functions). The package ldbounds is based on the Fortran code ld98 by Reboussin et al. (2000). Statistical packages can be downloaded from the Comprehensive R Archive Network from a multitude of mirror websites within R. This is done by selecting **Packages > Install package(s)...** and then the mirror site closest to the user (figure 2 outlines the steps). The USA(MD) Comprehensive R Archive Network mirror highlighted in figure 2 is at the United States National Cancer Institute (http://watson.nci.nih.gov/cran_mirror/).

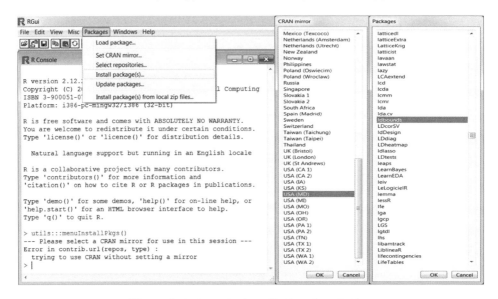

Figure 2. ldbounds installation description

Note that R does not have to be running when Stata is executing the metacumbounds command. The Stata program rsource is used to run R from inside Stata. It works by running the Rterm.exe program and may be downloaded from within Stata by typing ssc install rsource.

4 The metacumbounds command

4.1 Syntax for metacumbounds

Our command metacumbounds assumes that Stata's metan command (Harris et al. 2008) has been installed. Because of the complexity of the syntax and to facilitate its implementation, we have included the dialog-box file metacumbounds.dlg, which should be placed in the active Stata directory.

metacumbounds *varlist* [*if*] [*in*], data(count | loghr) <u>eff</u>ect(f | r)
   <u>sp</u>ending(*string*) <u>rdir</u>(*string*) is(ais | apis | lbis | lbhis) [id(*strvar*)
   surv(*#*) loss(*#*) lbid(*varname*) stat(rr | or | rd) <u>wkdir</u>(*string*)
   <u>kprsrce</u>(*string*) alpha(*#*) beta(*#*) <u>graph</u> rrr(*#*) listRout listRin keepR
   graph_options]

where *varlist* contains either count data or log hazard-ratios, their standard errors, and trial sample size.

4.2 Options

data(count | loghr) specifies whether the analysis is done for count data or on the log-hazard scale for time-to-event outcomes. Under the data(count) option, the user can specify effect size based on risk ratio, odds ratio, or risk difference. For both data(count) and data(loghr), the output is on the natural scale. logrr or logor may equally be used under the loghr option in the unlikely event that the count data are unavailable, in which case the survival rate S and loss to follow-up are both equal to 0. data() is required.

effect(f | r) specifies whether fixed- or random-effects estimates are used in the output and graph. If the fixed-effects model is chosen and heterogeneity I^2 is greater than 30%, then a warning message is displayed. The pooling method used is the inverse variance method (fixedi and randomi in metan). effect() is required.

spending(*string*) specifies the spending function that is calculated by ldbounds in R. spending(1) computes O'Brien–Fleming type bounds. spending(2) computes Pocock type bounds. spending(3) computes bounds of type αt. spending(4) computes bounds of type $\alpha t^{1.5}$. spending(5) computes bounds of type αt^2. spending() is required.

rdir(*string*) lists the path of the directory where the binary files for R can be found. rdir() is required.

is(ais | apis | lbis | lbhis) specifies the method to be used for calculation size. is() is required.

   ais represents the simple accrued information size—the fraction of the total number of participants in the meta-analysis used up to that point. The assumed a priori RRR (RRR = rrr()) is used to determine the power of the test for given alpha and given (actual) sample size.

   apis represents the a priori information size and means that the total sample size will be calculated so that the trial has the a priori intervention effect (RRR = rrr()) on the incidence rate in the control group (which is calculated from the provided trial data). The incidence rate for the experimental group is calculated using this RRR. The RRR is given by the user, as are alpha and beta. These variables are then used to determine the sample size (APIS).

lbis represents the low-bias information size and means that the total sample size will be calculated using the incidence rate of only those trials for which the low-bias ID variable is greater than 0. If LBIS = 1, then the trial has low bias. If LBIS = 0, then the trial does not have low bias, it has high bias. The intervention effect (RRR) is now calculated from the incidence rates of both control and experimental groups for only those trials for which the low-bias ID variable is greater than 0. For this RRR and for user-specified alpha and beta, we calculate the required sample size and call it LBIS.

lbhis (low-bias heterogeneity-adjusted information size) is the same as lbis except adjusted for heterogeneity; that is, LBHIS = LBIS/$(1 - I^2)$, where I^2 is the heterogeneity index of this group of trials for the given statistic.

id(*strvar*) is a character variable used to label the studies. If the data contain a labeled numeric variable, then the **decode** command can be used to create a character variable.

surv(*#*) for hazard-ratio data specifies the overall average survival rate and is defined on [0, 1).

loss(*#*) for hazard-ratio data specifies the percent of patients lost to follow-up and is defined on [0, 1).

lbid(*varname*) specifies whether each study is low risk for bias (coded 1) or high risk for bias (coded 0) under is(lbis) or is(lbhis).

stat(rr | or | rd) for count data specifies the effect size (risk ratio, odds ratio, or risk difference) to be pooled.

wkdir(*string*) is the directory where all the files should be saved.

kprsrce(*string*) saves the R source file after the program is completed.

alpha(*#*) specifies the type I error. *#* must be between 0 and 1.

beta(*#*) specifies the type II error. *#* must be between 0 and 1.

graph requests a graph.

rrr(*#*) specifies the trial a priori intervention effect size (RRR) to calculate APIS. For LBIS and LBHIS, **rrr()** is calculated from low-bias trials only.

listRout lists the R output on the Stata screen.

listRin lists the R source file on the Stata screen.

keepR keeps the R source file.

graph_options are overall graph options. shwRRR and pos() allow for the addition and position of the RRR, α, and power on the graph; xtitle(*string*) and ytitle(*string*) add labels to the *x* and *y* axes; title(*string*) and subtitle(*string*) add the title and subtitle to the graph. The dialog box makes performing TSA easier.

5   Examples

5.1   Example 1: Effects of artery catheter tip position in the newborn

Wetterslev et al. (2008) performed TSA with data from a systematic review by Barrington (2000). One of the review's aims was to determine whether the position (high versus low) of the tip of an umbilical arterial catheter led to clinical vascular compromise. Out of five total trials, only one was found to have adequate allocation concealment and was considered low bias (table 1). The author reported that high-placed catheters were found to produce a significantly lower incidence of clinical vascular complications with RRR = 47% (95% confidence interval (CI); [37%–56%]).

Table 1. High versus low catheter position for clinical vascular compromise

Study	High (n/N)	Low (n/N)	Low bias
Harris (1978)	3/18	12/18	no
Mokrohisky (1978)	9/33	26/40	no
Stork (1984)	12/85	25/97	no
Kempley (1992)	34/162	66/146	no
UACTSG (1992)	77/481	130/489	yes

For LBIS and LBHIS to be calculated, the low-bias ID variable needs to be specified. In their analysis, Wetterslev et al. (2008) assumed RRR = 15% based on clinical significance. Figure 3 provides a screenshot of the dialog box used to perform the TSA analysis, which confirms the results from the systematic review in figure 4(a)–(c). The figure also displays the actual power achieved given the information size. Trial sequential monitoring boundary (TSMB) for AIS and APIS detected three potentially spurious p-values; TSMB for LBIS and LBHIS detected two potentially spurious levels.

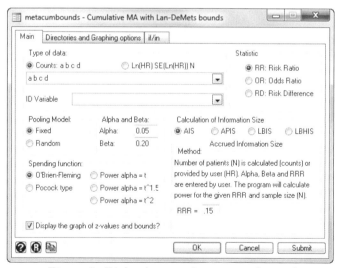

Figure 3. Dialog box used to create figure 4

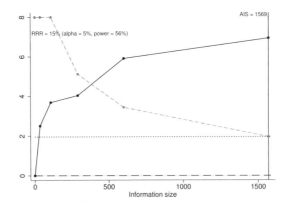

(a) Results showing three potentially spurious *p*-values
for AIS of 1,569 patients.

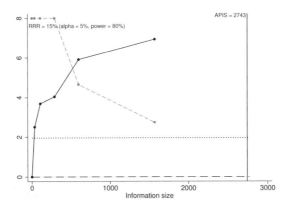

(b) Results showing three potentially spurious *p*-values
for APIS of 2,743 patients.

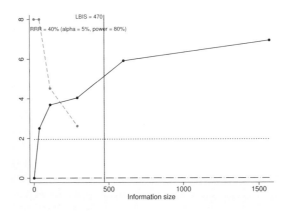

(c) Results showing three potentially spurious *p*-values
for LBIS of 470 patients. Note that because LBIS equals
LBHIS, results for the latter are the same.

Figure 4. TSA on the effects of umbilical artery catheter position in newborns

```
. use example1

. metacumbounds a b c d, data(count) effect(f) id(study) alpha(0.05) beta(0.20)
> is(AIS) stat(rr) graph spending(1) rrr(.15) kprsrce(StataRsource.R)
> rdir(C:\Program Files\R\R-2.12.2\bin\i386) shwRRR pos(10)
> xtitle(Information size)
Isquare =  0.00%

Cumulative fixed-effects meta-analysis of 5 studies  with Lan-DeMets bounds
-------------------------------------------------------------------------
```

```
                    Cumulative
Trial             estimate(rr)     z     P val       partN       UB
Harris_1978             0.250   2.508   0.012          36     8.000
Mokrohisky_1978         0.371   3.691   0.000         109     8.000
Stork_1984              0.436   4.041   0.000         291     5.128
Kempley_1992            0.452   5.911   0.000         599     3.445
UACTSG_1992             0.525   6.936   0.000        1569     1.962
```

5.2   Example 2: Neoadjuvant chemotherapy for invasive bladder cancer

Advanced Bladder Cancer Meta-analysis Collaboration (2011) conducted individual patient data meta-analysis to study whether neoadjuvant chemotherapy improves survival in patients with invasive bladder cancer. They concluded that the hazard ratio for all trials, including single-agent cisplatin, tended to favor neoadjuvant chemotherapy with RRR = 11% (95% CI; [2%–19%]) (the results were reported on the hazard scale as HR = 0.89; 95% CI; [81%–98%]). All 10 trials were found to have adequate allocation concealment and were considered low bias (see table 2). Because $I^2 = 0\%$, fixed- and random-effects meta-analyses produce identical TSMBs, and LBIS equals LBHIS.

Table 2. Neoadjuvant chemotherapy for invasive bladder cancer

Study	Neoadjuvant (n/N)	Local (n/N)	HR [95% CI]	Low bias
Raghavan (1991)	34/41	37/55	1.43 [0.88, 2.31]	yes
Wallace (1991)	59/83	50/76	1.11 [0.76, 1.61]	yes
Martinez (1995)	43/62	38/59	1.02 [0.66, 1.57]	yes
Malmstrom (1996)	68/151	84/160	0.77 [0.56, 1.06]	yes
Cortesi (unpub)	43/82	41/71	0.91 [0.6, 1.40]	yes
Bassi (1999)	53/102	60/104	0.93 [0.64, 1.35]	yes
MRC/EORTC (1999)	275/491	301/485	0.85 [0.72, 1]	yes
Sherif (2002)	79/158	90/159	0.86 [0.64, 1.16]	yes
Sengelov (2002)	70/78	60/75	1.06 [0.75, 1.50]	yes
Grossman (2003)	98/158	108/159	0.77 [0.58, 1.01]	yes

Figure 5 provides a screenshot of the dialog box used to perform the analysis. Using the estimated average survival rate of $S = 40\%$ and assuming $w = 0\%$ loss to follow-up, we found that TSA confirms the results from the systematic review for AIS [figure 6(a)–

(c)]. TSMB crosses the z curve for AIS of 2,809 patients. The TSA confirms the results for the systematic review of APIS = 1,990 under assumed RRR = 15%, $\alpha = 0.05$, and power$(1 - \beta) = 0.8$. However, the results of the systematic review do not hold under estimated LBIS = LBHIS = 4,418. There was one spurious p-value (Grossman trial) under LBIS and LBHIS estimates.

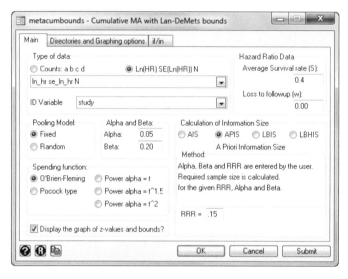

Figure 5. Dialog box used to create figure 6

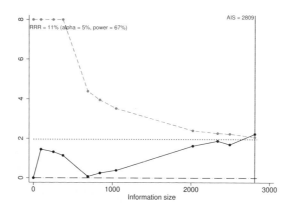

(a) Results showing that TSMB crosses z curve for AIS of 2,809 patients.

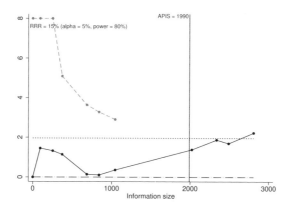

(b) Results for APIS of 1,990 patients.

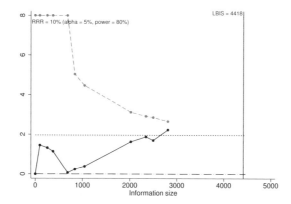

(c) Results for LBIS of 4,418 patients. Note that because LBIS equals LBHIS, results for the latter are the same.

Figure 6. TSA on the effects of neoadjuvant chemotherapy for invasive bladder cancer

```
. use example2

. metacumbounds ln_hr se_ln_hr N, data(loghr) effect(r) id(study) surv(0.40)
> loss(0.00) alpha(0.05) beta(0.20) is(APIS) graph spending(1) rrr(.15)
> kprsrce(StataRsource.R) rdir(C:\Program Files\R\R-2.12.2\bin\i386\)
> shwRRR pos(10) xtitle(Information size)
Isquare =  0.00%

Cumulative random-effects meta-analysis of 10 studies  with Lan-DeMets bounds
--------------------------------------------------------------
```

	Cumulative				
Trial	estimate()	z	P val	partN	UB
Raghavan_1991	1.430	1.453	0.146	96	8.000
Wallace_1991	1.221	1.322	0.186	255	8.000
Martinez_1995	1.153	1.142	0.253	376	5.087
Malmstrom_1996	1.019	0.143	0.887	687	3.640
Cortesi_1997	0.989	0.106	0.915	840	3.281
Bassi_1999	0.971	0.360	0.719	1046	2.910
MRC_EORTC_1999	0.917	1.384	0.166	2022	.
Sherif_2002	0.903	1.876	0.061	2339	.
Sengelov_2002	0.915	1.696	0.090	2492	.
Grossman_2003b	0.897	2.229	0.026	2809	.

6   Discussion

We presented a command, `metacumbounds`, for the implementation of TSA in Stata, which we recommend to minimize the risk of random error when performing cumulative meta-analyses. This way, the risk of finding a difference in treatment effects where no difference exists is minimized. The command uses a package for constructing Lan–Demets bounds in an open-source R statistical software.

`metacumbounds` can be implemented by using either fixed- or random-effects meta-analysis. It can incorporate heterogeneity in the calculation of boundaries. The method can be applied with count data or on the hazard scale for time-to-event data; TSA for both has not been available in the domain of public software. In addition to the subgroup analysis, funnel plots and meta-regression, the plot of the cumulative z curve, and monitoring boundaries, APIS and LBIS (or LBHIS in the presence of heterogeneity) should be a standard supplement to any meta-analysis.

7   Acknowledgment

The `rsource` program we used to run the R statistical software through Stata was developed by Roger Newson of the Imperial College London.

8   References

Advanced Bladder Cancer Meta-analysis Collaboration. 2011. Neoadjuvant cisplatin for advanced bladder cancer. *Cochrane Database of Systematic Reviews* 6: CD001426.

Armitage, P. 1969. Sequential analysis in therapeutic trials. *Annual Review of Medicine* 20: 425–430.

Barrington, K. J. 2000. Umbilical artery catheters in the newborn: Effects of position of the catheter tip. *Cochrane Database of Systematic Reviews* 2: CD000505.

Brok, J., K. Thorlund, C. Gluud, and J. Wetterslev. 2008. Trial sequential analysis reveals insufficient information size and potentially false positive results in many meta-analyses. *Journal of Clinical Epidemiology* 61: 763–769.

DeMets, D. L., and K. K. Gordon Lan. 1994. Interim analysis: The alpha spending function approach. *Statistics in Medicine* 13: 1341–1352.

Gordon Lan, K. K., and D. L. DeMets. 1983. Discrete sequential boundaries for clinical trials. *Biometrika* 70: 659–663.

Harris, R. J., M. J. Bradburn, J. J. Deeks, R. M. Harbord, D. G. Altman, and J. A. C. Sterne. 2008. metan: Fixed- and random-effects meta-analysis. *Stata Journal* 8: 3–28. (Reprinted in this collection on pp. 29–54.)

Lachin, J. M., and M. A. Foulkes. 1986. Evaluation of sample size and power for analyses of survival with allowance for nonuniform patient entry, losses to follow-up, noncompliance, and stratification. *Biometrics* 42: 507–519.

Pocock, S. J. 1977. Group sequential methods in the design and analysis of clinical trials. *Biometrika* 64: 191–199.

Pogue, J. M., and S. Yusuf. 1997. Cumulating evidence from randomized trials: Utilizing sequential monitoring boundaries for cumulative meta-analysis. *Controlled Clinical Trials* 18: 580–593.

———. 1998. Overcoming the limitations of current meta-analysis of randomised controlled trials. *Lancet* 351: 47–52.

Reboussin, D. M., D. L. DeMets, K. Kim, and K. K. G. Lan. 2000. Computations for group sequential boundaries using the Lan–DeMets spending function method. *Controlled Clinical Trials* 21: 190–207.

Tierney, J. F., L. A. Stewart, D. Ghersi, S. Burdett, and M. R. Sydes. 2007. Practical methods for incorporating summary time-to-event data into meta-analysis. *Trials* 8: 16.

Wetterslev, J., K. Thorlund, J. Brok, and C. Gluud. 2008. Trial sequential analysis may establish when firm evidence is reached in cumulative meta-analysis. *Journal of Clinical Epidemiology* 61: 64–75.

The Stata Journal (2013)
13, Number 3, st0304, pp. 451–473

Simulation-based sample-size calculation for designing new clinical trials and diagnostic test accuracy studies to update an existing meta-analysis

Michael J. Crowther
Department of Health Sciences
University of Leicester
Leicester, UK
michael.crowther@le.ac.uk

Sally R. Hinchliffe
Department of Health Sciences
University of Leicester
Leicester, UK

Alison Donald
Department of Health Sciences
University of Leicester
Leicester, UK

Alex J. Sutton
Department of Health Sciences
University of Leicester
Leicester, UK

Abstract. In this article, we describe a suite of commands that enable the user to estimate the probability that the conclusions of a meta-analysis will change with the inclusion of a new study, as described previously by Sutton et al. (2007, *Statistics in Medicine* 26: 2479–2500). Using the `metasim` command, we take a simulation approach to estimating the effects in future studies. The method assumes that the effect sizes of future studies are consistent with those observed previously, as represented by the current meta-analysis. Two-arm randomized controlled trials and studies of diagnostic test accuracy are considered for a variety of outcome measures. Calculations are possible under both fixed- and random-effects assumptions, and several approaches to inference, including statistical significance and limits of clinical significance, are possible. Calculations for specific sample sizes can be conducted (by using `metapow`). Plots, akin to traditional power curves, can be produced (by using `metapowplot`) to indicate the probability that a new study will change inferences for a range of sample sizes. Finally, plots of the simulation results are overlaid on extended funnel plots by using `extfunnel`, described in Crowther, Langan, and Sutton (2012, *Stata Journal* 12: 605–622), which can help to intuitively explain the results of such calculations of sample size. We hope the command will be useful to trialists who want to assess the potential impact new trials will have on the overall evidence base and to meta-analysts who want to assess the robustness of the current meta-analysis to the inclusion of future data.

Keywords: st0304, metasim, metapow, metapowplot, meta-analysis, diagnostic test, sample size, evidence-based medicine

1   Introduction

Sutton et al. (2007) argued that following the completion of a new randomized trial, the updated meta-analysis containing the new study would potentially be of more interest where multiple studies of the same topic exist in certain contexts. However, this goes against findings that many trialists do not consider previous trials, formally or informally, when designing new trials (Cooper, Jones, and Sutton 2005).

Relatively recently, formal methodology to assess the ability of new trials to affect the conclusions of an updated meta-analysis were developed (Sutton et al. 2009b). These methodologies were piloted in a study that applied them retrospectively to several clinical contexts (Goudie et al. 2010). A coherent framework for designing, analyzing, and reporting evidence that used such methods has also been described in Sutton, Cooper, and Jones (2009a). Very recently, the methods have been adapted for diagnostic test accuracy by Hinchliffe et al. (2013) and for cluster-randomized controlled trials by Rotondi and Donner (2012). Both Bayesian (Sutton et al. 2007) and frequentist (Goudie et al. 2010) implementations of the general approach have been considered.

Here we describe a collection of three commands to implement the frequentist version of the methodology in the contexts of (two-arm) randomized controlled trials and diagnostic test accuracy. `metasim` simulates data for future studies of a specified sample size by using predictions based on a meta-analysis of the existing evidence. Although this command can be used on its own, as described in sections 2.2 and 3, it is primarily designed to be used as a subroutine that is called by `metapow` (sections 2.2 and 4). `metapow` calculates the probability that a future study with a sample size specified by the user will change the inferences of an existing meta-analysis.[1] Several alternative approaches to inference can be specified, including both statistical significance and limits of clinical significance. `metapowplot` (sections 2.2 and 5) presents a graph of power for a range of sample sizes for the new study by repeatedly invoking `metapow`.

The structure of the remainder of this article is as follows: Section 2 describes the methods implemented in the three commands. Sections 3, 4, and 5 describe the syntax for `metasim`, `metapow`, and `metapowplot`, respectively. In section 6, plots of the simulated study results are overlaid on the previously described command `extfunnel` (Crowther, Langan, and Sutton 2012). This can help to intuitively explain the results of the power calculations through the use of boundary contours where inferences of the meta-analysis will change, indicating the effect size and precision combinations of future studies that would change inferences of the meta-analysis (Langan et al. 2012). Section 7, the discussion, concludes the article.

---

1. We refer to the probability that inferences of the meta-analysis will change as "power" throughout the remainder of the article, although we acknowledge this is not what is usually referred to as "power" in a single-study context.

2   Methods

2.1   Overview of methods

Simulation methods to establish appropriate sample sizes are often used as an alternative to closed-form solutions when complex analyses need to be carried out (Feiveson 2002) (for example, for analyses that include models with random effects). The approach suggested by Sutton et al. (2007) focuses not on the simulated study itself but on the modified meta-analysis, including the simulated study, because in some contexts, the results of the updated meta-analysis will be of more interest than those of the study on its own. Below is the nontechnical summary of the process as described by Sutton et al. (2007).

1. From a meta-analysis of the existing studies, a distribution for the chosen outcome measures in a new clinical trial or diagnostic test accuracy study is derived. An estimate for the outcome measure from this distribution is then sampled, representing the underlying effect in the new (simulated) study.

2. Data representing the new study are generated stochastically according to the estimate sampled in step 1 for a specified sample size.

3. These simulated study data are then added to the existing meta-analysis, which is then re-meta-analyzed.

4. The hypothesis test, on which decisions are to be based, is then considered. Whether the null is retained or rejected in favor of the alternative hypothesis at a specified level of statistical significance is recorded.

5. Steps 2–4 are repeated a large number of times (N), and the outcome of the hypothesis test is noted each time.

6. Power is estimated by calculating the proportion of the N simulations in which the null hypothesis is rejected.

7. The procedure is iterative: the sample size for a new study specified in step 2 is modified and steps 2–6 repeated until the desired level of power is achieved.

2.2   Overview of software

Figure 1 presents a schematic representation of the relationship between the Stata commands described in this article and previously described commands. It has already been explained that `metapowplot` calls `metapow`, which in turn calls `metasim`. Additionally, however, because others have written excellent routines for conducting meta-analysis in Stata, the preexisting `metan` command (Harris et al. 2008) is called by `metapow` to conduct meta-analyses of two-arm comparative study data, such as data from randomized controlled trials. Similarly, `metandi` (Harbord and Whiting 2009) and `midas` (Dwamena

2007) are both called by `metapow` for conducting meta-analyses of diagnostic test accuracy studies. Two commands are used because convergence issues with the bivariate diagnostic model are well documented (Rabe-Hesketh, Skrondal, and Pickles 2005). In addition, because both routines use different estimation algorithms, `midas` is invoked if `metandi` fails to converge. The command `extfunnel`, described in Crowther, Langan, and Sutton (2012), can be used to further illustrate the simulation results, and we show in section 6 how output from `metapow` can be overlaid on the plots produced by `extfunnel`.

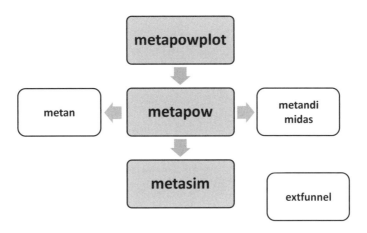

Figure 1. Software relationship diagram: arrows denote the calling of a command

metasim

`metasim` simulates a specified number of new studies based on the estimate obtained from a preexisting meta-analysis, assuming the effect size seen in the new study will be consistent with the existing studies in the meta-analysis. The command can be used independently, but it was designed to be used in conjunction with `metapow` (see section 4).

`metasim` will simulate data for a new study represented by the values for each of the variables entered in the variable list. These are saved in a Stata data file, `temppow.dta`, in the specified working directory.

metapow

`metapow` implements an approach to estimating the power of a study based on the evidence-based approach to sample-size determination for adding new studies to a meta-analysis of two-arm randomized controlled trials and diagnostic accuracy studies described in Sutton et al. (2007) and Hinchliffe et al. (2013), respectively. Power is determined through simulation, with data for new studies being generated with `metasim`.

As well as estimating the power of the updated meta-analysis including the new study, `metapow` can also estimate the power of the new study when analyzed on its own. The results of individual simulations are stored in a file, `temppow2.dta`, located in the specified working directory. While this function can be used directly to estimate the power for particular sample sizes, the higher-level command `metapowplot` uses this command to construct power curves across different sample sizes.

metapowplot

`metapowplot` produces a plot of the power values for a range of sample sizes. The command calls on `metapow`, which calculates power for a single sample size. `metapow` in turn calls on `metasim`, which simulates new studies by using the results of the existing meta-analysis.

Users need to input a minimum and maximum sample size for which they want to calculate a power estimate. The power estimates are stored with their confidence intervals (CIs) in a file called `temppow3.dta` within the working directory.

3 The metasim command

3.1 Syntax

`metasim` *varlist*, n(*integer*) es(*numlist*) var(*numlist*)

    <u>type</u>(clinical | diagnostic) [<u>measure</u>(or | rr | rd | nostandard | dor | ss)

    p(*real*) r(*real*) <u>studies</u>(*integer*)

    <u>mod</u>el(fixed | fixedi | random | randomi | bivariate) <u>tau</u>sq(*numlist*)

    dist(normal | t) corr(*real*) path(*string*)]

The dataset should contain the data for the existing studies with variable names that are consistent with those entered in *varlist*. The user should input a maximum of six variables. For trials with a binary outcome, four variables are required: these correspond to the number of events and nonevents in the experimental group followed by those of the control group. And for continuous outcomes, six variables should be entered: sample size, mean, and standard deviation of the experimental group followed by those of the control group. For diagnostic studies, four variables are required: the true positives, false positives, false negatives, and true negatives.

3.2 Options

n(*integer*) relates to the number of patients in the new study. If users simulate a new clinical trial, then n() specifies the number of patients in the control group. If users simulate a new diagnostic accuracy study with sensitivity and specificity as the outcome measure of accuracy, then n() is the number of diseased patients. If users simulate a new diagnostic accuracy study with the diagnostic odds ratio (DOR), then n() is the number of positive test results. n() is required.

es(*numlist*) specifies the pooled estimates from the meta-analysis of existing studies. If using the odds ratio (OR), the DOR, or relative risk (RR), then users need to specify ln(OR), ln(DOR), or ln(RR) estimates, respectively. If using sensitivity and specificity, then users need to specify logit(sensitivity) and logit(specificity), in that order. es() is required.

var(*numlist*) specifies the variances for es(). Two values should be entered when using sensitivity and specificity. var() is required.

type(clinical | diagnostic) specifies the type of new study that the user would like to simulate: a two-arm clinical trial or a diagnostic test accuracy study. type() is required.

measure(or | rr | rd | nostandard | dor | ss) specifies the outcome measure used in the meta-analysis to pool the results. The OR (or), RR (rr), risk difference (rd), and unstandardized mean difference (nostandard) can only be used when simulating a new clinical study. The DOR (dor) and sensitivity and specificity (ss) can only be used when simulating a new diagnostic accuracy study. The default for a type(clinical) study with four variables entered into the *varlist* is rr; the default for a type(clinical) study with six variables entered into the *varlist* is nostandard; and the default for a type(diagnostic) study is ss.

p(*real*) is the estimated event rate in the control group in a simulated, new clinical study. When users simulate a new diagnostic accuracy study, this is the estimated probability of being diseased given a positive result in the new study. When this option is not specified, metasim will calculate this value by averaging the probabilities across the studies included in the dataset in memory. Note that p() is only relevant in the diagnostic framework when dor is used as the option in measure().

r(*real*) is the ratio of patients in the control group to the treatment group in a simulated, new clinical study. When users simulate a new diagnostic accuracy study, this is the ratio of diseased to healthy people if using sensitivity and specificity and is the ratio of positive to negative results if using the DOR. The default is r(1).

studies(*integer*) specifies the number of new studies to be simulated and included in the updated meta-analysis. The default is studies(1). When more than one study is specified, each is assumed to have the same sample size.

model(fixed|fixedi|random|randomi|bivariate) defines the type of model used to meta-analyze the preexisting data. The default is model(fixed) unless the outcome measure is the nonstandardized mean difference, in which case the default is model(fixedi). The model(fixedi) option specifies a fixed-effects model by using the inverse-variance method. The model(random) option uses the random-effects DerSimonian and Laird method, taking the estimate for heterogeneity from the Mantel–Haenszel method. The model(randomi) option specifies a random-effects model by using the method of DerSimonian and Laird, with the estimate of heterogeneity being taken from the inverse-variance fixed-effects model. All the above options call on the metan command within metasim. The final option is the bivariate random-effects model (model(bivariate)). This method calls on a combination of the metandi and midas commands (a variable is created to indicate which has been used for each simulation). It may only be specified when simulating a new diagnostic accuracy study.

tausq(*numlist*) is the measure of between-study variance taken from the preexisting meta-analysis. The default is tausq(0). If measure(ss) is specified, then two values must be entered for tausq(). If a random-effects model is selected and the value for tausq() is still 0, then a warning message will appear to notify the user, but the command will continue to run.

dist(normal|t) specifies the distribution of effect sizes used to sample a value to simulate a new study. The default for model(random) and model(randomi) is a predictive distribution based on the *t* distribution (dist(t)), allowing for heterogeneity between studies (and the uncertainty in the heterogeneity). The default for all other models is dist(normal), based on the mean and variance entered in es() and var().

corr(*real*) is the correlation between the sensitivity and specificity. The default is corr(0). This option is only needed if the user chooses the bivariate model.

path(*string*) specifies the directory in which to save files created by metasim. This overrides the default of the working directory.

3.3 Example

We illustrate metasim with a systematic review of antibiotic use for the common cold from the Cochrane database of systematic reviews (Arroll and Kenealy 1999). We return to this example in sections 4.3 and 5.3 to illustrate metapow and metapowplot.

Six trials were conducted to compare antibiotics versus placebo for outcome symptoms persisting beyond seven days and labeled as "event" in table 1. A total of 1,147 subjects participated: 664 in the treatment group and 483 in the control group. The trials are summarized in table 1.

Table 1. Six trials included in antibiotics for the common cold and acute purulent rhinitis meta-analysis

Study	Year	a (event/trt)	b (no event/trt)	c (event/ctrl)	d (no event/ctrl)
Herne	1980	7	39	10	12
Hoaglund	1950	39	115	51	104
Kaiser	1996	97	49	94	48
Lexomboon	1971	8	166	4	83
McKerrow	1961	5	10	8	10
Taylor	1977	12	117	3	56

The review concluded that "there was insufficient evidence of benefit to warrant the use of antibiotics" (Arroll and Kenealy 1999). Further trials could be potentially beneficial. A fixed-effects meta-analysis using the inverse-variance method was carried out on the six trials with the OR. The command line is given below, and the results are presented in figure 2.

```
. metan event_t noevent_t event_c noevent_c, or fixedi
> label(namevar=Study, yearvar=Year) textsize(150)
> xlabel(0.125, 0.25, 0.5, 1, 2, 4, 8) scheme(sj)
> title("Forest plot") favours("Favours treatment" # "Favours control")
> xtitle("Odds ratio")
              Study  |   OR    [95% Conf. Interval]    % Weight
---------------------+-----------------------------------------------
Herne (1980)         |  0.215    0.067     0.689         6.87
Hoaglund (1950)      |  0.692    0.422     1.134        38.04
Kaiser (1996)        |  1.011    0.620     1.648        38.88
Lexomboon (1971)     |  1.000    0.293     3.417         6.15
McKerrow (1961)      |  0.625    0.151     2.586         4.61
Taylor (1977)        |  1.915    0.519     7.058         5.46
---------------------+-----------------------------------------------
I-V pooled OR        |  0.796    0.587     1.080       100.00
---------------------+-----------------------------------------------

  Heterogeneity chi-squared =   8.07 (d.f. = 5) p = 0.153
  I-squared (variation in OR attributable to heterogeneity) =  38.0%

  Test of OR=1 : z=   1.47 p = 0.143
```

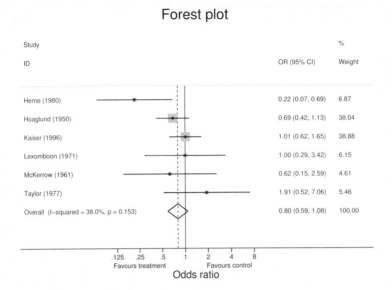

Figure 2. Forest plot of common cold data using fixed-effects meta-analysis with the inverse-variance method

The results suggest a slight treatment benefit, but this is not significant at the 5% level (OR = 0.80, 95% CI: [0.59,1.08]). It is possible that additional information in the form of another trial could lead to this result becoming statistically significant. The command `metasim` allows the user to simulate a new trial based on the above results. By inputting the pooled log OR from the meta-analysis as the estimate, along with the variance of the pooled log OR, users can write the command as follows:

```
. metasim event_t noevent_t event_c noevent_c, es(-0.228) var(0.155) n(100)
> type(clinical) measure(or) model(fixedi)
New study/studies simulated are saved in file called E:\Meta-analysis\
> metapow\temppow
. use temppow, clear
. list
```

	event_t	noeven~t	event_c	noeven~c
1.	21	79	31	69

A new trial has been generated with 100 patients in the treatment arm and 100 patients in the control arm. The trial is saved in the working directory in a file named `temppow`. The file will contain four variables with the same names as those in the current dataset. In this case, these will be `event_t`, `noevent_t`, `event_c`, and `noevent_c`.

4 The metapow command

4.1 Syntax

metapow *varlist*, n(*integer*) nit(*integer*) <u>type</u>(clinical|diagnostic)

    pow(*numlist*) [<u>meas</u>ure(or|rr|rd|nostandard|dor|ss)

    <u>infer</u>ence(ciwidth|pvalue|lci|uci) p(*real*) r(*real*) <u>stud</u>ies(*integer*)

    <u>mod</u>el(fixed|fixedi|random|randomi|bivariate) npow(*numlist*) ci(*real*)

    dist(normal|t) ind nip(*integer*) sos(sens|spec) path(*string*)

    level(*integer*)]

4.2 Options

n(*integer*); see the metasim options in section 3.2.

nit(*integer*) is the number of simulations on which the estimated power is based. The larger the number specified, the more accurate the estimate will be, but the longer the analysis will take. nit() is required.

type(clinical|diagnostic); see the metasim options in section 3.2.

pow(*numlist*) specifies the value used as a cutoff in determining the power. One or two values may be input. The value represents different things, depending on the option chosen for inference(). pow() is required.

measure(or|rr|rd|nostandard|dor|ss); see the metasim options in section 3.2.

inference(ciwidth|pvalue|lci|uci) defines the approach to inference used to calculate power. The default is inference(ciwidth). This counts the number of times that the CI width of the estimate from the updated meta-analysis (that is, with the simulated study included) is less than the specified value. This option can be used regardless of the measure of accuracy. Two other approaches to inference are inference(lci) and inference(uci). These will count the number of times that the lower or upper CI is higher or lower than a given value, respectively. The inference(lci) option can be used regardless of the measure of accuracy. The inference(uci) option is currently only available when working with clinical trial data and not diagnostic data. A final option only available when using clinical trial data is inference(pvalue). This counts the number of times that a *p*-value is significant to a specified level. When you use sensitivity and specificity, two values may be input into pow() for inference(ciwidth) and inference(lci). These will instruct the command to count the number of times that the CI widths for both sensitivity and specificity are less than their respective specified values. Sensitivity must be given first followed by specificity for the calculation to be correct. To use the inference(ciwidth) or inference(lci) option for just sensitivity or just specificity, you should also use the sos() option (described below).

p(*real*); see the `metasim` options in section 3.2.

r(*real*); see the `metasim` options in section 3.2.

studies(*integer*); see the `metasim` options in section 3.2.

model(fixed|fixedi|random|randomi|bivariate); see the `metasim` options in section 3.2.

npow(*numlist*) recalculates the power with a new value for the same `inference()` without having to rerun the whole command. Instead, it uses the data stored in `temppow2` and allows alternative approaches to inference to be explored. This is particularly valuable when the required simulation time is lengthy.

ci(*real*) specifies the width of the CI for the corresponding power estimate. The default is `ci(95)`.

dist(normal|t); see the `metasim` options in section 3.2.

ind instructs the command to calculate the power for the newly simulated study on its own in addition to the newly updated meta-analysis.

nip(*integer*) specifies the number of integration points used for quadrature when the bivariate model is selected. Higher values should result in greater accuracy but typically at the expense of longer execution times (see Harbord and Whiting [2009]).

sos(sens|spec) is used in addition to the `inference()` option and specifies whether inferences are focused on sensitivity or specificity when using `inference(ciwidth)` or `inference(lci)`. The default is `sos(sens)`. If `sos()` is not specified, then the inferences are based on both sensitivity and specificity, and two values should be entered for `pow()`.

path(*string*); see the `metasim` options in section 3.2.

level(*integer*) specifies the confidence level, as a percentage, for the individual study and pooled CIs. This is the level given in the `metan`, `metandi`, and `midas` commands when called on to meta-analyze the current dataset. The default is `level(95)`.

4.3   Example

The same example described in section 3.3 is used here to demonstrate the command `metapow`. This command allows the user to estimate the power that a new trial of a specified sample size would give to the meta-analysis. In this example, the inference is the *p*-value. `metapow` is told to estimate the power that a new trial with 100 patients in the treatment arm and 100 patients in the control arm would have at detecting a *p*-value less than 0.05 in the updated meta-analysis.

```
. metapow event_t noevent_t event_c noevent_c, n(100) type(clinical)
> measure(or) model(fixedi) nit(100) inference(pvalue) pow(0.05)
.....................................................................
> .....................
Fixed effect inverse variance-weighted model
Statistic used was odds ratio

n= 100 (in control group)
m= 100 (in treatment group)

Power of meta-analysis is: 31.00 (95% CI: 22.13, 41.03)

Level of significance used to estimate power = 0.05
Simulation estimates are saved in file called E:\Meta-analysis\metapow\temppow2
```

The output from the command describes the type of meta-analysis model specified and the inference used. It also gives the power estimate. In this case, the power estimate is 31.0% (95% CI: [22.1, 41.0]), meaning that the p-value was below 0.05 in 31 of the 100 iterations. It is possible to recalculate the power with a different cutoff value without having to rerun the whole analysis. The option npow() can be specified to do this as shown below. Notice that the dots are not displayed, which is because the analysis is not being run. The output also informs the user that the level used to estimate power has changed.

```
. metapow event_t noevent_t event_c noevent_c, n(100) type(clinical)
> measure(or) model(fixedi) nit(100) inference(pvalue) pow(0.05) npow(0.1)

Level used to estimate power has changed
Simulated data has not changed

Fixed effect inverse variance-weighted model
Statistic used was odds ratio

n= 100 (in control group)
m= 100 (in treatment group)

Power of meta-analysis is: 49.00 (95% CI: 38.86, 59.20)

Simulation estimates are saved in file called E:\Meta-analysis\metapow\temppow2
```

metapow stores the estimates from each of the 100 iterations in a file called temppow2. Because the command also calls on metasim, the last newly simulated study will also be saved in a file called temppow. These will both be found within the working directory in Stata.

In this example, only 100 simulations were run, resulting in quite a wide CI for power. This could be reduced by increasing the number of simulations.

5   The metapowplot command

5.1   Syntax

metapowplot *varlist*, start(*#*) stop(*#*) step(*#*) nit(*integer*)
    type(clinical | diagnostic) pow(*numlist*)
    [measure(or | rr | rd | nostandard | dor | ss)
    inference(ciwidth | pvalue | lci | uci) p(*real*) r(*real*) studies(*integer*)
    model(fixed | fixedi | random | randomi | bivariate) npow(*numlist*) ci(*real*)
    dist(normal | t) ind nip(*integer*) sos(sens | spec) path(*string*)
    graph(lowess | connected | overlay) noci regraph level(*integer*)]

5.2   Options

start(*#*) is the smallest total sample size of a new study for which the user wishes to
    calculate a power value. start() is required.

stop(*#*) is the largest total sample size of a new study for which the user wishes to
    calculate a power value. stop() is required.

step(*#*) is the step size to be used within the range of total sample sizes specified by
    start() and stop(). A step size of 10 between the range of 10 to 30 would mean
    that the power would be estimated for sample sizes of 10, 20, and 30. step() is
    required.

nit(*integer*); see the metapow options in section 4.2.

type(clinical | diagnostic); see the metasim options in section 3.2.

pow(*numlist*); see the metapow options in section 4.2.

measure(or | rr | rd | nostandard | dor | ss); see the metasim options in section 3.2.

inference(ciwidth | pvalue | lci | uci); see the metasim options in section 3.2.

p(*real*); see the metasim options in section 3.2.

r(*real*); see the metasim options in section 3.2.

studies(*integer*); see the metasim options in section 3.2.

model(fixed | fixedi | random | randomi | bivariate); see the metasim options in sec-
    tion 3.2.

npow(*numlist*); see the metapow options in section 4.2.

ci(*real*); see the metapow options in section 4.2.

dist(normal | t); see the metasim options in section 3.2.

ind; see the metapow options in section 4.2.

nip(*integer*); see the metapow options in section 4.2.

sos(sens | spec); see the metapow options in section 4.2.

path(*string*); see the metasim options in section 3.2.

graph(lowess | connected | overlay) allows the user to choose the type of line used to connect the specific estimates of power at the specified sample sizes. The default is graph(connected), which plots each point and connects them with a line. The other options are a lowess plot, which plots a smoothed line to the specific points, and an overlay plot, which plots both the points and the lowess curve. Because power is estimated through simulation, there is sampling error in each estimate that will decrease with the number of simulations specified (but also increase evaluation time). Thus smoothing may be desirable if several different but inaccurate estimates are considered. The lowess line should be similar to the connected option for larger simulations.

noci prevents the command from plotting CIs (indicating the sampling error in the estimation of power at specified sample sizes) on the graph.

regraph allows the user to regraph the power curves with alternative graph options without having to rerun the simulations for the specified range of sample sizes.

level(*integer*); see the metapow options in section 4.2.

5.3   Example

The command metapowplot is used to calculate the power value at various sample sizes by calling on metapow. The command then plots the power values against sample size. In the command below, the range of sample sizes has been specified as 100 to 1,000 with steps of 100; the results are shown in figure 3. All other options remain the same as those in section 4.3.

```
. metapowplot event_t noevent_t event_c noevent_c, start(100) step(100)
> stop(1000) type(clinical) measure(or) model(fixedi) nit(100)
> inference(pvalue) pow(0.05)
Sample size
t =      100      Treatment/Control = 50/50
t =      200      Treatment/Control = 100/100
t =      300      Treatment/Control = 150/150
t =      400      Treatment/Control = 200/200
t =      500      Treatment/Control = 250/250
t =      600      Treatment/Control = 300/300
t =      700      Treatment/Control = 350/350
t =      800      Treatment/Control = 400/400
t =      900      Treatment/Control = 450/450
t =      1000     Treatment/Control = 500/500
```

```
Fixed effect inverse variance-weighted model
Statistic used was odds ratio
Level of significance used to estimate power = 0.05
Power estimates used to plot the graph are saved in file called E:\Meta-analysis\
> metapow\temppow3
```

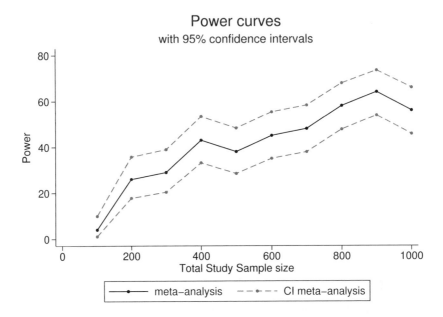

Figure 3. Power curve for common cold data based on the OR using a fixed-effects model with the inverse-variance method

metapowplot has stored the power values and corresponding sample sizes in a file called temppow3. Because the command calls on both metapow and metasim, the estimates from the last sample size are stored in temppow2, and the final newly simulated study is stored in temppow. All these files can be found in the working directory.

The output describes the options chosen by the user. Figure 3 shows the power curve generated by metapowplot. The power is estimated to reach 60% with a total sample size of 800: 400 patients in the treatment arm and 400 patients in the control arm. This implies that when updated with more information, the current meta-analysis from the Cochrane database could provide significant evidence to suggest a benefit in the use of antibiotics for the common cold.

If you are designing a new trial, we would recommend running a minimum of 1,000 simulations at each sample size and perhaps over a narrower, targeted range of sample sizes of interest.

5.4 Diagnostic example

This example focuses on the diagnostic test accuracy options within the commands. A meta-analysis was carried out in 1999 to assess the diagnostic value of the digital rectal examination (DRE) in detecting prostate cancer (Hoogendam, Buntinx, and de Vet 1999). Studies were included if they compared DRE with biopsy or surgery as the reference standard. A total of 14 studies met the inclusion criteria, giving a total of 21,839 patients. Table 2 gives the results from each study.

Table 2. Fourteen studies included in DRE as screening test for prostate cancer

Study	Year	TP	FP	FN	TN
Kirky	1994	8	6	6	541
Vihko	1985	6	21	3	741
Chodak	1989	32	112	13	1974
Ciatto	1994	17	8	9	1391
Lee	1989	10	19	12	743
Pode	1995	22	93	9	876
Dalkin	1993	9	33	15	695
Palken	1991	17	28	6	264
Teillac	1990	8	18	10	546
Catalona	1994	146	836	118	5530
Menor	1990	59	48	16	1389
Richie	1994	16	194	8	426
Gustafsson	1992	42	153	23	1564
Littrup	1994	77	287	95	2471

A separate DerSimonian and Laird random-effects meta-analysis of sensitivity and specificity was carried out on the 14 studies. Figure 4 gives the results of the random-effects meta-analysis of sensitivity. The results give a pooled estimate for sensitivity of 0.60 (95% CI: [0.53, 0.67]). This suggests that the test correctly identifies only 60% of the diseased patients. The other 40% would be given false negative results. The results of the random-effects meta-analysis of specificity are shown in figure 5. The pooled estimate for specificity was 0.95 (95% CI: [0.92, 0.96]). This suggests that 95% of the healthy patients are correctly identified by the test. This result is fairly good because it means that only 5% of the healthy patients would receive false positive results.

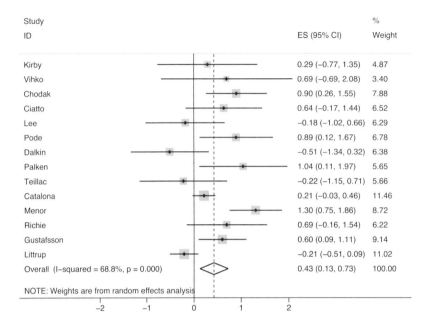

Figure 4. Forest plot of prostate data using DerSimonian and Laird random-effects meta-analysis of logit sensitivity

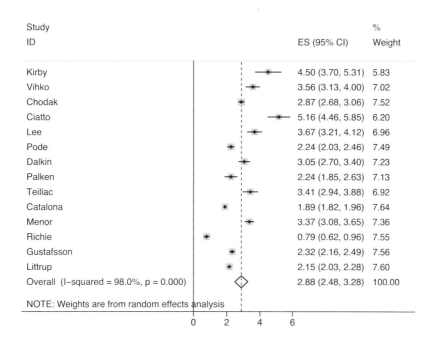

Figure 5. Forest plot of prostate data using DerSimonian and Laird random-effects meta-analysis of logit specificity

The command line given below estimates power values for sample sizes ranging from 100 to 1,000 in steps of 100 based on the lower confidence interval value for the pooled sensitivity estimate only. The cutoff for this has been set to 0.53, which is the same as the lower confidence interval value in the current meta-analysis for sensitivity (see figure 4). `metapowplot` will count how many times the lower confidence interval value for the pooled sensitivity is greater than or equal to 0.53 and base the power value on this.

```
. metapowplot TP FP FN TN, start(100) step(100) stop(1000) type(diagnostic)
> measure(ss) model(randomi) nit(200) sos(sens) inference(lci) pow(0.53)

Sample size
t =      100      Diseased/Healthy = 50/50
t =      200      Diseased/Healthy = 100/100
t =      300      Diseased/Healthy = 150/150
t =      400      Diseased/Healthy = 200/200
t =      500      Diseased/Healthy = 250/250
t =      600      Diseased/Healthy = 300/300
t =      700      Diseased/Healthy = 350/350
t =      800      Diseased/Healthy = 400/400
t =      900      Diseased/Healthy = 450/450
t =     1000      Diseased/Healthy = 500/500

Random effects model with inverse variance-weighted estimates of heterogeneity
Statistics used were sensitivity and specificity
Lower confidence interval value for sens used to estimate power = 0.53
Power estimates used to plot the graph are saved in file called E:\Meta-analysis\
> metapow\temppow3
```

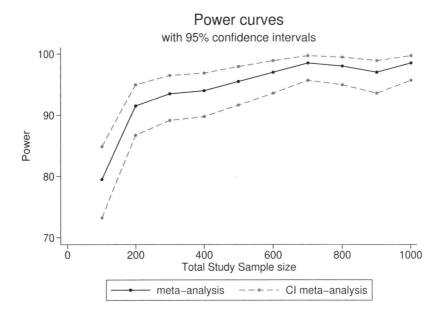

Figure 6. Power curve for prostate data based on sensitivity using DerSimonian and Laird random-effects model

Figure 6 shows the power curve obtained from the above command line. The power reaches about 98% for a total sample size of 1,000. This means that when a study with 500 diseased patients and 500 healthy patients was added to the current meta-analysis for sensitivity, the lower CI value for the pooled sensitivity was greater than or equal to 0.53 in about 98 of the 100 iterations.

6   Other uses

An intuitive way to visualize this process is to plot all the results of the individual simulations, at a specified sample size, on an extended funnel plot (Langan et al. 2012; Crowther, Langan, and Sutton 2012). Extended funnel plots illustrate how the conclusions of a meta-analysis would be impacted by the addition of a single new trial across a range of effect estimates and standard errors. By directly overlaying the simulation results at a specific sample size, stored in `temppow2.dta`, we can draw direct conclusions about the area where a new study would likely lie and its impact on hypothesis tests.

In figure 7, we overlay the simulated individual studies from the example in section 4.3. The majority of points lies in the region of the plot where a new study, when added to the existing meta-analysis, would produce a statistically significant result, with the updated effect estimate and 95% CI less than the null. This process can be repeated as desired for different sample sizes. In this example, we can directly relate the 28% power to 28 of the 100 simulated studies lying in the left-hand region of the plot, indicating a change in conclusions for the updated meta-analysis.

```
. merge 1:1 _n using "temppow2.dta", nogen noreport

. metan event_t noevent_t event_c noevent_c, or fixedi nograph
  (output omitted)

. gen logor=log(_ES)
(94 missing values generated)

. gen t1 = log(indes)

. extfunnel logor _selogES, fixedi eform
> xlabel(0.1 0.2 0.5 1 2 5, format(%2.1f)) yrange(0 1)
> addplot(scatter indse_es indes, msize(tiny) msym(T) mcol(black) xscale(log))
> ylabel(,format(%2.1f)) sumd sumdpos(0.9) pred
> legend(order(1 "Non-sig. effect (5% level)" 2 "Sig. effect > NULL (5% level)"
> 3 "Sig. effect < NULL (5% level)" 4 "Prediction interval" 6 "Null effect"
> 7 "Pooled effect" 8 "Original studies" 9 "Simulated studies"))
Original meta-analysis results:
             Study     |     ES     [95% Conf. Interval]     % Weight
---------------------+---------------------------------------------------
1                      |   0.215     0.067       0.689           6.87
2                      |   0.692     0.422       1.134          38.04
3                      |   1.011     0.620       1.648          38.88
4                      |   1.000     0.293       3.417           6.15
5                      |   0.625     0.151       2.586           4.61
6                      |   1.915     0.519       7.058           5.46
---------------------+---------------------------------------------------
I-V pooled ES          |   0.796     0.587       1.080         100.00
---------------------+---------------------------------------------------

  Heterogeneity chi-squared =   8.07 (d.f. = 5) p = 0.153
  I-squared (variation in ES attributable to heterogeneity) =  38.0%

  Test of ES=1 : z=   1.47 p = 0.143
Building graph:
```

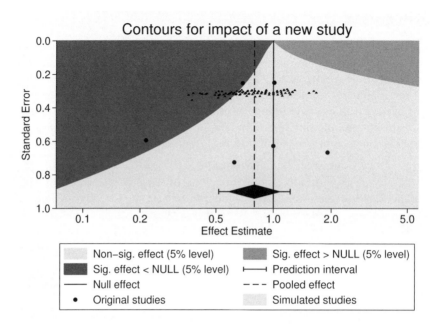

Figure 7. Extended funnel plot with simulated studies overlaid

7 Discussion

We hope the commands will be useful to 1) trialists who want to assess the potential impact new trials will have on the overall evidence base and those involved in funding new trials; and 2) meta-analysts who want to assess the robustness of the current meta-analysis to the inclusion of future data.

Finally, we thought it may be helpful to outline ongoing and potential future work. We have created a prototype set of commands that conducts the same calculations as the commands described here but that uses a Bayesian approach to all meta-analyses estimation. This is done through the use of the WinBUGS software, which links with Stata through a previously written command (Thompson, Palmer, and Moreno 2006). We hope to develop these to a point where they can be released in the future because a Bayesian approach to meta-analysis offers several advantages, as described elsewhere (Sutton et al. 2007).

A further Stata command, with the specific purpose of prioritizing a portfolio of meta-analyses for updating and which adapts much of the methodology described herein, is also very near completion.

Finally, others have extended the approaches described here to the context of cluster randomized controlled trials (Rotondi and Donner 2012) and written software in R to implement them; we hope this extension of the methodology can also be coded in a Stata command.

8 Acknowledgments

The authors would like to thank an anonymous reviewer for suggestions that improved the suite of commands and the manuscript. Michael Crowther was funded by a National Institute for Health Research methodology fellowship (RP-PG-0407-10314).

9 References

Arroll, B., and T. Kenealy. 1999. Antibiotics versus placebo in the common cold (Cochrane Review). *Cochrane Library Oxford* Issue 2.

Cooper, N. J., D. R. Jones, and A. J. Sutton. 2005. The use of systematic reviews when designing studies. *Clinical Trials* 2: 260–264.

Crowther, M. J., D. Langan, and A. J. Sutton. 2012. Graphical augmentations to the funnel plot to assess the impact of a new study on an existing meta-analysis. *Stata Journal* 12: 605–622. (Reprinted in this collection on pp. 193–210.)

Dwamena, B. 2007. midas: Stata module for meta-analytical integration of diagnostic test accuracy studies. Statistical Software Components S456880, Department of Economics, Boston College. https://ideas.repec.org/c/boc/bocode/s456880.html.

Feiveson, A. H. 2002. Power by simulation. *Stata Journal* 2: 107–124.

Goudie, A. C., A. J. Sutton, D. R. Jones, and A. Donald. 2010. Empirical assessment suggests that existing evidence could be used more fully in designing randomized controlled trials. *Journal of Clinical Epidemiology* 63: 983–991.

Harbord, R. M., and P. Whiting. 2009. metandi: Meta-analysis of diagnostic accuracy using hierarchical logistic regression. *Stata Journal* 9: 211–229. (Reprinted in this collection on pp. 213–231.)

Harris, R. J., M. J. Bradburn, J. J. Deeks, R. M. Harbord, D. G. Altman, and J. A. C. Sterne. 2008. metan: Fixed- and random-effects meta-analysis. *Stata Journal* 8: 3–28. (Reprinted in this collection on pp. 29–54.)

Hinchliffe, S. R., M. J. Crowther, R. S. Phillips, and A. J. Sutton. 2013. Using meta-analysis to inform the design of subsequent studies of diagnostic test accuracy. *Research Synthesis Methods* 4: 156–168.

Hoogendam, A., F. Buntinx, and H. C. de Vet. 1999. The diagnostic value of digital rectal examination in primary care screening for prostate cancer: A meta-analysis. *Family Practice* 16: 621–626.

Langan, D., J. P. T. Higgins, W. Gregory, and A. J. Sutton. 2012. Graphical augmentations to the funnel plot assess the impact of additional evidence on a meta-analysis. *Journal of Clinical Epidemiology* 65: 511–519.

Rabe-Hesketh, S., A. Skrondal, and A. Pickles. 2005. Maximum likelihood estimation of limited and discrete dependent variable models with nested random effects. *Journal of Econometrics* 128: 301–323.

Rotondi, M., and A. Donner. 2012. Sample size estimation in cluster randomized trials: An evidence-based perspective. *Computational Statistics and Data Analysis* 56: 1174–1187.

Sutton, A. J., N. J. Cooper, and D. R. Jones. 2009a. Evidence synthesis as the key to more coherent and efficient research. *BMC Medical Research Methodology* 9: 29.

Sutton, A. J., N. J. Cooper, D. R. Jones, P. C. Lambert, J. R. Thompson, and K. R. Abrams. 2007. Evidence-based sample size calculations based upon updated meta-analysis. *Statistics in Medicine* 26: 2479–2500.

Sutton, A. J., S. Donegan, Y. Takwoingi, P. Garner, C. Gamble, and A. Donald. 2009b. An encouraging assessment of methods to inform priorities for updating systematic reviews. *Journal of Clinical Epidemiology* 62: 241–251.

Thompson, J. R., T. M. Palmer, and S. G. Moreno. 2006. Bayesian analysis in Stata with WinBUGS. *Stata Journal* 6: 530–549.

Appendix: Further Stata meta-analysis commands

Stata users have written meta-analysis commands that have not, so far, been accepted for publication in the *Stata Journal*. Here are brief descriptions of commands known to the editors at the time of publishing this collection. Readers should note that these commands have not undergone the review process required for publication in the *Stata Journal*. This list is likely to be incomplete, and the editors apologizes to authors of any commands that have been overlooked. For the most up-to-date information on these and other meta-analysis commands, readers are encouraged to check the Stata frequently asked question on meta-analysis:

http://www.stata.com/support/faqs/statistics/meta-analysis/

- `metacurve` models a response as a function of a continuous covariate (optionally) adjusting for other variables. The estimated functions are averaged across the studies. The variables used to adjust for confounding can differ for each study. The methodology implemented in the command is described by Sauerbrei and Royston (2011). To install the program, type `net install metacurve,` `from(http://www.homepages.ucl.ac.uk/ ucakjpr/stata)` in Stata.

- `metannt` is intended to aid interpretation of meta-analyses of binary data by presenting intervention effect sizes in absolute terms, as the number needed to treat (NNT) and the number of events avoided (or added) per 1,000. The user inputs design parameters, and `metannt` uses the `metan` command to calculate the required statistics. This command is available as part of the `metan` package.

The NNT is the number of individuals required to experience the intervention in order to expect there to be one additional event to be observed. It is defined as the reciprocal of the absolute value of the risk difference (risk of the outcome in the intervention group minus risk in control).

$$\text{NNT} = \frac{1}{|\text{risk difference}|}$$

Assuming the event is undesirable, this is termed the *number needed to treat to benefit*. If the intervention arm experiences more events, this is commonly referred to as the *number needed to treat to harm*. Because most meta-analyses are based on ratio measures, the risk difference is calculated based on an assumed value of the risk in the control group. The `metannt` command calculates this by deriving an estimate of the intervention effect (for example, a risk ratio), applying it to a population with a given outcome event risk, and deriving from this a projected event risk if the population were to receive the intervention. The number of avoided or excess events (respectively) per 1,000 population is the difference between the two event risks multiplied by 1,000. Optionally, a confidence interval is also presented, using the confidence limits for the estimated intervention effect applied to the control group event rate.

- `metaninf` investigates the influence of one study on the overall meta-analysis estimate and shows graphically the results when the meta-analysis estimates are computed, omitting one study in each turn. This command makes repeated calls to the `metan` command for its analyses. It was released in 2001 and was last updated in 2004. It requires the user to provide input in the form needed by `metan`. To install the package, type `ssc install metaninf` in Stata. Articles describing `metainf`, a previous version of the command, were published in the *Stata Technical Bulletin* (Tobias 1999, 2000).

- `midas` provides statistical and graphical routines for undertaking meta-analysis of diagnostic test performance in Stata. Primary data synthesis is performed within the bivariate mixed-effects binary regression modeling framework. Model specification, estimation, and prediction are carried out with `xtmelogit` in Stata 10 or the `gllamm` command in Stata 9 by adaptive quadrature. Using the estimated coefficients and variance–covariance matrices, `midas` calculates summary operating sensitivity and specificity (with confidence and prediction contours in summary receiver operating characteristic space), summary likelihood, and odds ratios. Global and relevant test performance metric-specific heterogeneity statistics are provided. `midas` facilitates extensive statistical and graphical data synthesis and exploratory analyses of heterogeneity, covariate effects, publication bias, and influence. Bayes' nomograms and likelihood-ratio matrices can be obtained and used to guide clinical decision making. The minimum required input data are variables containing the elements of the 2×2 contingency tables (true positives, false positives, false negatives, and true negatives) of test results from each study. To install the package, type `ssc install midas` in Stata.

  Further information on the comprehensive suite of facilities provided by `midas` is available at http://www.sitemaker.umich.edu/metadiagnosis/midas_home. In particular, two presentations given at Stata Users Group meetings are available at http://www.sitemaker.umich.edu/metadiagnosis/presentations and via RePEc at http://econpapers.repec.org/paper/bocasug07/4.htm and http://ideas.repec.org/p/boc/wsug07/1.html.

- `meta_lr` graphs positive and negative likelihood ratios in diagnostic tests. It can do stratified meta-analysis of individual estimates. The user must provide the effect estimates (log positive likelihood ratio and log negative likelihood ratio) and their standard errors. Commands `meta` and `metareg` are used for internal calculations. This is a version 8 command released in 2004. To install the package, type `ssc install meta_lr` in Stata.

- `metaparm` performs meta-analyses and calculates confidence intervals and p-values for differences or ratios between parameters for different subpopulations, for data stored in the `parmest` format (Newson 2003). To install the package, type `ssc install metaparm` in Stata.

- `metaeff` is a pre-processing command for meta-analysis and a companion to `metaan` (Kontopantelis and Reeves 2010; Kontopantelis et al. 2013) which calculates effect sizes and their standard errors. To install the program, type `ssc install metaeff` in Stata.

10   References

Kontopantelis, E., and D. Reeves. 2010. metaan: Random-effects meta-analysis. *Stata Journal* 10: 395–407. (Reprinted in this collection on pp. 55–67.)

Kontopantelis, E., D. A. Springate, and D. Reeves. 2013. A re-analysis of the Cochrane Library data: The dangers of unobserved heterogeneity in meta-analyses. *PLOS ONE* 8: e69930.

Newson, R. B. 2003. Confidence intervals and p-values for delivery to the end user. *Stata Journal* 3: 245–269.

Sauerbrei, W., and P. Royston. 2011. A new strategy for meta-analysis of continuous covariates in observational studies. *Statistics in Medicine* 30: 3341–3360.

Tobias, A. 1999. sbe26: Assessing the influence of a single study in the meta-analysis estimate. *Stata Technical Bulletin* 47: 15–17. Reprinted in *Stata Technical Bulletin Reprints*, vol. 8, pp. 108–110. College Station, TX: Stata Press.

———. 2000. sbe26.1: Update of metainf. *Stata Technical Bulletin* 56: 15. Reprinted in *Stata Technical Bulletin Reprints*, vol. 10, p. 72. College Station, TX: Stata Press.

Author index

Command index

S